GROUP THEORY

A Physicist's Survey

Group theory has long been an important computational tool for physicists, but, with the advent of the Standard Model, it has become a powerful conceptual tool as well. This book introduces physicists to many of the fascinating mathematical aspects of group theory, and mathematicians to its physics applications.

Designed for advanced undergraduate and graduate students, this book gives a comprehensive overview of the main aspects of both finite and continuous group theory, with an emphasis on applications to fundamental physics. Finite groups are extensively discussed, highlighting their irreducible representations and invariants. Lie algebras, and to a lesser extent Kac–Moody algebras, are treated in detail, including Dynkin diagrams. Special emphasis is given to their representations and embeddings. The group theory underlying the Standard Model is discussed, along with its importance in model building. Applications of group theory to the classification of elementary particles are treated in detail.

PIERRE RAMOND is Distinguished Professor of Physics at the University of Florida and Director of the Institute for Fundamental Theory. He has held positions at FermiLab, Yale University and Caltech. He has made seminal contributions to supersymmetry, superstring theory and the theory of neutrino masses, and is a Member of the American Academy of Arts & Sciences.

Group Theory

A Physicist's Survey

PIERRE RAMOND

Institute for Fundamental Theory, Physics Department
University of Florida

CAMBRIDGE UNIVERSITY PRESS

CAMBRIDGE
UNIVERSITY PRESS

University Printing House, Cambridge CB2 8BS, United Kingdom

Cambridge University Press is part of the University of Cambridge.

It furthers the University's mission by disseminating knowledge in the pursuit of education, learning and research at the highest international levels of excellence.

www.cambridge.org
Information on this title: www.cambridge.org/9780521896030

First published 2010
Reprinted 2010

A catalogue record for this publication is available from the British Library

ISBN 978-0-521-89603-0 Hardback

To Richard Slansky
Who would have written this book

Contents

1

Preface: the pursuit of symmetries

Symmetric objects are so singular in the natural world that our ancestors must have noticed them very early. Indeed, symmetrical structures were given special magical status. The Greeks' obsession with geometrical shapes led them to the enumeration of platonic solids, and to adorn their edifices with various symmetrical patterns. In the ancient world, symmetry was synonymous with perfection. What could be better than a circle or a sphere? The Sun and the planets were supposed to circle the Earth. It took a long time to get to the apparently less than perfect ellipses!

Of course most shapes in the natural world display little or no symmetry, but many are almost symmetric. An orange is close to a perfect sphere; humans are almost symmetric about their vertical axis, but not quite, and ancient man must have been aware of this. Could this lack of exact symmetry have been viewed as a sign of imperfection, imperfection that humans need to atone for?

It must have been clear that highly symmetric objects were special, but it is a curious fact that the mathematical structures which generate symmetrical patterns were not systematically studied until the nineteenth century. That is not to say that symmetry patterns were unknown or neglected, witness the Moors in Spain who displayed the seventeen different ways to tile a plane on the walls of their palaces!

Évariste Galois in his study of the roots of polynomials of degree larger than four, equated the problem to that of a set of substitutions which form that mathematical structure we call a group. In physics, the study of crystals elicited wonderfully regular patterns which were described in terms of their symmetries. In the twentieth century, with the advent of Quantum Mechanics, symmetries have assumed a central role in the study of Nature.

The importance of symmetries is reinforced by the Standard Model of elementary particle physics, which indicates that Nature displays more symmetries in the small than in the large. In cosmological terms, this means that our Universe emerged from the Big Bang as a highly symmetrical structure, although most of its symmetries are no longer evident today. Like an ancient piece of pottery, some of its parts may not have survived the eons, leaving us today with its shards. This is a very pleasing concept that resonates with the old Greek ideal of perfection. Did our Universe emerge at the Big Bang with perfect symmetry that was progressively shattered by cosmological evolution, or was it born with internal defects that generated the breaking of its symmetries? It is a profound question which some physicists try to answer today by using conceptual models of a perfectly symmetric universe, e.g. superstrings.

Some symmetries of the natural world are so commonplace, that they are difficult to identify. The outcome of an experiment performed by undergraduates should not depend on the time and location of the bench on which it was performed. Their results should be impervious to shifts in time and space, as consequences of time and space translation invariances, respectively. But there are more subtle manifestations of symmetries. The great Galileo Galilei made something of a "trivial" observation: when your ship glides on a smooth sea with a steady wind, you can close your eyes and not "feel" that you are moving. Better yet, you can perform experiments whose outcomes are the same as if you were standing still! Today, you can leave your glass of wine while on an airplane at cruising altitude without fear of spilling. The great genius that he was elevated this to his principle of relativity: the laws of physics do not depend on whether you are at rest or move with constant velocity! However, if the velocity changes, you can feel it (a little turbulence will spill your wine). Our experience of the everyday world appears complicated by the fact that it is dominated by frictional forces; in a situation where their effect can be neglected, simplicity and symmetries (in some sense analogous concepts) are revealed.

According to Quantum Mechanics, physics takes place in Hilbert spaces. Bizarre as this notion might be, we have learned to live with it as it continues to be verified whenever experimentally tested. Surely, this abstract identification of a physical system with a state vector in Hilbert space will eventually be found to be incomplete, but in a presently unimaginable way, which will involve some other weird mathematical structure. That Nature uses the same mathematical structures invented by mathematicians is a profound mystery hinting at the way our brains are wired. Whatever the root cause, mathematical structures which find natural representations in Hilbert spaces have assumed enormous physical interest. Prominent among them are *groups* which, subject to specific axioms, describe transformations in these spaces.

Since physicists are mainly interested in how groups operate in Hilbert spaces, we will focus mostly on the study of their representations. Mathematical concepts will be introduced as we go along in the form of *scholia* sprinkled throughout the text. Our approach will be short on proofs, which can be found in many excellent textbooks. From representations, we will focus on their products and show how to build group invariants for possible physical applications. We will also discuss the embeddings of the representations of a subgroup inside those of the group. Numerous tables will be included.

This book begins with the study of *finite* groups, which as the name indicates, have a finite number of symmetry operations. The smallest finite group has only two elements, but there is no limit as to their number of elements: the permutations on n letters form a finite group with $n!$ elements. Finite groups have found numerous applications in physics, mostly in crystallography and in the behavior of new materials. In elementary particle physics, only small finite groups have found applications, but in a world with extra dimensions, and three mysterious families of elementary particles, this situation is bound to change. Notably, the sporadic groups, an exceptional set of twenty-six finite groups, stand mostly as mathematical curiosities waiting for an application.

We then consider *continuous* symmetry transformations, such as rotations by arbitrary angles, or open-ended time translation, to name a few. Continuous transformations can be thought of as repeated applications of infinitesimal steps, stemming from generators. Typically these generators form algebraic structures called Lie algebras. Our approach will be to present the simplest continuous groups and their associated Lie algebras, and build from them to the more complicated cases. Lie algebras will be treated *à la Dynkin*, using both Dynkin notation and diagrams. Special attention will be devoted to exceptional groups and their representations. In particular, the Magic Square will be discussed. We will link back to finite groups, as most can be understood as subgroups of continuous groups.

Some non-compact symmetries are discussed, especially the representations of space-time symmetries, such as the Poincaré and conformal groups. Group-theoretic aspects of the Standard Model and Grand Unification are presented as well. The algebraic construction of the five exceptional Lie algebras is treated in detail. Two generalizations of Lie algebras are also discussed, super-Lie algebras and their classification, and infinite-dimensional affine Kac–Moody algebras.

I would like to express my gratitude to the Institute for Advanced Study and the Aspen Center for Physics for their hospitality, where a good part of this book was written. I would like to thank Professors L. Brink and J. Patera, as well as Drs. D. Belyaev, Sung-Soo Kim, and C. Luhn, for their critical reading of the manuscript, and many useful suggestions. Finally, I owe much to my wife Lillian, whose patience, encouragements, and understanding made this book possible.

2

Finite groups: an introduction

Symmetry operations can be discrete or continuous. Easiest to describe are the discrete symmetry transformations. In this chapter we lay out notation and introduce basic concepts in the context of finite groups of low order. The simplest symmetry operation is *reflection* (parity):

$$P: \quad x \to x' = -x.$$

Doing it twice produces the identity transformation

$$I: \quad x \to x' = x,$$

symbolically

$$PP = I.$$

There are many manifestations of this symmetry operation. Consider the isosceles triangle. It is left the same by a reflection about its vertical axis. In geometry this operation is often denoted as σ. It takes place in the x–y plane, and you will see it written as σ_z which denotes a reflection in the x–y plane.

The second simplest symmetry operation is *rotation*. In two dimensions, it is performed about a point, but in three dimensions it is performed about an axis. A square is clearly left invariant by an anti-clockwise rotation about its center by 90°. The inverse operation is a clockwise rotation. Four such rotations are akin to the identity operation. Generally, anti-clockwise rotations by $(2\pi/n)$ generate the cyclic group $\mathcal{Z}_n$ when n is an integer. Repeated application of n such rotations is the identity operation. It is only in two dimensions that a reflection is a 180° rotation.

A third symmetry operation is *inversion* about a point, denoted by i. Geometrically, it involves a reflection about a plane together with a 180° rotation in the plane. Symbolically

$$i = \sigma_h \, \mathsf{Rot}(180°).$$

Reflections and inversions are sometimes referred to as *improper* rotations. They are rotations since they require a center of symmetry.

The symmetry operations which leave a given physical system or shape invariant must satisfy a number of properties: (1) a symmetry operation followed by another must be itself a symmetry operation; (2) although the order in which the symmetry operations are performed is important, the symmetry operations must associate; (3) there must be an identity transformation, which does nothing; and (4) whatever operation transforms a shape into itself, must have an inverse operation. These intuitive considerations lead to the group axioms.

2.1 Group axioms

A **group** $\mathcal{G}$ is a collection of operators,

$$\mathcal{G}: \quad \{a_1, a_2, \ldots, a_k, \ldots\}$$

with a "$\star$" operation with the following properties.

Closure. For every ordered pair of elements, a_i and a_j, there exists a unique element

$$a_i \star a_j = a_k, \tag{2.1}$$

for any three $i,\ j,\ k$.

Associativity. The $\star$ operation is associative

$$(a_i \star a_j) \star a_k = a_i \star (a_j \star a_k). \tag{2.2}$$

Unit element. The set $\mathcal{G}$ contains a unique element e such that

$$e \star a_i = a_i \star e = a_i, \tag{2.3}$$

for all i. In particular, this means that

$$e \star e = e.$$

Inverse element. Corresponding to every element a_i, there exists a unique element of $\mathcal{G}$, the inverse $(a_i)^{-1}$ such that

$$a_i \star (a_i)^{-1} = (a_i)^{-1} \star a_i = e. \tag{2.4}$$

When $\mathcal{G}$ contains a finite number of elements

$$(\mathcal{G}: \quad a_1, a_2, \ldots, a_k, \ldots, a_n)$$

it is called a finite group, and n is called the *order* of the group.

In the following we will discuss groups with a finite number of elements, but we should note that there are many examples of groups with an infinite number of elements; we now name a few.

- The real numbers, including zero, constitute an infinite group under addition ($\star \to +$). Its elements are the zero, the positive and the negative real numbers. Closure is satisfied: if x and y are real numbers, so is their sum $x + y$. Each x has an inverse $-x$, such that
$$x + (-x) = 0, \qquad x + 0 = 0 + x,$$
and we see that the zero plays the role of the unit element.
- The real numbers also form a group under multiplication ($\star \to \times$). Indeed xy is a real number. The inverse of x is $1/x$, and the unit element is 1. In this case, zero is excluded.
- The rational numbers of the form $\frac{n}{m}$, where m and n are non-zero integers also form a group under multiplication, as can easily be checked.

2.2 Finite groups of low order

We begin by discussing the finite groups of order less than thirteen (see Ledermann [14]). In the process we will introduce much notation, acquaint ourselves with many useful mathematical concepts, and be introduced to several ubiquitous groups and to the different aspects of their realizations.

Group of order 2

We have already encountered the unique group, called $\mathcal{Z}_2$, with two elements. One element is the identity operation, e, and the second element a must be its own inverse, leading to the following multiplication table.

$\mathcal{Z}_2$	e	a
e	e	a
a	a	e

It can serve many functions, depending on the physical situation: a can be the parity operation, the reflection about an axis, a 180° rotation, etc.

Group of order 3

There is only one group of order 3. This is easy to see: one element is the identity e. Let a_1 be its second element. The group element $a_1 \star a_1$ must be the third element of the group. Otherwise $a_1 \star a_1 = a_1$ leads to $a_1 = e$, or the second possibility $a_1 \star a_1 = e$ closes the group to only two elements. Hence it must be that $a_1 \star a_1 = a_2$, the third element, so that $a_1 \star a_2 = a_2 \star a_1$. It is $\mathcal{Z}_3$, the *cyclic group* of order three, defined by its multiplication table.

$\mathcal{Z}_3$	e	a_1	a_2
e	e	a_1	a_2
a_1	a_1	a_2	e
a_2	a_2	e	a_1

It follows that

$$a_1 \star a_1 \star a_1 = e,$$

which means that a_1 is an element of *order* three, and represents a 120° rotation.

It should be obvious that these generalize to arbitrary n, $\mathcal{Z}_n$, the cyclic group of order n. It is generated by the repeated action of one element a of order n

$$\mathcal{Z}_n : \quad \{ e \,,\, a \,,\, a \star a \,,\, a \star a \star a \,,\, \dots \,,\, (a \star a \star \cdots a \star a)_{n-1} \},$$

with

$$(a \star a \star \cdots a \star a)_k \equiv a^k, \qquad a^n = e.$$

If we write its different elements as $a_j = a^{j-1}$, we deduce that $a_i \star a_j = a_j \star a_i$. A group for which any two of its elements commute with one another is called *Abelian*. Groups which do not have this property are called *non-Abelian*.

Groups of order 4

It is equally easy to construct all possible groups with four elements $\{ e, a_1, a_2, a_3 \}$. There are only two possibilities.

The first is our friend $\mathcal{Z}_4$, the cyclic group of order four, generated by 90° rotations. We note that the generator of this cyclic group has the amusing realization

$$a : \quad z \rightarrow z' = i\, z, \tag{2.5}$$

where z is a complex number. It is easy to see that a is an element of order four that generates $\mathcal{Z}_4$.

The second group of order four, is the *dihedral group* $\mathcal{D}_2$, with the following multiplication table.

$\mathcal{D}_2$	e	a_1	a_2	a_3
e	e	a_1	a_2	a_3
a_1	a_1	e	a_3	a_2
a_2	a_2	a_3	e	a_1
a_3	a_3	a_2	a_1	e

It is Abelian, since the multiplication table is symmetrical about its diagonal. It is sometimes called V (*Vierergruppe*) or Klein's four-group.

It is the first of an infinite family of groups called the dihedral groups of order $2n$, $\mathcal{D}_n$. They have the simple geometrical interpretation of mapping a plane polygon with n vertices into itself. The case $n = 2$ corresponds to the invariance group of a line: a line is left invariant by two 180° rotations about its midpoint, one about the axis perpendicular to the plane of the line, the other about the axis in the plane of the line. It is trivially invariant under a rotation about the line itself. The multiplication table shows three elements of order two, corresponding to these three 180° rotations about any three orthogonal axes in three dimensions. It has many other realizations, for instance in terms of (2×2) matrices

$$\begin{pmatrix} 1 & 0 \\ 0 & 1 \end{pmatrix}, \quad \begin{pmatrix} 1 & 0 \\ 0 & -1 \end{pmatrix}, \quad \begin{pmatrix} -1 & 0 \\ 0 & 1 \end{pmatrix}, \quad \begin{pmatrix} -1 & 0 \\ 0 & -1 \end{pmatrix}. \tag{2.6}$$

An interesting realization involves functional dependence. Consider the four mappings (functions)

$$f_1(x) = x, \quad f_2(x) = -x, \quad f_3(x) = \frac{1}{x}, \quad f_4(x) = -\frac{1}{x}. \tag{2.7}$$

It is easily verified that they close on $\mathcal{D}_2$. These two groups are distinct since $\mathcal{Z}_4$ has an element of order four and $\mathcal{D}_2$ does not.

Groups of order four share one feature that we have not yet encountered: a subset of their elements forms a group, in this case $\mathcal{Z}_2$. The subgroup within $\mathcal{Z}_4$ is $\mathcal{Z}_2$, generated by $(e,\, a^2)$, expressed as

$$\mathcal{Z}_4 \supset \mathcal{Z}_2.$$

As for $\mathcal{D}_2$, it contains three $\mathcal{Z}_2$ subgroups, generated by $(e,\, a_1)$, $(e,\, a_2)$, and $(e,\, a_3)$, respectively. Since $a_2 \star a_1 = a_1 \star a_2$, the elements of the first two $\mathcal{Z}_2$ commute with one another, and we can express $\mathcal{D}_2$ as the *direct product* of the first two commuting subgroups

$$\mathcal{D}_2 = \mathcal{Z}_2 \times \mathcal{Z}_2. \tag{2.8}$$

This is the first and simplest example of a general mathematical construction.

Scholium. *Direct product*

Let $\mathcal{G}$ and $\mathcal{K}$ be two groups with elements $\{g_a\}, a = 1, \ldots, n_g$ and $\{k_i\}$, $i = 1, \ldots, n_k$, respectively. We assemble new elements $(g_a,\, k_i)$, with multiplication rule

$$(g_a,\, k_i)\,(g_b,\, k_j) = (g_a \star g_b,\, k_i \star k_j). \tag{2.9}$$

They clearly satisfy the group axioms, forming a group of order $n_g n_k$ called the *direct (Kronecker) product group* $\mathcal{G} \times \mathcal{K}$. Since $\mathcal{G}$ and $\mathcal{K}$ operate in different spaces, they can always be taken to be commuting subgroups, and the elements of the direct product can be written simply as $g_a k_i$. This construction provides a simple way to generate new groups of higher order.

Scholium. *Lagrange's theorem*

This theorem addresses the conditions for the existence of subgroups. Consider a group $\mathcal{G}$ with elements $\{g_a\}$, $a = 1, 2, \ldots, N$, that contains a subgroup $\mathcal{H}$ with n elements $(h_1, h_2, \ldots, h_n) \equiv \{h_i\}$, $i = 1, 2, \ldots, n < N$,

$$\mathcal{G} \supset \mathcal{H}.$$

Pick an element g_1 of $\mathcal{G}$ that is not in the subgroup $\mathcal{H}$. The n elements of the form $g_1 \star h_i$ are of course elements of $\mathcal{G}$, but *not* of the subgroup $\mathcal{H}$. If any of them were in $\mathcal{H}$, we would have for some i and j

$$g_1 \star h_i = h_j,$$

but this would imply that

$$g_1 = h_j \star (h_i)^{-1}$$

is an element of $\mathcal{H}$, contradicting our hypothesis: the two sets $\{h_i\}$ and $\{g_1 \star h_i\}$ have no element in common. Now we repeat the procedure with another element of $\mathcal{G}$, g_2, not in $\mathcal{H}$, nor in $g_1\mathcal{H}$. The new set $\{g_2 \star h_i\}$ is distinct from both $\{h_i\}$ and $\{g_1 \star h_i\}$, for if they overlap, we would have for some i and j

$$g_1 \star h_i = g_2 \star h_j.$$

This would in turn imply that $g_2 = g_1 \star h_k$, contradicting our hypothesis. We proceed in this way until we run out of group elements after forming the last set $\{g_k \star h_i\}$. Hence we can write the full $\mathcal{G}$ as a *(right) coset decomposition*

$$\begin{aligned} \mathcal{G} &= \{h_i\} + \{g_1 \star h_i\} + \cdots + \{g_k \star h_i\} \\ &\equiv \mathcal{H} + g_1 \star \mathcal{H} + \cdots + g_k \star \mathcal{H}, \end{aligned} \tag{2.10}$$

in which none of the sets overlap: the order of $\mathcal{G}$ must therefore be a multiple of the order of its subgroup. Hence we have Lagrange's theorem.

If a group $\mathcal{G}$ of order N has a subgroup $\mathcal{H}$ of order n, then N is necessarily an integer multiple of n

The integer ratio $k = N/n$ is called the *index* of $\mathcal{H}$ in $\mathcal{G}$. This theorem is about to save us a lot of work.

Let a be an element of a finite group $\mathcal{G}$, and form the sequence

$$a,\ a^2,\ a^3,\ \ldots,\ a^k,\ \ldots,$$

all of which are elements of $\mathcal{G}$. Since it is finite not all can be different, and we must have for some $k > l$

$$a^k = a^l \quad \rightarrow \quad a^{k-l} = e,$$

that is some power of any element of a finite group is equal to the identity element. When $a^n = e$, we say that a is an element of *order* n. Let k be the order of any element b of $\mathcal{G}$; it generates the cyclic subgroup $\mathcal{Z}_k$ of $\mathcal{G}$. By Lagrange's theorem, k must be a multiple of n, the order of the group $\mathcal{G}$.

As a second application, let $\mathcal{G}$ be a group of *prime* order p. Since a prime has no divisor, the order of any of its elements must be either one or p itself. Hence it must be that $\mathcal{G} = \mathcal{Z}_p$: we do not need to construct groups of order 5, 7, 11, ..., they are all cyclic, and Lagrange's theorem tells us that the cyclic groups of prime order have no subgroup.

Groups of order 6

From what we have just learned, we know that there are at least two groups of order six: the cyclic group $\mathcal{Z}_6$, generated by 60° rotations, and the direct product group $\mathcal{Z}_2 \times \mathcal{Z}_3$, but they are the same. To see this, let a and b be the generators of $\mathcal{Z}_3$ and $\mathcal{Z}_2$, respectively, that is $a^3 = b^2 = e$, and $ab = ba$. Consider the element ab of $\mathcal{Z}_2 \times \mathcal{Z}_3$. Clearly, $(ab)^3 = b$, so that $(ab)^6 = (b^2) = (e)$, and ab is of order six. Hence both have an element of order six, and necessarily the two groups must be *isomorphic* to one another

$$\mathcal{Z}_6 = \mathcal{Z}_2 \times \mathcal{Z}_3. \tag{2.11}$$

This is true only because the two factors are relatively primes.

Any other group of order six must contain an order-three element a $(a^3 = e)$. If b is a different element, we find six elements $(e,\ a,\ a^2,\ b,\ ab,\ a^2 b)$. It is easily seen that all must be distinct: they must form a group of order six.

In particular the element b^2 must be e, a or a^2. The latter two choices imply that b is of order three, and lead to a contradiction. Hence b must be an element of order two: $b^2 = e$. Now we look at the element ba. It can be either ab or a^2b. If $ba = ab$, we find that ab must be of order six, a contradiction. Hence this group is non-Abelian. By default it must be that $ba = a^2b$; the multiplication table is now fixed to yield the following dihedral group.

$\mathcal{D}_3$	e	a_1	a_2	a_3	a_4	a_5
e	e	a_1	a_2	a_3	a_4	a_5
a_1	a_1	a_2	e	a_4	a_5	a_3
a_2	a_2	e	a_1	a_5	a_3	a_4
a_3	a_3	a_5	a_4	e	a_2	a_1
a_4	a_4	a_3	a_5	a_1	e	a_2
a_5	a_5	a_4	a_3	a_2	a_1	e

This is the last multiplication table you will see! There is a much better way to present the information it contains. We note that all group elements can be obtained from one order-three element $a_1 \equiv a$, and one order-two element $a_3 \equiv b$; a and b are the *generators* of the group.

Scholium. *Presentation*

A *presentation* of a finite group is a list of its generators, their order, and whatever further relations are needed to specify the group. The same group may have several presentations. For example, $\mathcal{D}_3$ has the simple presentation

$$< a, b \,|\, a^3 = b^2 = e \; ; \; bab^{-1} = a^{-1} > . \qquad (2.12)$$

With two generators, $\mathcal{D}_3$ is said to have *rank* two. $\mathcal{D}_3$ is also the symmetry group of the equilateral triangle.

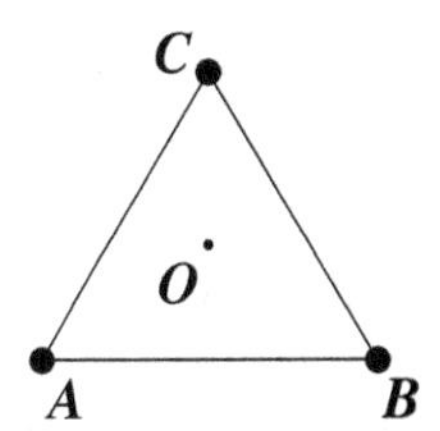

Its element of order three describes the 60° rotation about O, the center of the triangle. Its three elements of order two are the reflections about the three axes OA, OB, and OC, from each vertex to its center.

There is yet a better way to represent this group, since mapping the triangle into itself amounts to permuting its three vertices: the group operations are permutations on the three letters A, B, C which label its vertices. The reflection about the vertical OC axis which interchanges the vertices A and B is the permutation on two letters, called a *transposition*, and denoted by the symbol $(A\ B)$. The other two reflections are simply $(B\ C)$ and $(A\ C)$. The 120° rotations about O are permutations on three letters, denoted by the symbol $(A\,B\,C)$, meaning the transformations $A \to\ B \to\ C \to A$. We call transpositions two-cycles, and we call permutations like $(A\,B\,C)$ three-cycles. The six elements of $\mathcal{D}_3$ break up into the unit element, three transpositions, and two three-cycles. This cycle notation will prove ubiquitously useful. We note that these operations generate *all* 3! permutations on three objects. There are in general $n!$ permutations on n objects; they obviously form a group called the permutation group $\mathcal{S}_n$. In this case we see that $\mathcal{D}_3 = \mathcal{S}_3$.

We can also think of the three vertices as components of a vector on which the cycles act. The group operations are then represented by (3×3) matrices acting on a column vector with entries labeled A, B, and C. We make the identification

$$b = (A\,B) \to \begin{pmatrix} 0 & 1 & 0 \\ 1 & 0 & 0 \\ 0 & 0 & 1 \end{pmatrix}, \qquad a = (A\,B\,C) \to \begin{pmatrix} 0 & 1 & 0 \\ 0 & 0 & 1 \\ 1 & 0 & 0 \end{pmatrix}. \tag{2.13}$$

Group action is simply matrix multiplication. For instance

$$ab = (A\,B\,C)(A\,B) = (B\,C) \to \begin{pmatrix} 0 & 1 & 0 \\ 0 & 0 & 1 \\ 1 & 0 & 0 \end{pmatrix} \begin{pmatrix} 0 & 1 & 0 \\ 1 & 0 & 0 \\ 0 & 0 & 1 \end{pmatrix} = \begin{pmatrix} 1 & 0 & 0 \\ 0 & 0 & 1 \\ 0 & 1 & 0 \end{pmatrix}, \tag{2.14}$$

and so on.

This geometrical interpretation suggests a generalization to the dihedral group $\mathcal{D}_n$ with $2n$ elements, as the symmetry group of n-sided plane polygons.

- The n-polygon is clearly left invariant by the n $(2\pi/n)$ rotations about its center.
- It is also left invariant by n mirror reflections. If n is odd as it is for the triangle, the reflections are about the polygon's n symmetry axes drawn from its center to each vertex. If n is even (as in the square shown below), the reflections are about the $(n/2)$ symmetry axes drawn from the center to the vertices, *and* $(n/2)$ axes drawn from its center to the middle of the opposite faces. Correspondingly, its presentation is

$$\mathcal{D}_n : \quad < a, b\,|\, a^n = b^2 = e\ ;\ bab^{-1} = a^{-1} > . \tag{2.15}$$

Groups of order 8

By now it should be obvious how to construct many groups with eight elements. First of all we have the cyclic group $\mathcal{Z}_8$, and the dihedral group $\mathcal{D}_4$, the symmetry group of the following rectangle.

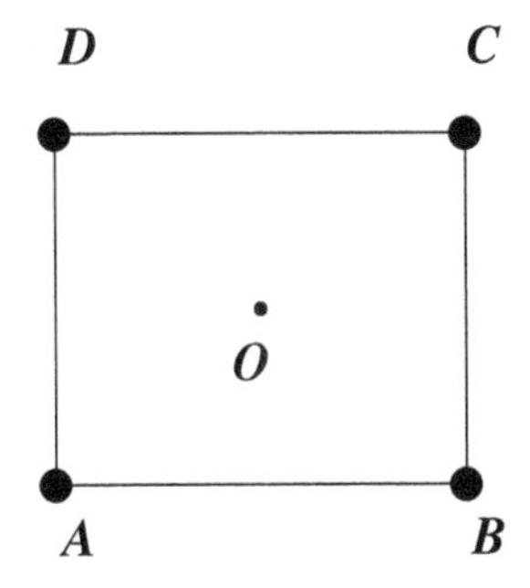

We can also form the direct products $\mathcal{Z}_2 \times \mathcal{Z}_2 \times \mathcal{Z}_2 = \mathcal{D}_2 \times \mathcal{Z}_2$, and $\mathcal{Z}_4 \times \mathcal{Z}_2$.

Are all these groups different? Clearly, the direct product $\mathcal{Z}_4 \times \mathcal{Z}_2$ which contains an element of order four, is different from $\mathcal{Z}_2 \times \mathcal{Z}_2 \times \mathcal{Z}_2$ which has only elements of order two. Furthermore it does not contain any order eight element and cannot be $\mathcal{Z}_8$.

Any other group of order eight must have at least one element of order four. Call it a, with $(a)^4 = e$. Let b be a different element, which is necessarily of order two or four. In this way we generate eight distinct elements

$$\{e,\ a,\ a^2,\ a^3,\ b,\ ab,\ a^2b,\ a^3b\}$$

which must form a group. We consider two possibilities.

The first is where $b^2 = e$. There are three possibilities here: $ba = a^2b$ leads to a contradiction; $ba = ab$ yields the Abelian $\mathcal{Z}_4 \times \mathcal{Z}_2$; and finally $ba = a^3b$, and the element ab is of order two, generating the dihedral group $\mathcal{D}_4$.

The second is where $b^4 = e$. In that case we must have $b^2 = a^2$ which leads also to three possibilities: again, $ba = a^2b$ leads to a contradiction, while $ba = ab$ yields the Abelian group $\mathcal{Z}_4 \times \mathcal{Z}_2$. The third possibility $ba = a^3b$ generates a new group called the quaternion group, $\mathcal{Q}$. Rather than showing its cumbersome multiplication table, it suffices to give its presentation in terms of two order-four generators:

$$\mathcal{Q}: \quad < a, b \,|\, a^4 = e, a^2 = b^2, bab^{-1} = a^{-1} > . \tag{2.16}$$

$\mathcal{Q}$ is a *rank* two group.

Why is it called the quaternion group? Quaternions are a generalization of real and complex numbers which have the property that the norm of a product of two real or complex numbers is equal to the product of their norm, the Hurwitz property. The norm of a complex number $z = x + i\,y$ is defined as

$$N(z) = \sqrt{z\,\bar{z}}, \tag{2.17}$$

where $\overline{z} = x - i\,y$. For any two complex numbers w and z, we have

$$N(z\,w) = N(z)N(w). \tag{2.18}$$

This property is trivial for the real numbers. Quaternions are generalizations of real and complex numbers which satisfy the Hurwitz property. A quaternion is like a complex number with three imaginary units

$$q = x_0 + e_1\,x_1 + e_2\,x_2 + e_3\,x_3, \qquad \overline{q} = x_0 - e_1\,x_1 - e_2\,x_2 - e_3\,x_3. \tag{2.19}$$

Defining

$$N(q) = \sqrt{q\,\overline{q}}, \tag{2.20}$$

it is easy to verify that, for any two quaternions,

$$N(q\,q') = N(q)N(q'), \tag{2.21}$$

as long as

$$(e_1)^2 = (e_2)^2 = (e_3)^2 = -1, \qquad e_1\,e_2 = -e_2\,e_1 = e_3, \tag{2.22}$$

plus cyclic combinations, producing Hamilton's famous quaternion algebra. It can be realized in terms of the (2×2) Pauli (Felix Klein) matrices, as

$$e_j = -i\,\sigma_j, \qquad \sigma_j\,\sigma_k = \delta_{jk} + i\,\epsilon_{jkl}\sigma_l, \tag{2.23}$$

with

$$\sigma_1 = \begin{pmatrix} 0 & 1 \\ 1 & 0 \end{pmatrix}, \qquad \sigma_2 = \begin{pmatrix} 0 & -i \\ i & 0 \end{pmatrix}, \qquad \sigma_3 = \begin{pmatrix} 1 & 0 \\ 0 & -1 \end{pmatrix}. \tag{2.24}$$

The generators of the quaternion group are the multiplication of quaternions by the imaginary units

$$a: \quad q \to q' = e_1\,q, \qquad b: \quad q \to q' = e_2\,q. \tag{2.25}$$

The order-eight quaternion group is, as we shall soon see, the first of the infinite family of the finite *dicyclic* (*binary dihedral*) groups.

Groups of order 9

Rather than dragging the reader through a laborious algebraic procedure, we rely on a theorem that a group (to be proved later) whose order is the square of a prime must be Abelian. There are two different groups of order nine

$$\mathcal{Z}_9\,; \qquad \mathcal{Z}_3 \times \mathcal{Z}_3. \tag{2.26}$$

Groups of order 10

There are no new types of groups at this order. One finds only one cyclic group

$$\mathcal{Z}_{10} = \mathcal{Z}_5 \times \mathcal{Z}_2,$$

since two and five are relative primes, and the dihedral group $\mathcal{D}_5$ which leaves the pentagon invariant.

Groups of order 12

First of all, we have the obvious possibilities

$$\mathcal{Z}_{12}; \qquad \mathcal{D}_6; \qquad \mathcal{Z}_6 \times \mathcal{Z}_2; \qquad \mathcal{Z}_4 \times \mathcal{Z}_3; \qquad \mathcal{D}_3 \times \mathcal{Z}_2; \qquad \mathcal{D}_2 \times \mathcal{Z}_3,$$

but not all are different: since four and three are relative primes, the isomorphism

$$\mathcal{Z}_{12} = \mathcal{Z}_4 \times \mathcal{Z}_3$$

obtains, and also

$$\mathcal{Z}_2 \times \mathcal{Z}_6 = \mathcal{Z}_2 \times \mathcal{Z}_2 \times \mathcal{Z}_3 = \mathcal{D}_2 \times \mathcal{Z}_3,$$

resulting in only two Abelian groups of order twelve, $\mathcal{Z}_4 \times \mathcal{Z}_3$, and $\mathcal{Z}_6 \times \mathcal{Z}_2$, and the non-Abelian

$$\mathcal{D}_6 = \mathcal{D}_3 \times \mathcal{Z}_2. \tag{2.27}$$

There are two new non-Abelian groups at this order. The first is the tetrahedral group, $\mathcal{T}$, the symmetry group of the regular tetrahedron.

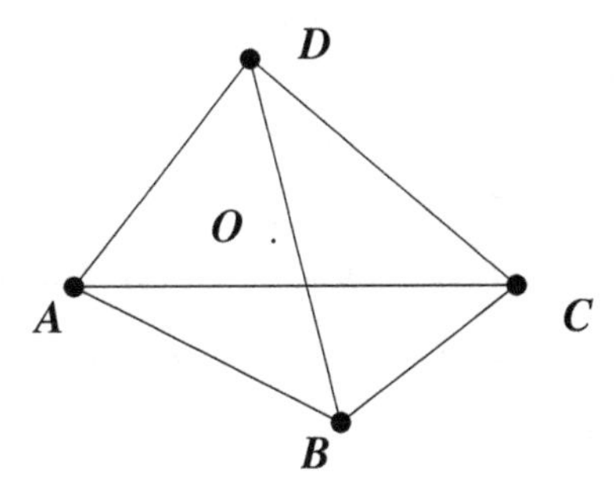

Each of its four faces is an equilateral triangle left invariant by two rotations around the four axes OA, OB, OC, and OD, for a total of eight symmetry operations of order three, as well as three order-two rotations about an axis linking opposite edges, e.g. AD and BC. With the identity, this makes for a size-twelve group with elements of order two and three.

An elegant way to describe this (and any other) finite group is to think of its operations as permutations of the four letters $A,\ B,\ C,\ D$ which label the vertices of the tetrahedron. We have already seen this type of construction applied to the equilateral triangle.

The twelve elements $\mathcal{T}$ neatly break down as the product of two transpositions

$$(A\,B)(C\,D), \qquad (A\,C)(B\,D), \qquad (A\,D)(B\,C), \tag{2.28}$$

and rotations about the axes from the center to each vertex, written as

$$(A\,B\,C), (A\,C\,B); \ \ (A\,B\,D), (A\,D\,B); \tag{2.29}$$

$$(A\,C\,D), (A\,D\,C), (B\,C\,D), (B\,D\,C). \tag{2.30}$$

Together with the unit element, these form the tetrahedral group, $\mathcal{T}$. We easily verify that it is generated by the two (4×4) matrices

$$a = (1\,2)(3\,4) \rightarrow \begin{pmatrix} 0 & 1 & 0 & 0 \\ 1 & 0 & 0 & 0 \\ 0 & 0 & 0 & 1 \\ 0 & 0 & 1 & 0 \end{pmatrix}; \quad b = (1\,2\,3) \rightarrow \begin{pmatrix} 0 & 1 & 0 & 0 \\ 0 & 0 & 1 & 0 \\ 1 & 0 & 0 & 0 \\ 0 & 0 & 0 & 1 \end{pmatrix}, \tag{2.31}$$

which generate the four-dimensional matrix representation of $\mathcal{T}$. We will see shortly that this procedure generalizes since any group operation can be viewed as a permutation. For completeness, we can write its presentation as

$$\mathcal{T}: \ \ < a, b \,|\, a^2 = b^3 = (ab)^3 = e > . \tag{2.32}$$

The tetrahedral group turns out to be isomorphic to a member of an infinite family of groups, called the *alternating* groups $\mathcal{A}_n$ ($\mathcal{T} = \mathcal{A}_4$), generated by the even permutations of n letters.

The second non-Abelian group of order twelve, Γ, is generated by two elements a and b with presentation

$$\Gamma: \ \ < a, b \,|\, a^6 = e; \ b^2 = a^3, bab^{-1} = a^{-1} > . \tag{2.33}$$

The generators can be written as Pauli spin matrices, with

$$a = \frac{1}{2}(\sigma_0 + i\sqrt{3}\sigma_3) = \frac{1}{2}\begin{pmatrix} 1 + i\sqrt{3} & 0 \\ 0 & 1 - i\sqrt{3} \end{pmatrix}; \quad b = i\sigma_1 = \begin{pmatrix} 0 & i \\ i & 0 \end{pmatrix}. \tag{2.34}$$

This group is part of an infinite family of dicyclic groups $\mathcal{Q}_{2n}$, with presentation

$$\mathcal{Q}_{2n}: \ \ < a, b \,|\, a^{2n} = e, b^2 = a^n, bab^{-1} = a^{-1} > . \tag{2.35}$$

Our limited exploration of groups of low order has produced the low-lying members of five infinite families of finite groups: the cyclic family, $\mathcal{Z}_n$ generated by one

Table 2.1. *Finite groups of lowest order*

Order	Group	Isomorphism	Type
2	$\mathcal{Z}_2$		Cyclic
3	$\mathcal{Z}_3$		Cyclic
4	$\mathcal{Z}_4$		Cyclic
4	$\mathcal{D}_2$	$\mathcal{Z}_2 \times \mathcal{Z}_2$	Dihedral, Klein's four-group
5	$\mathcal{Z}_5$		Cyclic
6	$\mathcal{Z}_6$	$\mathcal{Z}_3 \times \mathcal{Z}_2$	Cyclic
6*	$\mathcal{D}_3$	$\mathcal{S}_3$	Dihedral, permutation
7	$\mathcal{Z}_7$		Cyclic
8	$\mathcal{Z}_8$		Cyclic
8	$\mathcal{Z}_4 \times \mathcal{Z}_2$		
8	$\mathcal{Z}_2 \times \mathcal{Z}_2 \times \mathcal{Z}_2$	$\mathcal{D}_2 \times \mathcal{Z}_2$	
8*	$\mathcal{D}_4$		Dihedral
8*	$\mathcal{Q}$	$\mathcal{Q}_4$	Quaternion, dicyclic
9	$\mathcal{Z}_9$		Cyclic
9	$\mathcal{Z}_3 \times \mathcal{Z}_3$		
10	$\mathcal{Z}_{10}$	$\mathcal{Z}_5 \times \mathcal{Z}_2$	Cyclic
10*	$\mathcal{D}_5$		Dihedral
11	$\mathcal{Z}_{11}$		Cyclic
12	$\mathcal{Z}_{12}$	$\mathcal{Z}_4 \times \mathcal{Z}_3$	Cyclic
12	$\mathcal{Z}_6 \times \mathcal{Z}_2$	$\mathcal{D}_2 \times \mathcal{Z}_3$	
12*	$\mathcal{D}_6$	$\mathcal{D}_3 \times \mathcal{Z}_2$	Dihedral
12*	$\mathcal{Q}_6$		Dicyclic, binary dihedral
12*	$\mathcal{T}$	$\mathcal{A}_4$	Tetrahedral, alternating

element of order n, the dihedral groups $\mathcal{D}_n$ of order $2n$, generated by one element of order n and one of order 2. $\mathcal{D}_n$ is non-Abelian for $n > 2$. At order six, we found the permutation group on three objects, $\mathcal{S}_3$, a member of the infinite family $\mathcal{S}_n$ of permutations on n objects. At order eight we found the quaternion group $\mathcal{Q}_4$, of the dicyclic group family $\mathcal{Q}_{2n}$. At order twelve, we found another dicyclic group $\mathcal{Q}_6$, as well as the tetrahedral group $\mathcal{T}$. Later we will identify it as $\mathcal{A}_4$, a member of the infinite family $\mathcal{A}_n$, the group of even permutations on n objects. At each order we encountered in addition groups that were direct products of groups of lesser order.

We summarize our results in two tables. In Table 2.1, non-Abelian groups are marked with a star. We show the names of the group as well as isomorphisms. Table 2.2 gives the total number of finite groups up to order 200, with the number of Abelian groups indicated in brackets, following Lomont [15].

Table 2.2 is pretty amazing. We note, as expected, that there is only one group of prime order. Another regularity is the existence of only two Abelian groups of squared prime order. What about the order equal to twice a prime? Then there are some stupendous numbers, where there are large numbers on non-Abelian groups.

Table 2.2. *Number of groups of order up to* 200

Order	$N[N_{Abel}]$	Order	$N[N_{Abel}]$	Order	$N[N_{Abel}]$	Order	$N[N_{Abel}]$	Order	$N[N_{Abel}]$
1	1 [1]	41	1 [1]	81	15 [5]	121	2 [2]	161	1 [1]
2	1 [1]	42	6 [1]	82	2 [1]	122	2 [1]	162	55 [5]
3	1 [1]	43	1 [1]	83	1 [1]	123	1 [1]	163	1 [1]
4	2 [2]	44	4 [2]	84	15 [2]	124	4 [2]	164	5 [2]
5	1 [1]	45	2 [2]	85	1 [1]	125	5 [3]	165	2 [1]
6	2 [1]	46	2 [1]	86	2 [1]	126	16 [2]	166	2 [1]
7	1 [1]	47	1 [1]	87	1 [1]	127	1 [1]	167	1 [1]
8	5 [3]	48	52 [5]	88	12 [3]	128	2328 [15]	168	57 [3]
9	2 [2]	49	2 [2]	89	1 [1]	129	2 [1]	169	2 [2]
10	2 [1]	50	2 [2]	90	10 [2]	130	4 [1]	170	4 [1]
11	1 [1]	51	1 [1]	91	1 [1]	131	1 [1]	171	5 [2]
12	5 [2]	52	5 [2]	92	4 [2]	132	10 [2]	172	4 [2]
13	1 [1]	53	1 [1]	93	2 [1]	133	1 [1]	173	1 [1]
14	2 [1]	54	15 [3]	94	2 [1]	134	2 [1]	174	4 [1]
15	1 [1]	55	2 [1]	95	1 [1]	135	5 [3]	175	2 [2]
16	14 [5]	56	13 [3]	96	230 [7]	136	15 [3]	176	42 [5]
17	1 [1]	57	2 [1]	97	1 [1]	137	1 [1]	177	1 [1]
18	5 [2]	58	2 [1]	98	5 [2]	138	4 [1]	178	2 [1]
19	1 [1]	59	1 [1]	99	2 [2]	139	1 [1]	179	1 [1]
20	5 [2]	60	13 [2]	100	16 [4]	140	11 [2]	180	37 [4]
21	2 [1]	61	1 [1]	101	1 [1]	141	1 [1]	181	1 [1]
22	2 [1]	62	2 [1]	102	4 [1]	142	2 [1]	182	4 [1]
23	1 [1]	63	4 [2]	103	1 [1]	143	1 [1]	183	2 [1]
24	15 [3]	64	267 [11]	104	14 [3]	144	197 [10]	184	12 [3]
25	2 [2]	65	1 [1]	105	2 [1]	145	1 [1]	185	1 [1]
26	2 [1]	66	4 [1]	106	2 [1]	146	2 [1]	186	6 [1]
27	5 [3]	67	1 [1]	107	1 [1]	147	6 [2]	187	1 [1]
28	4 [2]	68	5 [2]	108	45 [6]	148	5 [2]	188	4 [2]
29	1 [1]	69	1 [1]	109	1 [1]	149	1 [1]	189	13 [3]
30	4 [1]	70	4 [1]	110	6 [1]	150	13 [2]	190	4 [1]
31	1 [1]	71	1 [1]	111	2 [1]	151	1 [1]	191	1 [1]
32	51 [7]	72	50 [6]	112	43 [5]	152	12 [3]	192	1543 [11]
33	1 [1]	73	1 [1]	113	1 [1]	153	2 [2]	193	1 [1]
34	2 [1]	74	2 [1]	114	6 [1]	154	4 [1]	194	2 [1]
35	1 [1]	75	3 [2]	115	1 [1]	155	2 [1]	195	2 [1]
36	14 [4]	76	4 [2]	116	5 [2]	156	18 [2]	196	17 [4]
37	1 [1]	77	1 [1]	117	4 [2]	157	1 [1]	197	1 [1]
38	2 [1]	78	6 [1]	118	2 [1]	158	2 [1]	198	10 [2]
39	2 [1]	79	1 [1]	119	1 [1]	159	1 [1]	199	1 [1]
40	14 [3]	80	52 [5]	120	47 [3]	160	238 [7]	200	52 [6]

It gets worse; for example, there are 20 154 non-Abelian groups of order 384, but only 11 of order 390.

To go further, more systematic methods need to be developed. Can groups be derived from basic building blocks? Are some groups more equal than others? Believe it or not, none of the groups we have constructed so far are viewed as "fundamental" by mathematicians: the smallest such fundamental non-Abelian group happens to have sixty elements, corresponding to the even permutations on five objects!

2.3 Permutations

The systematic study of groups was started by Évariste Galois as permutations of roots of polynomial equations (more on this later). Permutations do play a central role in the study of finite groups because any finite group of order n can be viewed as permutations over n letters. We have seen several examples of this. In this section we develop the idea systematically.

A permutation on n identified objects, $1, 2, 3, \ldots, k, \ldots, n$, shuffles them into a different order, say, $a_1, a_2, a_3, \ldots, a_k, \ldots, a_n$. This transformation $k \to a_k$, $k = 1, 2, \ldots, n$, is represented by the symbol

$$\begin{pmatrix} 1 & 2 & \cdots & n \\ a_1 & a_2 & \cdots & a_n \end{pmatrix}. \tag{2.36}$$

It clearly satisfies all the group axioms, for instance

$$\begin{pmatrix} 1 & 2 & \cdots & n \\ a_1 & a_2 & \cdots & a_n \end{pmatrix}\begin{pmatrix} a_1 & a_2 & \cdots & a_n \\ b_1 & b_2 & \cdots & b_n \end{pmatrix} = \begin{pmatrix} 1 & 2 & \cdots & n \\ b_1 & b_2 & \cdots & b_n \end{pmatrix}, \tag{2.37}$$

and so on. Every permutation has a unique inverse, and the identity permutation leaves the array invariant. Clearly, the $n!$ permutations on n letters form a group, the *symmetric group*, denoted by $\mathcal{S}_n$.

We use the more concise *cycle notation* for permutations, which we have already encountered. A permutation that shuffles $k < n$ objects into themselves, leaving the rest untouched, is called a *k-cycle*. Using permutations on four objects as examples, we write

$$\begin{pmatrix} 1 & 2 & 3 & 4 \\ 2 & 3 & 4 & 1 \end{pmatrix} \sim (1\,2\,3\,4). \tag{2.38}$$

Reading from left to right it reads $1 \to 2$, $2 \to 3$, $3 \to 4$, $4 \to 1$. Another four-cycle element is

$$\begin{pmatrix} 1 & 2 & 3 & 4 \\ 3 & 4 & 2 & 1 \end{pmatrix} \sim (1\,3\,2\,4). \tag{2.39}$$

On the other hand, we have

$$\begin{pmatrix} 1 & 2 & 3 & 4 \\ 2 & 1 & 3 & 4 \end{pmatrix} \sim (1\,2)(3)(4), \tag{2.40}$$

$$\begin{pmatrix} 1 & 2 & 3 & 4 \\ 2 & 1 & 4 & 3 \end{pmatrix} \sim (1\,2)(3\,4), \tag{2.41}$$

as well as

$$\begin{pmatrix} 1 & 2 & 3 & 4 \\ 3 & 1 & 2 & 4 \end{pmatrix} \sim (1\,3\,2)(4). \tag{2.42}$$

One-cycles, included here for clarity, will hereon be omitted.

The *degree* of a permutation is the number of objects on which it acts. Permutations can be organized into several types called *classes* (see later for a formal definition). On these four objects the classes are the transpositions on two objects, (xx); the product of two transpositions, (xx)(xx), three-cycles, (xxx); and four-cycles, (xxxx).

Any permutation can be decomposed in this manner. Consider an arbitrary permutation on n objects. Pick one object, say k, and follow what happens to the sequence $k \to a_k^{[1]} \to a_k^{[2]} \to \cdots \to k$ as we go through the permutation. The sequence terminates and k goes back into itself since the number of objects is finite. One of two things can happen. Either we need to go over all n objects, and the permutation is written as an n-cycle

$$(k\, a_k^{[1]}\, a_k^{[2]}\, \cdots\, a_k^{[n-1]}), \tag{2.43}$$

or it terminates earlier, after only r iterations. In that case, let m be one of $(n-r)$ objects unaccounted for, and follows its fate through the permutation $m \to a_m^{[1]} \to a_m^{[2]} \to \cdots \to a_m^{[s-1]} \to m$, such that it returns to itself after s iterations. Again one of two things can happen. If $s = (n-r)$, all objects are accounted for and the permutation is written as

$$(k\, a_k^{[1]}\, a_k^{[2]}\, \cdots\, a_k^{[r-1]})(m\, a_m^{[1]}\, a_m^{[2]}\, \cdots\, a_m^{[n-r-1]}). \tag{2.44}$$

If $s < (n-r)$, we need to account for the remaining $(n-r-s)$ objects. By repeating this procedure until we run out of objects, we arrive at the cycle decomposition

$$(k\, a_k^{[1]}\, a_k^{[2]}\, \cdots\, a_k^{[r-1]})(m\, a_m^{[1]}\, a_m^{[2]}\, \cdots\, a_m^{[s-1]})\cdots(t\, a_t^{[1]}\, a_t^{[2]}\, \cdots\, a_t^{[u-1]}), \tag{2.45}$$

with $r+s+\cdots+u = n$. This decomposition is unique because it does not depend on the element we started from, nor on the choices we made at each stage. We just proved the following.

Every permutation can be uniquely resolved into cycles which operate on mutually exclusive sets

Any permutation can also be decomposed as a (non-commutative) product of transpositions

$$(a_1\, a_2\, \cdots\, a_n) = (a_1\, a_2)(a_1\, a_3)\cdots(a_1\, a_n). \tag{2.46}$$

If the number of transpositions in this decomposition is *even (odd)*, the permutation is said to be *even (odd)*. The product of two even permutations is an even permutation, and even permutations form a subgroup of order $n!/2$, called the *alternating group* $\mathcal{A}_n$.

In our example, $\mathcal{T} = \mathcal{A}_4$, the *even* permutations are the product of two transpositions like $(1\,2)(3\,4)$, and three-cycles like $(1\,2\,3) = (1\,2)(1\,3)$, also the product of two transpositions. The elements of $\mathcal{S}_4$ not in the subgroup $\mathcal{A}_4$ are the odd permutations, that is those containing one transposition (xx), and one four-cycle (xxxx). These do not form a group since the product of two odd permutations is an even permutation.

Consider a permutation on n objects described by α_1 one-cycles, α_2 two-cycles, ..., α_k k-cycles, so that

$$n = \sum_{j=1}^{k} j\alpha_j. \tag{2.47}$$

We want to find out the number of permutations with this cycle structure. This is like finding the number of ways of putting n distinct objects into α_1 one-dimensional boxes, α_2 two-dimensional boxes, ... There will be two types of redundancies: with α_j boxes of the same type, we expect $\alpha_j!$ redundancies from shuffling the boxes. In addition, there are j ways to put the objects in each j-dimensional box, and since there are α_j such boxes this will produce a further redundancy of j^{α_j}. It follows that the number of permutations with this cycle structure is

$$\frac{n!}{\prod_j^k j^{\alpha_j}\alpha_j!}. \tag{2.48}$$

For example, the twenty-four elements of $\mathcal{S}_4$ break up into the identity, six two-cycles, three double transpositions, eight three-cycles, and six four-cycles.

Let $\mathcal{G}$ be a group with elements g_a, $a = 1, 2, \ldots, n$. Pick an element g_b, and form the n elements $g_1\,g_b$, $g_2\,g_b$, ..., $g_n\,g_b$, and associate to g_b the permutation

$$P_b = \begin{pmatrix} g_1 & g_2 & g_3 & \cdots & g_n \\ g_1 g_b & g_2 g_b & g_3 g_b & \cdots & g_n g_b \end{pmatrix}. \tag{2.49}$$

This can be done for every element of $\mathcal{G}$, to produce n *distinct* permutations. In addition, this construction leaves the multiplication table invariant,

$$g_a g_b = g_c \quad \rightarrow P_a P_b = P_c, \tag{2.50}$$

and the $\{P_a\}$ form a group representation of $\mathcal{G}$. Since the P_a satisfy the group axioms, they form a subgroup of the permutations on n objects. Thus we have Cayley's theorem.

Every group of finite order n is isomorphic to a subgroup of the permutation group $\mathcal{S}_n$.

The permutations P_a of degree n form a representation called the *regular* representation, where each P_a is a $(n \times n)$ matrix acting on the n objects arranged as a column matrix. These are very large matrices. The regular representation of the order-eight quaternion group is formed by (8×8) matrices, yet we have constructed a representation of the same group in terms of (2×2) matrices. Clearly the regular representation is overkill. Such a representation is said to be *reducible*. We will soon develop systematics to find the *irreducible* representations of the finite groups, but first we need to introduce more concepts that will prove useful in the characterization of finite groups.

2.4 Basic concepts

In this section, we introduce many of the concepts and tools which have proven crucial for a systematic study of finite groups (see Carmichael's [2], and Hall's [9] books). In addition we provide *scholia* (marginal explanatory notes), of the type found in ancient scientific books and texts.

2.4.1 Conjugation

There is a very important operation which has the property of preserving any multiplication table. Let $\mathcal{G}$ be a group with elements g_a. We define the *conjugate* of any element g_a with respect to any other group element g as

$$\tilde{g}_a \equiv g\, g_a\, g^{-1}. \tag{2.51}$$

Clearly, such a transformation maps the multiplication table into itself

$$g_a\, g_b = g_c \;\rightarrow\; \tilde{g}_a\, \tilde{g}_b = \tilde{g}_c, \tag{2.52}$$

since

$$g\, g_a\, g^{-1}\, g\, g_b\, g^{-1} = g(g_a\, g_b)g^{-1} = g\, g_c\, g^{-1}. \tag{2.53}$$

This is an example of a mapping which leaves the multiplication table invariant, called a *homomorphism*. In addition, since it is one-to-one, it is a special homomorphism called an *automorphism*. This particular mapping is called *inner* since it is generated by one of its elements. If two elements g_a and g_b commute, then g_a is self-conjugate with respect to g_b and vice versa.

Conjugacy has the important property that it does not change the cycle structure of a group element. To see this, consider a group element, written as a permutation

$P = (k\,l\,m\,p\,q)$ that cycles through five elements of a group of order n. To verify that

$$g\,P\,g^{-1} = (g(k)\,g(l)\,g(m)\,g(p)\,g(q)), \tag{2.54}$$

where $g(k)$ is the action of group operation g on group element k, etc., it suffices to note that since

$$g\,P\,g^{-1}g(k) = g\,P(k) = g(l), \tag{2.55}$$

conjugacy by g replaces $k \to l$ by $g(k) \to g(l)$. We conclude that conjugacy does not alter the cycle decomposition of the permutation: conjugate elements have the same cycle structure.

Scholium. *Classes*

We can use conjugacy to organize the elements of any group into non-overlapping sets. Pick any element g_b, and conjugate it with respect to every element of $\mathcal{G}$

$$C_b : \qquad \widetilde{g}_b = g_a g_b g_a^{-1}, \;\; \forall\, g_a \in \mathcal{G}, \tag{2.56}$$

thus generating a set of elements C_b, called a *class*. Now pick a second element g_c of $\mathcal{G}$ that is not in C_b, and form a new class

$$C_c : \qquad \widetilde{g}_c = g_a g_c g_a^{-1}, \;\; \forall\, g_a \in \mathcal{G}. \tag{2.57}$$

The classes C_b and C_c have no elements in common, for if they did, it would imply that g_c is in C_b, which contradicts our hypothesis. The procedure continues until we have exhausted all the group elements. We conclude that any finite group of order n can be decomposed into $k < n$ classes.

Since this construction does not depend on our choice of elements $g_b, g_c, \ldots$, it is unique, and k, the number of classes is a characteristic of the group. Any Abelian group of order n has n classes, each containing one element. The identity element is always in a class by itself, traditionally denoted as C_1.

Elements of a given class have the same cycle structure, but permutations with the same cycle structure may belong to different classes.

Scholium. *Normal subgroup*

Let $\mathcal{H}$ with elements h_i be a subgroup of $\mathcal{G}$

$$\mathcal{G} \supset \mathcal{H}.$$

Some subgroups have a very special property. If the *conjugate* of h_i with respect to every element g of $\mathcal{G}$ remains in $\mathcal{H}$, that is

$$g\,h_i\,g^{-1} = h_j \in \mathcal{H}, \quad \forall\; g \in \mathcal{G}, \tag{2.58}$$

the subgroup $\mathcal{H}$ is said to be a *normal subgroup*. In that case, mathematicians use the symbol

$$\mathcal{H} \lhd \mathcal{G}, \qquad (\mathcal{G} \rhd \mathcal{H}). \tag{2.59}$$

Most groups have normal subgroups; those without normal subgroups, called *simple* groups are very special, and form the building blocks for all the others.

Scholium. *Quotient group*

To understand what is so special about a normal subgroup, consider the homomorphic mapping of the group $\mathcal{G}$ into a smaller group $\mathcal{K}$,

$$\mathcal{G} \to \mathcal{K}, \qquad g_a \to k_a,$$

which preserves the multiplication table: $g_a g_b = g_c$ implies that $k_a k_b = k_c$. Let $\mathcal{H}$ be the subset of elements of $\mathcal{G}$ that are mapped into the identity (the *kernel* of the mapping)

$$\mathcal{H} \to 1.$$

The set $h_i \in \mathcal{H}$ clearly is a subgroup of $\mathcal{G}$, and furthermore, since $g_a h_i g_a^{-1} \to k_a k_a^{-1} = 1$, it is a normal subgroup.

Now take the argument the other way. Start from $\mathcal{G}$ and assume it has a normal subgroup $\mathcal{H}$. Consider any elements $g_a h_i$, and $g_b h_j$ of two different cosets $g_a \mathcal{H}$ and $g_b \mathcal{H}$, and form their product

$$(g_a h_i)(g_b h_j) = g_a g_b g_b^{-1} h_i g_b h_j = g_a g_b (g_b^{-1} h_i g_b) h_j = g_a g_b \tilde{h}_i h_j, \tag{2.60}$$

where the conjugate $\tilde{h}_i$ is an element of the normal subgroup $\mathcal{H}$. Hence the product of cosets

$$g_a \mathcal{H} \otimes g_b \mathcal{H} = g_c \mathcal{H}, \tag{2.61}$$

satisfies the group property. To complete the group structure, the unit element is the subgroup $\mathcal{H}$ itself, and if $g_a g_b = h_i$, the cosets $g_a \mathcal{H}$ and $g_b \mathcal{H}$ are inverses of one another. The group axioms are satisfied (associativity follows easily), and we have constructed a group with n_g / n_h elements. This group is called the *quotient (factor) group*,

$$\mathcal{G}/\mathcal{H}.$$

It is a homomorphic image of $\mathcal{G}$, but *not* a subgroup of $\mathcal{G}$.

Suppose $\mathcal{H}$ is *not* the largest subgroup of $\mathcal{G}$. We split its generators in the form

$$\mathcal{G}: \ \{ g_a \,, \, g_\alpha \,, \, h_i \}, \tag{2.62}$$

with

$$\mathcal{G} \supset \mathcal{H}' \rhd \mathcal{H}, \qquad \mathcal{H}': \ \{ g_\alpha \, , \, h_i \}, \tag{2.63}$$

since $\mathcal{H}$ is a normal subgroup of $\mathcal{H}'$. We can form two factor groups

$$\mathcal{G}/\mathcal{H}: \ \{ g_a \mathcal{H} \, , \, g_\alpha \mathcal{H} \, , \, \mathcal{H}(=1) \}, \qquad \mathcal{K} \equiv \mathcal{H}'/\mathcal{H}: \ \{ g_\alpha \mathcal{H} \, , \, \mathcal{H}(=1) \}. \tag{2.64}$$

Clearly $\mathcal{K}$ is a subgroup of $\mathcal{G}/\mathcal{H}$

$$\mathcal{K} \subset \mathcal{G}/\mathcal{H}. \tag{2.65}$$

Since $\mathcal{H}$ is a normal subgroup, conjugation satisfies

$$\widetilde{g_\alpha \mathcal{H}} = \widetilde{g}_\alpha \mathcal{H}, \qquad \widetilde{g}_\alpha = g_a g_\alpha g_a^{-1}. \tag{2.66}$$

Thus if $\widetilde{g}_\alpha$ is an element of $\mathcal{H}'$ ($\in \{g_\alpha\}$), $\mathcal{K}$ is a normal subgroup of the quotient group; otherwise it is not. But then, $\mathcal{H}'$ must itself be a normal subgroup of $\mathcal{G}$: $\mathcal{G}/\mathcal{H}$ has a normal subgroup only if $\mathcal{H}$ is not maximal.

The quotient group $\mathcal{G}/\mathcal{H}$ has no normal subgroup as long as $\mathcal{H}$ is the maximal normal subgroup of $\mathcal{G}$. This fact opens the way for a systematic decomposition of all groups in terms of simple groups.

2.4.2 Simple groups

Finite groups with normal subgroups are dime-a-dozen, and not viewed as fundamental by mathematicians. Galois, the founder of group theory, split groups into two types: *simple groups* which have *no* normal subgroup, and the rest.

This was a pretty important distinction to Galois, since he related the simplicity of $\mathcal{A}_5$, the alternating group on five objects, to the impossibility of finding a formula that solves the quintic equation by radicals (using square roots, cube roots, ...)!

So far we have encountered only one family of simple groups, the cyclic groups of prime order $\mathcal{Z}_p$, which have no subgroups. One of the great triumphs of modern mathematics has been the comparatively recent classification of *all* simple groups. Typically simple groups are very large: for example, $\mathcal{A}_5$ contains 60 elements, and the next largest (not counting the cyclic groups) has 168 elements! We shall return to their classification once we have covered the theory of Lie algebras. For the moment we just list the types of simple groups.

- Cyclic groups of prime order, $\mathcal{Z}_p$.
- Alternating groups on five or more objects, $\mathcal{A}_n$, $n \geq 5$.
- Infinite families of *groups of Lie type* (to be defined later).
- Twenty-six *sporadic groups*.

All finite groups can be systematically constructed from simple groups, by means of a procedure we now proceed to describe.

Scholium. *Composition series*

We can decompose any group in terms of its normal subgroups by seeking its *largest* normal subgroup $\mathcal{H}_1$. If it has a normal subgroup, the group elements split into $\mathcal{H}_1$ and the quotient group $\mathcal{G}/\mathcal{H}_1$. We repeat the procedure applied to $\mathcal{H}_1$: we seek its maximal normal subgroup $\mathcal{H}_2$, and splits its elements into $\mathcal{H}_2$, and $\mathcal{H}_1/\mathcal{H}_2$, and so on. This procedure yields the *composition series* of the group. It always terminates on the identity element

$$\mathcal{G} \rhd \mathcal{H}_1 \rhd \mathcal{H}_2 \rhd \cdots \rhd \mathcal{H}_k \text{ (simple subgroup)} \rhd e, \tag{2.67}$$

generating the quotient subgroups

$$\mathcal{G}/\mathcal{H}_1, \mathcal{H}_1/\mathcal{H}_2, \ldots, \mathcal{H}_k. \tag{2.68}$$

In general quotient groups need not be simple, but those in the composition series are because they are constructed at each step out of the *maximal* normal subgroup. This procedure dissects a group down in term of its simple constituents. That is the reason mathematicians consider the simple groups to be the fundamental entities out of which all other finite groups can be built. In the special case where all the quotient groups are the cyclic groups of prime order, the group is said to be *soluble* or *solvable* (Galois's terminology).

The wonderful thing is that, although a group can have more than one composition series, i.e. several normal subgroups to choose from, composition series have some invariant features: the number of steps to the identity is always the same, and the order of the quotient groups, called the *composition indices* are also the same! They are therefore characteristics of the group.

The quaternion group has several subgroups, and they are all normal and Abelian; the largest is $\mathcal{Z}_4$, and three different $\mathcal{Z}_2$. Its three-step composition series is

$$\mathcal{Q} \rhd \mathcal{Z}_4 \rhd \mathcal{Z}_2 \supset e, \tag{2.69}$$

with quotient groups

$$\mathcal{Q}/\mathcal{Z}_4 = \mathcal{Z}_2; \qquad \mathcal{Z}_4/\mathcal{Z}_2 = \mathcal{Z}_2; \qquad \mathcal{Z}_2. \tag{2.70}$$

Its composition indices are 2, 2, 2, and the length is three.

We encourage the reader to determine the composition series and normal subgroups of $\mathcal{S}_4$ and $\mathcal{A}_4$.

𝔖𝔠𝔥𝔬𝔩𝔦𝔲𝔪. *Derived (commutator) subgroup*

Let a and b be any two elements of a group $\mathcal{G}$. We define their *commutator* as

$$[a,\, b] \;\equiv\; a^{-1}b^{-1}ab, \tag{2.71}$$

which has the virtue to be equal to the unit element if they commute. Furthermore, since $[b,\, a] = ([a,\, b])^{-1}$ the product of all possible commutators form a group, $\mathcal{G}'$ called the *derived or commutator subgroup*. It is easy to see that the derived subgroup is a normal subgroup. We have

$$\widetilde{[a,\, b]} = g\left(a^{-1}b^{-1}ab\right)g^{-1} = [\widetilde{a},\, \widetilde{b}], \tag{2.72}$$

where

$$\widetilde{a} = g\,ag^{-1}, b = g\,bg^{-1}, \ldots, \tag{2.73}$$

so that conjugation maps the derived subgroup into itself. As an example, to compute $\mathcal{D}'_n$, the derived subgroup of the dihedral group with elements of the form $a^m,\ a^m b,\ m = 1, 2, \ldots, n$ with $b^2 = e$ and $aba = b$. Explicitly, we find

$$[a^m,\, a^k b] = a^{-2m}, \qquad [a^m b,\, a^k b] = a^{2(m-k)}, \tag{2.74}$$

all other commutators being either the unit element or the inverse of the above. Hence the commutator subgroup of $\mathcal{D}_n$ is the cyclic group $\mathcal{Z}_{n/2}$ for n even, and $\mathcal{Z}_n$ if n is odd.

There are three possibilities: $\mathcal{G}'$ is the same as $\mathcal{G}$, then $\mathcal{G}$ is a *perfect group*; $\mathcal{G}' = e$, $\mathcal{G}$ is Abelian; otherwise $\mathcal{G}' \lhd \mathcal{G}$, and $\mathcal{G}$ is not a simple group. With a distinct commutator subgroup, $\mathcal{D}_n$ is not a simple group. Hence a group can be perfect (the fact that it is its own commutator subgroup does not mean it does not have normal subgroups), while a non-Abelian simple group must be perfect!

The derived subgroup of $\mathcal{S}_n$ is $\mathcal{A}_n$. That of $\mathcal{A}_4$ is the Klein four-group, and the derived subgroup of the quaternion group $\mathcal{Q}$ is $\mathcal{Z}_2$. None of these groups is simple. We leave it to the reader to find the derived subgroup of Γ, the other group of order twelve.

2.4.3 Sylow's criteria

Let p be a prime number. A *p-group* is a group for which the order of every element is a power of p. They can be Abelian, such as the cyclic groups, or non-Abelian: $\mathcal{D}_4$ whose elements are of order two and four, is a Sylow two-group.

Let $\mathcal{G}$ be a group of order n. Decompose its order in powers of primes, say, $n = p^m r$, where p is a prime, and r is integer not multiple of p. Then, there is a series of theorems, due to Ludwig Sylow that state the following.

- $\mathcal{G}$ contains n_p (Sylow) p-subgroups of order p^m, $\mathcal{G}_p^i$, $i = 1, 2 \ldots, n_p$.
- All $\mathcal{G}_p^i$ are isomorphic to one another, related by $\mathcal{G}_p^j = g\mathcal{G}_p^k g^{-1}$, $g \in \mathcal{G}$.
- n_p is a divisor of r.
- $n_p = 1 \bmod(p)$.

These theorems put very strong restrictions of the possible groups of a given order.

Consider groups of order $n = pq$ where both p and q are prime. Sylow tells us these groups have n_p subgroups $\mathcal{G}_p$ of order p, and n_q subgroups $\mathcal{G}_q$ of order q. These are necessarily cyclic groups. Sylow requires that n_p divide q, that is $n_p = 1$ or q. In addition, $n_p = 1, 1+p, \ldots$, so that if $p > q$, the unique solution is $n_p = 1$. On the other hand, $n_q = 1, 1+q, 1+2q, \ldots$ are possible, as long as n_q divides p. There is always one solution with $n_q = 1$, corresponding to

$$\mathcal{Z}_{pq} = \mathcal{Z}_p \times \mathcal{Z}_q.$$

We have already seen this case: if a, b are elements of $\mathcal{Z}_p$ and $\mathcal{Z}_q$ respectively, the element (a, b) is clearly of order pq.

The second possibility is when $n_q = p$, but this is possible only if $p = 1 \bmod(q)$. This proves that there are at most two groups of order pq, if $p = 1 \bmod(q)$, and only one otherwise. If there is only one Sylow subgroup ($n_p = 1$), it is a normal subgroup.

Let us consider a few applications. There is only one group of order $15 = 3 \cdot 5$, since $5 \neq 1 \bmod(3)$, $\mathcal{Z}_{15} = \mathcal{Z}_3 \times \mathcal{Z}_5$.

On the other hand, there are two groups of order $21 = 7 \cdot 3$, since $7 = 1 \bmod(3)$. The second group is called the Frobenius group. It is a subgroup of the permutation group on seven objects, with one Sylow seven-subgroup generated by $(1\,2\,3\,4\,5\,6\,7)$, and seven Sylow three-subgroups generated by $(2\,3\,5)$, and $(4\,7\,6)$.

Sylow analysis was very useful in the hunt for simple groups. The reason is that a group with only *one* Sylow subgroup cannot be simple, since conjugation maps it into itself. For instance, any group of order $84 = 2^2 \cdot 3 \cdot 7$ contains n_7 Sylow seven-subgroups, where n_7 is required to divide 12 and be $1, 8, 15, \ldots$. The only solution is $n_7 = 1$, and there are no simple groups of order 84.

2.4.4 Semi-direct product

There is another way to construct a group out of two groups. It is more subtle than the direct product construction. It requires two groups $\mathcal{G}$ and $\mathcal{K}$, *and* the action of one on the other.

Assume that $\mathcal{G}$ acts on $\mathcal{K}$ in the sense that it sends any of its elements $k \in \mathcal{K}$ into another one

$$\mathcal{G}: \quad k \to k' \equiv k^{g}, \tag{2.75}$$

where $g \in \mathcal{G}$. For any two group elements

$$(k^{g_1})^{g_2} = k^{g_1 g_2} = k^{g'}. \tag{2.76}$$

For any two elements $k_a, k_b \in \mathcal{K}$,

$$(k_a k_b)^{g} = k_a^{g} k_b^{g}, \tag{2.77}$$

so that the $\mathcal{K}$-multiplication table is preserved (homomorphism)

$$k_a\, k_b = k_c \;\to\; (k_a)^{g} (k_b)^{g} = k_c^{g}. \tag{2.78}$$

Then we can verify that the elements (k_a, g_i) form a group with composition law

$$(k_a\,, g_i\,) \cdot (k_b\,,\, g_j\,) \;\equiv\; (\,(k_a)^{g_j}\, k_b\,,\; g_i\, g_j). \tag{2.79}$$

This new group of order $(n_k\, n_g)$ is called the *semi-direct product* of $\mathcal{G}$ with $\mathcal{K}$, and denoted by

$$\mathcal{K} \rtimes \mathcal{G}. \tag{2.80}$$

The funny symbol is to distinguish it from the direct product construction.

For finite groups, one can think of several possibilities: a natural choice is to take $\mathcal{G}$ to be the automorphism group of $\mathcal{K}$, sometimes called $\mathcal{A}ut\ \mathcal{K}$, in which case it is called the *holomorph* of the group

$$Hol(\mathcal{K}) \;\equiv\; \mathcal{K} \rtimes \mathcal{A}ut\ \mathcal{K}. \tag{2.81}$$

A second possibility is to take $\mathcal{K}$ to be a normal subgroup of $\mathcal{G}$. Then $\mathcal{G}$ acts on $\mathcal{K}$ by conjugation.

Examples can illustrate the construction. Consider $\mathcal{D}_2$, with $\mathcal{A}ut\ \mathcal{D}_2 = \mathcal{S}_3$. This allows us to form the semi-direct product of $\mathcal{D}_2$ with $\mathcal{S}_3$ or any of its subgroups. In general, we have

$$\mathcal{D}_n = \mathcal{Z}_n \;\rtimes\; \mathcal{Z}_2. \tag{2.82}$$

The presentation

$$\mathcal{D}_n: \quad (\,a^n = e;\; b^2 = e;\; bab^{-1} = a^{-1}\,)$$

shows that $\mathcal{Z}_2$ generated by b acts on $\mathcal{Z}_n$ by conjugation.

Scholium. *Transitivity*

The notion of transitivity plays an important role in the classification of finite groups. Permutations are described by cycles acting on sets of objects $[a_1, a_2, \ldots, a_{n-1}, a_n]$. The $n!$ permutations in $\mathcal{S}_n$ generate $n!$ distinct arrangements $[b_1, b_2, \ldots, b_{n-1}, b_n]$. Label the objects by number, and start from one arrangement, $[1, 2, \ldots, (n-1), n]$. Clearly any other arrangement can be attained by a permutation in $\mathcal{S}_n$: $\mathcal{S}_n$ is said to be (singly) *transitive*. Groups without this property are called *intransitive*. For example, the permutation group on four objects, generated by the cycles $(1\,2)$, $(3\,4)$ is intransitive because it does not contain the permutation that maps $[1, 2, 3, 4]$ to $[3, 2, 1, 4]$. Since $\mathcal{S}_n$ maps any set of n objects into another, it is called *n-ply transitive*. Put in another way, if we define the *orbit* of an element g as the set of all elements generated from g by group operation, a transitive group has only one orbit. Obviously, any permutation in $\mathcal{S}_n$ is *a fortiori* k-ply transitive with $k < n$. The correspondence

$$\text{even permutations} \;\rightarrow\; 1, \qquad \text{odd permutations} \;\rightarrow\; -1, \tag{2.83}$$

maps $\mathcal{S}_n$ into $\mathcal{Z}_2$, which is a subgroup of $\mathcal{S}_n$. As we have seen, $\mathcal{A}_n$ the group of even permutations form the group the *kernel* of this mapping. It is a normal subgroup, since conjugation by an odd permutation leaves the parity unaltered. Hence $\mathcal{S}_n$ is not a simple group. Since $\mathcal{A}_n$ is the largest subgroup of $\mathcal{S}_n$, we verify that its quotient group is indeed simple, as

$$\mathcal{S}_n/\mathcal{A}_n = \mathcal{Z}_2, \tag{2.84}$$

Consider the array of $(n-1)$ objects $[a_2, a_3, \ldots, a_{n-1}, a_n]$ which does not contain a_1. The only way to generate the $(n-1)$ array $[a_1, a_3, \ldots, a_{n-1}, a_n]$, where a_2 has been replaced by a_1 is by the transposition $(a_1\, a_2)$:

$$[a_2, a_3, \ldots, a_{n-1}, a_n] \;\underset{(a_1\, a_2)}{\longrightarrow}\; [a_1, a_3, \ldots, a_{n-1}, a_n]. \tag{2.85}$$

This is an odd permutation, and we conclude that $\mathcal{A}_n$ is **not** $(n-1)$-ply transitive. Now look at the set of $(n-2)$ objects $[a_3, a_4, \ldots, a_{n-1}, a_n]$ which contains neither a_1 nor a_2. The array $[a_1, a_4, \ldots, a_{n-1}, a_n]$ where a_3 has been replaced by a_1 can also be reached by a transposition $(a_1\, a_3)$ as before, but since we have one "idle" object a_2, we can also reach it by means of the *even* permutation $(a_3\, a_1\, a_2)$

$$[a_3, a_4, \ldots, a_{n-1}, a_n] \;\underset{(a_3\, a_1\, a_2)}{\longrightarrow}\; [a_1, a_4, \ldots, a_{n-1}, a_n]. \tag{2.86}$$

The set where both a_3 and a_4 are replaced by a_1 and a_2 is clearly obtained by the (even) product of two transpositions. Hence $\mathcal{A}_n$ is $(n-2)$-ply transitive.

Other than $\mathcal{S}_n$ and $\mathcal{A}_n$, there are relatively few multiply transitive groups: several infinite families of doubly transitive groups, and one infinite family of triply transitive groups. All have interesting interpretations in terms of discrete geometries. In addition, there are several isolated doubly transitive groups, one triply transitive, two quadruply transitive, and two quintuply transitive groups. All but one of these are sporadic groups.

2.4.5 *Young Tableaux*

Finally, using $\mathcal{S}_5$, the first meaty permutation group on five objects as exemplar, we present a general method to organize its $5! = 120$ permutations. We begin with the following cycle decomposition.

(x x x x x)	5	24	+
(x x x x)(x)	4+1	30	−
(x x x)(x x)	3+2	20	−
(x x x)(x)(x)	3+1+1	20	+
(x x)(x x)(x)	2+2+1	15	+
(x x)(x)(x)(x)	2+1+1+1	10	−
(x)(x)(x)(x)(x)	1+1+1+1+1	1	+

Here, (x x x x x) denotes a cycle with arbitrary entries. The number of different cycles is in one-to-one correspondence with the partitions of 5, as indicated in the above table, where we also list the number of permutations in each partition, as well as their parity. To each partition of n corresponds an irreducible representation of $\mathcal{S}_n$. It is common practice to express the partitions in terms of an ordered set of n integers $\{\lambda_1 \geq \lambda_2 \geq \cdots \geq \lambda_n\}$, defined in terms of α_k, the number of cycles of order k as

$$\lambda_1 = \sum_{i=1}^{n} \alpha_i, \ \lambda_2 = \sum_{i=2}^{n} \alpha_i, \ldots, \ \lambda_n = \alpha_n. \tag{2.87}$$

For instance, (x x x x)(x) with $\alpha_1 = 1$, $\alpha_4 = 1$, yields the five integers $\{2, 1, 1, 1, 0\}$, in short $\{2, 1^3\}$.

Partitions can also be represented graphically, by means of *Young Tableaux*. The Young Tableau associated with the partition $\{\lambda_1, \lambda_2, \ldots, \lambda_n\}$ is obtained by drawing horizontally λ_1 boxes side to side, then just below λ_2 boxes, down to the lowest value of λ_k. This rule associates a unique Young Tableau to each partition and class of $\mathcal{S}_5$, as follows.

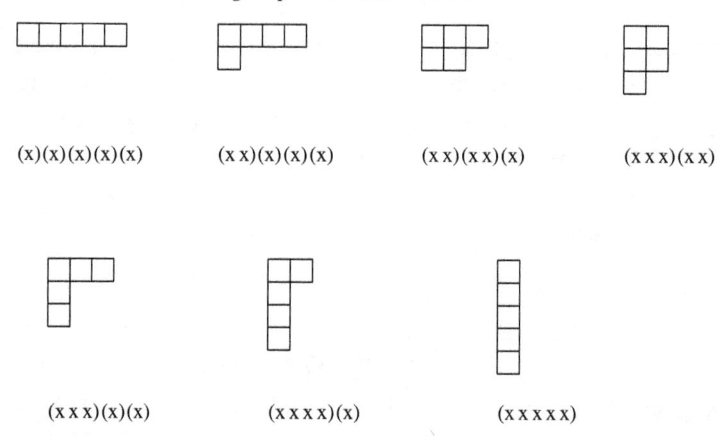

This representation of the partitions in terms of Young Tableaux turns out to be very useful in other contexts, especially in determining tensor symmetries, and transformation properties of products of representations.

3

Finite groups: representations

The representation theory of finite groups addresses the number of ways the operations of a certain group can be written in terms of matrices acting on some vector space. It answers questions such as how many distinct representations does a group have, and how to distinguish between them? We have already seen that the group actions of any finite group $\mathcal{G}$ of order n can be represented in terms of $(n \times n)$ matrices, forming the *regular representation*. In the following, we show how to break up the regular representation in terms of irreducible blocks of *irreducible representations* (irreps). Their number and type characterize the group, and are the starting point for physical applications. For more details, the reader is directed to classic books by Carmichael [2], Hall [9], and Hammermesh [10].

3.1 Introduction

Consider a finite vector space V, spanned by an orthonormal set of N bras and kets,

$$\langle i \mid j \rangle = \delta_{ij}, \qquad \sum_{i=1}^{N} \mid i \rangle \langle i \mid = 1. \tag{3.1}$$

We wish to study the action of an order-n group $\mathcal{G}$ on this vector space. We assume that the action of $g \in \mathcal{G}$ in V is represented by $(N \times N)$ non-singular matrices acting as

$$\mid i \rangle \rightarrow \mid i(g) \rangle = \mathcal{M}_{ij}(g) \mid j \rangle, \tag{3.2}$$

with

$$\mathcal{M}_{ij}(g^{-1}) = \mathcal{M}^{-1}_{ij}(g), \tag{3.3}$$

forming an N-dimensional representation $\mathfrak{R}$ of $\mathcal{G}$. The regular representation would correspond to $n = N$. Under what conditions can this representation be broken down into smaller *irreducible* representations? We already know there is at least one, the trivial one-dimensional representation where every group action is multiplication by one.

Assume that $\mathfrak{R}$ is reducible, with a d_1-dimensional matrix representation $\mathfrak{R}_1$ of $\mathcal{G}$ in V, acting on a smaller set $|\, a\,\rangle$, $a = 1, 2, \ldots, d_1 < N$, spanning the subspace V_1. The remaining $|\, m\,\rangle$, $m = 1, 2, \ldots, (N - d_1)$ span its orthogonal complement $\mathsf{V}_1^{\perp}$. Assemble the kets in an N-dimensional column, with the group matrices acting as

$$\begin{pmatrix} \mathcal{M}^{[1]}(g) & 0 \\ \mathcal{N}(g) & \mathcal{M}^{[\perp]}(g) \end{pmatrix}. \tag{3.4}$$

These matrices keep their form under multiplication, with

$$\mathcal{M}^{[1]}(gg') = \mathcal{M}^{[1]}(g)\,\mathcal{M}^{[1]}(g'), \qquad \mathcal{M}^{[\perp]}(gg') = \mathcal{M}^{[\perp]}(g)\,\mathcal{M}^{[\perp]}(g'), \tag{3.5}$$

and

$$\mathcal{N}(gg') = \mathcal{N}(g)\,\mathcal{M}^{[1]}(g') + \mathcal{M}^{[\perp]}(g)\,\mathcal{N}(g'). \tag{3.6}$$

A clever change of basis,

$$\begin{pmatrix} 1 & 0 \\ \mathcal{V} & 1 \end{pmatrix}, \tag{3.7}$$

where

$$\mathcal{V} = \frac{1}{n}\sum_{g} \mathcal{M}^{[\perp]}(g^{-1})\mathcal{N}(g), \tag{3.8}$$

brings the representation matrices to block-diagonal form

$$\begin{pmatrix} \mathcal{M}^{[1]} & 0 \\ 0 & \mathcal{M}^{[\perp]} \end{pmatrix}. \tag{3.9}$$

In this basis, the original representation $\mathfrak{R}$ is said to be *fully reducible*, that is

$$\mathfrak{R} = \mathfrak{R}_1 \oplus \mathfrak{R}_1^{\perp}.$$

Clearly, the process can be continued if $\mathfrak{R}_1^{\perp}$ is itself reducible, until one runs out of subspaces. In general, a representation $\mathfrak{R}$ on V will be expressed as a sum of irreducible representations

$$\mathfrak{R} = \mathfrak{R}_1 \oplus \mathfrak{R}_2 \oplus \cdots \oplus \mathfrak{R}_k, \tag{3.10}$$

corresponding to

$$\mathsf{V} = \mathsf{V}_1 \oplus \mathsf{V}_2 \oplus \cdots \oplus \mathsf{V}_k. \tag{3.11}$$

The same irreducible representation may appear several times in this decomposition. Application of this decomposition to the regular representation is the key of representation theory.

3.2 Schur's lemmas

Express $\mathfrak{R}_1$ of $\mathcal{G}$ in matrix form

$$|\,a\,\rangle \to |\,a(g)\,\rangle = \mathcal{M}^{[1]}_{ab}(g)|\,b\,\rangle. \tag{3.12}$$

Since both $|\,a\,\rangle$ and $|\,i\,\rangle$ live in V, they are linearly related

$$|\,a\,\rangle = \mathcal{S}_{ai}|\,i\,\rangle, \tag{3.13}$$

where $\mathcal{S}$ is a $(d_1 \times N)$ array. This enables us to view the group action in two ways:

$$|\,a\,\rangle \to |\,a(g)\,\rangle = \mathcal{M}^{[1]}_{ab}(g)|\,b\,\rangle = \mathcal{M}^{[1]}_{ab}(g)\mathcal{S}_{bi}|\,i\,\rangle, \tag{3.14}$$

or

$$|\,a\,\rangle = \mathcal{S}_{ai}|\,i\,\rangle \to \mathcal{S}_{ai}|\,i(g)\,\rangle = \mathcal{S}_{ai}\mathcal{M}_{ij}(g)|\,j\,\rangle. \tag{3.15}$$

Comparing the two and using the completeness of the $\{|\,i\,\rangle\}$, we deduce that the reducibility of $\mathfrak{R}$ implies

$$\mathcal{S}_{ai}\mathcal{M}_{ij}(g) = \mathcal{M}^{[1]}_{ab}(g)\mathcal{S}_{bj}. \tag{3.16}$$

Conversely, starting from this equation, we see that the $\{|\,a\,\rangle\}$ form a smaller representation of $\mathcal{G}$: $\mathfrak{R}$ is reducible. Thus if $\mathfrak{R}$ is *irreducible*, there is no such $\mathcal{S}$. We can state the result as *Schur's first lemma.*

If the matrices of two irreducible *representations of different dimensions are related as* $\mathcal{S}_{ai}\mathcal{M}_{ij}(g) = \mathcal{M}^{[1]}_{ab}(g)\mathcal{S}_{bj}$, *then* $\mathcal{S} = 0$.

In the special case when $d_1 = N$, the $\mathcal{S}$ array becomes a $(N \times N)$ matrix. If both sets $|\,i\,\rangle$ and $|\,a\,\rangle$ span the same Hilbert space, then $\mathcal{S}$ must have an inverse, and $\mathfrak{R}$ is related to $\mathfrak{R}_1$ by a similarity transformation

$$\mathcal{M}^{[1]} = \mathcal{S}\mathcal{M}\mathcal{S}^{-1}. \tag{3.17}$$

The two representations are said to be equivalent. It could also be that the two representations have the same dimension, but live in different Hilbert spaces; in that case $\mathcal{S} = 0$. We conclude that a non-vanishing $\mathcal{S}$ implies either reducibility or similarity.

Now assume that $\mathfrak{R}$ is irreducible. Any matrix $\mathcal{S}$ which satisfies

$$\mathcal{M}(g)\mathcal{S} = \mathcal{S}\mathcal{M}(g), \quad \forall\, g \in \mathcal{G}, \tag{3.18}$$

must be proportional to the unit matrix: if $|\,i\,\rangle$ is an eigenket of $\mathcal{S}$, then every $|\,i(g)\,\rangle$ is also an eigenket with the *same* eigenvalue, reducing $\mathcal{S}$ to a multiple of the unit matrix (the eigenvalue cannot be zero). This is *Schur's second lemma.*

These two lemmas have important consequences. To start, consider $\mathfrak{R}_\alpha$ and $\mathfrak{R}_\beta$, two *irreps* of $(d_\alpha \times d_\alpha)$ matrices $\mathcal{M}^{[\alpha]}(g)$ in V_α, and $(d_\beta \times d_\beta)$ matrices $\mathcal{M}^{[\beta]}(g)$ in V_β. Construct the mapping $\mathcal{S}$

$$\mathcal{S} = \sum_g \mathcal{M}^{[\alpha]}(g)\, \mathcal{N}\, \mathcal{M}^{[\beta]}(g^{-1}), \tag{3.19}$$

where $\mathcal{N}$ is *any* $(d_\alpha \times d_\beta)$ array. It satisfies

$$\mathcal{M}^{[\beta]}(g)\mathcal{S} = \mathcal{S}\mathcal{M}^{[\alpha]}(g). \tag{3.20}$$

If the two representations are different, $d_\alpha \neq d_\beta$, and by Schur's first lemma, $\mathcal{S}$ necessarily vanishes. If on the other hand, $\mathcal{S} \neq 0$, then one of the representations must be reducible.

Thus for any two *different* irreps, $\mathfrak{R}_\alpha$ and $\mathfrak{R}_\beta$, an immediate consequence of Schur's first lemma is

$$\frac{1}{n}\sum_g \mathcal{M}^{[\alpha]}_{ij}(g)\mathcal{M}^{[\beta]}_{pq}(g^{-1}) = 0, \quad \alpha \neq \beta, \tag{3.21}$$

so that for fixed $i, j = 1, 2 \ldots d_\alpha$ and $p, q = 1, 2 \ldots d_\beta$, the matrix elements of two different irreps are orthogonal over an n-dimensional vector space.

Setting $\mathfrak{R}_\beta = \mathfrak{R}_\alpha$, we find

$$\frac{1}{n}\sum_g \mathcal{M}^{[\alpha]}_{ij}(g)\mathcal{M}^{[\alpha]}_{pq}(g^{-1}) = \frac{1}{d_\alpha}\delta_{iq}\delta_{jp}. \tag{3.22}$$

The (properly normalized) matrix elements of an irrep form an orthonormal set. This equation can be verified by multiplying both sides by an arbitrary matrix t_{jp}, and summing over j and p, reproducing the result of Schur's second lemma. The proper normalization is determined by summing over i and q.

Let us now introduce the group-averaged inner product

$$\left(i,\, j\right) \equiv \frac{1}{n}\sum_g \langle i(g) \mid j(g) \rangle, \tag{3.23}$$

where the sum is over all elements of $\mathcal{G}$. It is group-invariant by construction, in the sense that for any element g_a,

$$\begin{aligned}(i(g_a),\, j(g_a)) &= \frac{1}{n}\sum_g \langle i(g_a g) \mid j(g_a g) \rangle \\ &= \frac{1}{n}\sum_{g'=g_a g} \langle i(g') \mid j(g') \rangle = (i,\, j).\end{aligned} \tag{3.24}$$

Then for any two vectors **u** and **v**, we have

$$(\mathcal{M}(g^{-1})\mathbf{v},\ \mathbf{u}) = (\mathbf{v},\ \mathcal{M}(g)\mathbf{u}), \tag{3.25}$$

since the inner product is group-invariant. On the other hand, for any representation,

$$(\mathcal{M}^{\dagger}(g)\mathbf{v},\ \mathbf{u}) = (\mathbf{v},\ \mathcal{M}(g)\mathbf{u}), \tag{3.26}$$

so that

$$\mathcal{M}(g^{-1}) = \mathcal{M}^{\dagger}(g), \qquad \mathcal{M}(g^{-1})\,\mathcal{M}(g) = 1. \tag{3.27}$$

It follows that *all representations of finite groups are unitary* with respect to the group-averaged inner product. This will not be true when we consider continuous groups, where the group-averaged inner product cannot in general be defined.

We summarize all these results in a master equation for different unitary irreps:

$$\frac{1}{n}\sum_{g}\mathcal{M}^{[\alpha]}_{ij}(g)\overline{\mathcal{M}}^{[\beta]}_{qp}(g) = \frac{1}{d_\alpha}\delta^{\alpha\beta}\delta_{iq}\delta_{jp}. \tag{3.28}$$

This equation can be rewritten in terms of the *character* of an element g in a representation $\mathfrak{R}_\alpha$, defined as the trace of its representation matrix

$$\chi^{[\alpha]}(g) \equiv \mathrm{Tr}\,\mathcal{M}^{[\alpha]}(g). \tag{3.29}$$

It is not a unique function of g; rather, the cyclic property of the trace

$$\chi^{[\alpha]}(g_a g g_a^{-1}) = \mathrm{Tr}\,\left(\mathcal{M}^{[\alpha]}(g_a)\mathcal{M}^{[\alpha]}(g)\mathcal{M}^{[\alpha]}(g_a^{-1})\right) = \mathrm{Tr}\,\mathcal{M}^{[\alpha]}(g), \tag{3.30}$$

shows that all elements of one class have the same character: character is a class attribute. In addition, for a unitary transformation,

$$\chi^{[\alpha]}(g^{-1}) = \mathrm{Tr}\,\mathcal{M}^{[\alpha]\dagger}(g) = \overline{\chi}^{[\alpha]}(g). \tag{3.31}$$

Applied to characters, our master formula becomes

$$\frac{1}{n}\sum_{g}\chi^{[\alpha]}(g)\overline{\chi}^{[\beta]}(g) = \delta^{\alpha\beta}. \tag{3.32}$$

In terms of n_i the number of elements in the ith class, with character $\chi_i^{[\alpha]}$, the sum over the group elements becomes

$$\{\chi^{[\alpha]},\ \chi^{[\beta]}\} \equiv \frac{1}{n}\sum_{i=1}^{n_C} n_i\,\chi_i^{[\alpha]}\overline{\chi}_i^{[\beta]} = \delta^{\alpha\beta}, \tag{3.33}$$

where n_C is the number of classes. Assuming n_R different irreps, $\alpha = 1, 2, \ldots, n_R$, this equation states that the n_R vectors $\chi^{[\alpha]}$ form an orthonormal set in an n_C-dimensional vector space. This can only be if

$$n_R \leq n_C : \tag{3.34}$$

the number of different irreps of a group is no larger than the number of its classes. Another consequence of this equation is that for any irrep,

$$\{\chi^{[\alpha]}, \chi^{[\alpha]}\} = 1. \tag{3.35}$$

Consider a reducible representation that is the sum of r_α irreps $\mathfrak{R}_\alpha$, with characters

$$\chi = \sum_\alpha r_\alpha \chi^{[\alpha]}. \tag{3.36}$$

The multiplicity of the $\mathfrak{R}_\alpha$ is easily computed from the scalar product

$$r_\alpha = \{\chi, \chi^{[\alpha]}\}, \tag{3.37}$$

which holds for *any* character χ. Also

$$\{\chi, \chi\} = \sum_{\alpha\,\beta} r_\alpha r_\beta \{\chi^{[\alpha]}, \chi^{[\beta]}\} = \sum_\alpha r_\alpha^2, \tag{3.38}$$

provides a simple test for irreducibility: if the scalar product of its characters is equal to one, the representation is irreducible.

These formulæ can be applied to the n-dimensional regular representation, which contains *all* the irreps $\mathfrak{R}_\alpha$,

$$\mathfrak{R}^{\rm reg} = \sum_{\alpha=1}^{n_R} r_\alpha \mathfrak{R}_\alpha, \tag{3.39}$$

so that

$$n = \sum_{\alpha=1}^{n_R} r_\alpha\, d_\alpha, \tag{3.40}$$

its characters are expressed as

$$\chi_i^{\rm reg} = \sum_{\alpha=1}^{n_R} r_\alpha \chi_i^{[\alpha]}. \tag{3.41}$$

In the regular representation, any group element shuffles the n-vector elements into different ones: they are represented by matrices which have zeroes on their diagonal. The only exception is the identity element. Therefore all the characters of the regular representation vanish except for the identity

$$\chi_1^{\rm reg} = n, \qquad \chi_i^{\rm reg} = 0, \quad i \neq 1. \tag{3.42}$$

As a consequence,

$$r_\alpha = \{\chi^{\text{reg}},\, \chi^{[\alpha]}\} = \{\chi_1^{\text{reg}},\, \chi^{[\alpha]}\} = \chi_1^{[\alpha]}. \tag{3.43}$$

Since the unit element of any representation is represented by the unit matrix, its character is the dimension of the representation

$$\chi_1^{[\alpha]} = d_\alpha, \tag{3.44}$$

so that

$$r_\alpha = d_\alpha. \tag{3.45}$$

Hence we arrive at the following important result.

The multiplicity of any irrep in the regular representation is equal to its dimension.

A relation between the order of the group and the dimensions of its irreps ensues,

$$n = \sum_{\alpha=1}^{n_R} (d_\alpha)^2\,. \tag{3.46}$$

This equation is quite useful in determining the various irreps of groups of low order.

Given the class C_i made up of n_i elements $(g_1^{(i)},\, g_2^{(i)}, \ldots, g_{n_i}^{(i)})$ obtained by conjugation, we define

$$A_i \equiv \frac{1}{n_i} \sum_{m=1}^{n_i} g_m^{(i)}, \tag{3.47}$$

which is self-conjugate by construction,

$$\widetilde{A}_i = A_i. \tag{3.48}$$

Furthermore, the A_i form a closed algebra. Their product is of course a sum of elements of the group, but it is a self-conjugate sum. Hence the elements must come in full classes. Their product is made up of A_k as well

$$A_i A_j = \sum_k c_{ijk} A_k, \tag{3.49}$$

where the c_{ijk}, sometimes called the *class coefficients*, depend on the group. In $\mathfrak{R}_\alpha$, they are matrices

$$\mathcal{N}_{(i)}^{[\alpha]} \equiv \mathcal{M}^{[\alpha]}(A_i) = \frac{1}{n_i} \sum_{g \in C_i} \mathcal{M}^{[\alpha]}(g), \tag{3.50}$$

with

$$\mathrm{Tr}\,\mathcal{N}^{[\alpha]}_{(i)} = \chi^{[\alpha]}_i. \tag{3.51}$$

Since

$$\mathcal{M}^{[\alpha]}(g)\mathcal{N}^{[\alpha]}_{(i)}\mathcal{M}^{[\alpha]}(g^{-1}) = \mathcal{N}^{[\alpha]}_{(i)}, \tag{3.52}$$

the $\mathcal{N}^{[\alpha]}_{(i)}$ commute with all the group matrices: when $\mathfrak{R}_\alpha$ is irreducible, they are multiples of the unit matrix in the representation

$$\mathcal{N}^{[\alpha]}_{(i)} = \frac{\chi^{[\alpha]}_i}{\chi^{[\alpha]}_1}\mathbf{1}. \tag{3.53}$$

Since

$$\mathcal{N}^{[\alpha]}_{(i)}\mathcal{N}^{[\alpha]}_{(j)} = \sum_k c_{ijk}\mathcal{N}^{[\alpha]}_{(k)}, \tag{3.54}$$

we arrive at

$$\chi^{[\alpha]}_i\chi^{[\alpha]}_j = \chi^{[\alpha]}_1\sum_k c_{ijk}\chi^{[\alpha]}_k. \tag{3.55}$$

We can rewrite the characters of the regular representation as

$$\chi^{\mathrm{reg}}_i = \sum_{\alpha=1}^{n_R} r_\alpha\chi^{[\alpha]}_i = \sum_{\alpha=1}^{n_R} d_\alpha\chi^{[\alpha]}_i = \sum_{\alpha=1}^{n_R}\chi^{[\alpha]}_1\chi^{[\alpha]}_i. \tag{3.56}$$

Hence we see that

$$\sum_{\alpha=1}^{n_R}\chi^{[\alpha]}_i\chi^{[\alpha]}_j = \sum_k c_{ijk}\chi^{\mathrm{reg}}_k. \tag{3.57}$$

Since all characters of the regular representation vanish except for the identity, it follows that

$$\sum_{\alpha=1}^{n_R}\chi^{[\alpha]}_i\chi^{[\alpha]}_j = nc_{ij1}. \tag{3.58}$$

Now let us consider $C_{\bar{i}}$ the class made of all the inverses of class C_i. In a unitary representation, their characters are related by complex conjugation,

$$\chi^{[\alpha]}_{\bar{i}} = \overline{\chi}^{[\alpha]}_i. \tag{3.59}$$

Thus if the character for a particular class is real, the class contains its own inverses; if it is complex, the class and its inverse class are different. The product A_iA_j

contains the unit element only if some member of the j class is the inverse of one in the i class, but this only happens if $j = \bar{i}$, when we have

$$A_i A_{\bar{i}} = \frac{1}{n_i} e + \cdots . \tag{3.60}$$

Thus we conclude that

$$c_{ij1} = \begin{cases} 0, & j \neq \bar{i} \\ \frac{1}{n_i}, & j = \bar{i} \end{cases} . \tag{3.61}$$

Putting it all together, we arrive at

$$\sum_{\alpha=1}^{n_R} \chi_i^{[\alpha]} \overline{\chi}_j^{[\alpha]} = \frac{n}{n_i} \delta_{ij} . \tag{3.62}$$

The n_C vectors χ_i form an orthogonal basis in the n_R-dimensional vector space, which implies that

$$n_C \leq n_R . \tag{3.63}$$

The number of classes is limited to the number of irreps, but we already saw that the number of irreps is limited by the number of classes. Hence we arrive at the fundamental result of representation theory.

The number of irreps of a finite group is equal to its number of classes.

Once we know the classes of a group, we can deduce its representations. The information is usually displayed in a *character table*, which lists the values of the characters for the different representations. A note of caution: in rare occasions two different groups (e.g. the quaternion group $\mathcal{Q}$ and the dihedral group $\mathcal{D}_4$) can have the same character table. In the following, we show how to derive the character table for $\mathcal{A}_4$, and list the character tables of the low-order groups we have already encountered.

3.3 The $\mathcal{A}_4$ character table

The alternating group $\mathcal{A}_4$ contains the even permutations on four letters, of the form (x x x) and (x x)(x x). A hasty conclusion would be that it has three classes, but there is a subtlety in the three-cycle: it contains elements that differ from one another by *one* transposition, such as (1 2 3) and (1 3 2), which cannot be mapped by $\mathcal{A}_4$ conjugation into one another. Hence the three-cycles break up into two classes which contain the inverses of one another. Their characters are related by complex conjugation.

If $a = (1\,2)(3\,4)$, and $b = (1\,2\,3)$, the four classes are $C_1(e)$ which contains the unit element, $C_2(b, ab, aba, ba)$ with four elements, its inverse class $C_3(b^2, ab^2, b^2a, ab^2a)$ also with four elements, and $C_4(a, b^2ab, bab^2)$ with three elements. It has four irreps, one of which is the trivial unit representation. The order is low enough that we can easily find their dimensions d_i. Indeed, they must satisfy

$$12 = 1 + d_2^2 + d_3^2 + d_4^2, \tag{3.64}$$

which has the unique solution $d_2 = d_3 = 1$, $d_4 = 3$. Let us denote the four representations by $\mathbf{1}$, $\mathbf{1_1}$, $\mathbf{1_2}$, and $\mathbf{3}$. Their multiplicity in the regular representation is equal to their dimension, so that (we dispense with the arcane $\otimes$ and $\oplus$, and simply use $\times$ and $+$ for the product and sum of representations)

$$\mathfrak{R}^{\text{reg}}(\mathcal{A}_4) = \mathbf{1} + \mathbf{1_1} + \mathbf{1_2} + \mathbf{3} + \mathbf{3} + \mathbf{3}. \tag{3.65}$$

Let $\chi^{[\alpha]}$ be the characters of the four irreps. We form a (4×4) *character table*. Its rows are labeled by the characters, and its columns by the classes. The first column is the unit matrix, the trace of which is the dimension of the irrep. The first row is the (1×1) unit irrep where all characters are equal to one. We start with the following.

$\mathcal{A}_4$	C_1	$4C_2$	$4C_3$	$3C_4$
$\chi^{[1]}$	1	1	1	1
$\chi^{[1_1]}$	1	z	$\bar{z}$	r
$\chi^{[1_2]}$	1	u	$\bar{u}$	s
$\chi^{[3]}$	3	w	$\bar{w}$	t

Since the class C_4 is its own inverse, its characters r, s and t are real numbers. To find the characters in these representations, we can use the orthogonality of characters, or that the weighted sum of the characters along any row (except for $\chi^{[1]}$) vanishes. The latter determines r, s, and t in terms of the real parts of z, u and w, but we need more information. It is encoded in the class coefficients c_{ijk} through

$$\chi_i^{[\alpha]}\chi_j^{[\alpha]} = d_\alpha \sum_k c_{ijk}\chi_k^{[\alpha]}. \tag{3.66}$$

The class coefficients of $\mathcal{A}_4$ are easily worked out from the commutative multiplication table of the A_i matrices (eq. (3.49))

$$A_4A_4 = \frac{1}{3}A_1 + \frac{2}{3}A_4, \qquad A_2A_3 = \frac{1}{4}A_1 + \frac{3}{4}A_4,$$

$$A_4A_2 = A_3A_3 = A_2; \qquad A_4A_3 = A_2A_2 = A_3. \tag{3.67}$$

These tell us that for $\alpha = 1, 2, 3$,

$$\chi_3^{[\alpha]}\chi_3^{[\alpha]} = \chi_2^{[\alpha]}\chi_4^{[\alpha]} = \chi_2^{[\alpha]}, \tag{3.68}$$

implying that

$$\bar{z}\,\bar{z} = zr = z, \qquad \bar{u}\,\bar{u} = us = u\,: \tag{3.69}$$

so that z and u are either zero or else cubic roots of unity. The orthogonality of the characters with the unit representation leads to the vanishing of the weighted row sum of $\chi^{[2]}$

$$1 + 4(z + \bar{z}) + 3r = 0, \tag{3.70}$$

together with $1 + z + \bar{z} = 0$, which yields $r = 1$, and $z = e^{\pm 2\pi i/3}$. We see that z cannot be zero; if it were, the class equation

$$0 = \chi_2^{[2]}\chi_3^{[2]} = \frac{1}{4} + \frac{3}{4}\,\chi_4^{[2]}, \tag{3.71}$$

would require $r = -1/3$, which would contradict the class relation

$$\chi_4^{[2]}\chi_4^{[2]} = \frac{1}{3} + \frac{2}{3}\,\chi_4^{[2]}. \tag{3.72}$$

The same method applied to the third row fixes

$$s = 1, \qquad u = \bar{z} = e^{\mp 2\pi i/3}, \tag{3.73}$$

taking into account the orthogonality between the second and third rows. For the three-dimensional irrep, the class relation,

$$\chi_3^{[4]}\chi_3^{[4]} = \chi_2^{[4]}\chi_4^{[4]} = 3\chi_2^{[4]}, \tag{3.74}$$

yields

$$\bar{w}\,\bar{w} = w\,t = 3w, \tag{3.75}$$

while

$$\chi_4^{[4]}\chi_4^{[4]} = 3\left(\frac{1}{3}\chi_1^{[4]} + \frac{2}{3}\chi_4^{[4]}\right), \tag{3.76}$$

leads to two solutions, $t = -1$ and $t = 3$. The solution $t = -1$ yields by orthogonality $w + \overline{w} = 0$. The second solution yields $w + \overline{w} = wz + \overline{wz} = w\overline{z} + \overline{w}z = -12$, which is not consistent unless $w + \overline{w}$ vanishes. Hence the one solution $t = -1$, $w = 0$ completes the character table. The characters of $\mathbf{1_1}$ and $\mathbf{1_2}$ are complex conjugates of one another: these two irreps are said to be conjugates as well, $\mathbf{1_2} = \overline{\mathbf{1}}_1$.

The three non-trivial classes are $C_2^{[3]}(b, ab, aba, ba)$, $C_3^{[3]}(b^2, ab^2, b^2a, ab^2a)$, and $C_4^{[2]}(a, b^2ab, bab^2)$, with the order of their elements in a square-bracketed superscript. The elements are written in term of the two generators with presentation $< a, b \,|\, a^2 = e, b^3 = e, (ba)^3 = e >$. The $\mathcal{A}_4$ character table is then as follows.

$\mathcal{A}_4$	C_1	$4C_2^{[3]}$	$4C_3^{[3]}$	$3C_4^{[2]}$
$\chi^{[1]}$	1	1	1	1
$\chi^{[1_1]}$	1	$e^{2\pi i/3}$	$e^{4\pi i/3}$	1
$\chi^{[\overline{1}_1]}$	1	$e^{4\pi i/3}$	$e^{2\pi i/3}$	1
$\chi^{[3]}$	3	0	0	-1

Since $\mathbf{3}$ is a real irrep, it is not hard to deduce the expression for its generators as (3×3) matrices:

$$\mathbf{3}: \quad a = \begin{pmatrix} -1 & 0 & 0 \\ 0 & 1 & 0 \\ 0 & 0 & -1 \end{pmatrix}; \qquad b = \begin{pmatrix} 0 & 1 & 0 \\ 0 & 0 & 1 \\ 1 & 0 & 0 \end{pmatrix}. \tag{3.77}$$

3.4 Kronecker products

In order to study how representations combine, consider two irreps of $\mathcal{G}$, $\mathfrak{R}_\alpha$ and $\mathfrak{R}_\beta$, represented on two Hilbert spaces spanned by d_α and d_β bras and kets, $\langle\, i \,|_\alpha$, $i, j = 1, 2 \ldots, d_\alpha$, and $\langle\, s \,|_\beta$, $s, t = 1, 2 \ldots, d_\beta$, respectively, that is

$$|\, i \,\rangle_\alpha \to |\, i(g) \,\rangle_\alpha = \mathcal{M}_{ij}^{[\alpha]}(g) |\, j \,\rangle_\alpha, \tag{3.78}$$

$$|\, s \,\rangle_\beta \to |\, s(g) \,\rangle_\beta = \mathcal{M}_{st}^{[\beta]}(g) |\, t \,\rangle_\beta. \tag{3.79}$$

We wish to study how $\mathcal{G}$ is represented on the product space spanned by

$$|\, i \,\rangle_\alpha |\, s \,\rangle_\beta \equiv |\, A \,\rangle, \tag{3.80}$$

where A, B take on $d_\alpha d_\beta$ values of the pairs $\{i, s\}$.

We define a new matrix for the representation $\mathfrak{R}_\alpha \times \mathfrak{R}_\beta$ by

$$\begin{aligned} |\,A\,\rangle \to |\,A(g)\,\rangle &\equiv \mathcal{M}^{[\alpha\times\beta]}_{AB}(g)|\,B\,\rangle, \\ &= \mathcal{M}^{[\alpha]}_{ij}(g)\mathcal{M}^{[\beta]}_{st}(g)|\,j\,\rangle_\alpha|\,t\,\rangle_\beta. \end{aligned} \tag{3.81}$$

This new reducible representation, called the Kronecker product, is like any other expressible as a sum of irreps of $\mathcal{G}$. We write the *Clebsch–Gordan series*

$$\mathfrak{R}_\alpha \times \mathfrak{R}_\beta = \sum_\gamma d(\alpha,\beta\,|\,\gamma)\mathfrak{R}_\gamma, \tag{3.82}$$

where the $d(\alpha,\beta\,|\,\gamma)$ are to be determined. In terms of characters,

$$\chi^{[\alpha\times\beta]} = \sum_\gamma d(\alpha,\beta\,|\,\gamma)\chi^{[\gamma]}. \tag{3.83}$$

Now, trace over the AB to obtain the characters of the Kronecker product

$$\chi^{[\alpha\times\beta]} = \sum_A \mathcal{M}^{[\alpha\times\beta]}_{AA}(g) = \sum_{i\,s} \mathcal{M}^{[\alpha]}_{ii}(g)\mathcal{M}^{[\beta]}_{ss}(g), \tag{3.84}$$

that is for any class

$$\chi^{[\alpha\times\beta]} = \chi^{[\alpha]}\,\chi^{[\beta]}. \tag{3.85}$$

Using the orthogonality property of the characters, we obtain the coefficients

$$d(\alpha,\beta\,|\,\gamma) = \{\,\chi^{[\alpha\times\beta]},\ \chi^{[\gamma]}\,\} = \frac{1}{n}\sum_{i=1}^{n_C} n_i\chi^{[\alpha]}_i\chi^{[\beta]}_i\overline{\chi}^{[\gamma]}_i. \tag{3.86}$$

This allows us to find the product of irreps from the character table. The product of the unit irrep **1** with any other irrep reproduces that irrep, using orthogonality of the characters. As an example, let us compute the Kronecker products of $\mathcal{A}_4$ irreps. From the character table, we find

$$d(\mathbf{3},\mathbf{3}\,|\,\mathbf{1}) = d(\mathbf{3},\mathbf{3}\,|\,\mathbf{1}_1) = d(\mathbf{3},\mathbf{3}\,|\,\mathbf{1}_2) = 1; \tag{3.87}$$

$$d(\mathbf{3},\mathbf{3}\,|\,\mathbf{3}) = \frac{3^3+3(-1)^3}{12} = 2, \tag{3.88}$$

whence

$$\mathbf{3}\times\mathbf{3} = \mathbf{1}+\mathbf{1}_1+\mathbf{1}_2+\mathbf{3}+\mathbf{3}. \tag{3.89}$$

Similarly

$$\mathbf{3}\times\mathbf{1}_1 = \mathbf{3}; \qquad \mathbf{3}\times\mathbf{1}_2 = \mathbf{3}; \qquad \mathbf{1}_1\times\mathbf{1}_2 = \mathbf{1}. \tag{3.90}$$

We can separate the product of the same representation into its symmetric and antisymmetric parts, which must transform separately. Thus we see that we have

$$(\mathbf{3}\times\mathbf{3})_{\text{sym}} = \mathbf{1}+\mathbf{1}_1+\mathbf{1}_2+\mathbf{3}, \qquad (\mathbf{3}\times\mathbf{3})_{\text{antisym}} = \mathbf{3}. \tag{3.91}$$

We want to find in detail the decomposition of the Kronecker product of two of its representations $\mathbf{r}$ and $\mathbf{s}$. In these representations, the group generators will be written as $a^{[\mathbf{r}]}$, $b^{[\mathbf{r}]}$, and $a^{[\mathbf{s}]}$ $b^{[\mathbf{s}]}$, assuming for simplicity that the group has two generators a and b (rank two). These act on the Hilbert spaces $|\,i\,\rangle_{\mathbf{r}}$ and $|\,m\,\rangle_{\mathbf{s}}$, respectively. The Kronecker product $\mathbf{r} \times \mathbf{s}$ lives in the direct product Hilbert space $|\,i\,\rangle_{\mathbf{r}}\,|\,m\,\rangle_{\mathbf{s}}$: the group acts on this space with generators

$$A^{[\mathbf{r}\times\mathbf{s}]} = a^{[\mathbf{r}]}a^{[\mathbf{s}]}, \qquad B^{[\mathbf{r}\times\mathbf{s}]} = b^{[\mathbf{r}]}b^{[\mathbf{s}]}, \tag{3.92}$$

in such a way that

$$A^{[\mathbf{r}\times\mathbf{s}]}|\,i\,\rangle_{\mathbf{r}}\,|\,m\,\rangle_{\mathbf{s}} = a^{[\mathbf{r}]}|\,i\,\rangle_{\mathbf{r}}\,a^{[\mathbf{s}]}|\,m\,\rangle_{\mathbf{s}}, \tag{3.93}$$

and

$$B^{[\mathbf{r}\times\mathbf{s}]}|\,i\,\rangle_{\mathbf{r}}\,|\,m\,\rangle_{\mathbf{s}} = b^{[\mathbf{r}]}|\,i\,\rangle_{\mathbf{r}}\,b^{[\mathbf{s}]}|\,m\,\rangle_{\mathbf{s}}. \tag{3.94}$$

Assuming that we know the action of the generators in the original representations, this very convenient formula allows us to explicitly decompose the direct product Hilbert space into irreducible subspaces.

Consider for example the antisymmetric product of two $\mathcal{A}_4$ triplets. The product Hilbert space contains the three antisymmetric states. Then by using the explicit form of a in the triplet, eq. (3.77), we find

$$A^{[(\mathbf{3}\times\mathbf{3})_a]}\left(|\,2\,\rangle_3\,|\,3\,\rangle_{3'} - |\,3\,\rangle_3\,|\,2\,\rangle_{3'}\right) = -\left(|\,2\,\rangle_3\,|\,3\,\rangle_{3'} - |\,3\,\rangle_3\,|\,2\,\rangle_{3'}\right)$$

$$A^{[(\mathbf{3}\times\mathbf{3})_a]}\left(|\,3\,\rangle_3\,|\,1\,\rangle_{3'} - |\,1\,\rangle_3\,|\,3\,\rangle_{3'}\right) = +\left(|\,3\,\rangle_3\,|\,1\,\rangle_{3'} - |\,1\,\rangle_3\,|\,3\,\rangle_{3'}\right)$$

$$A^{[(\mathbf{3}\times\mathbf{3})_a]}\left(|\,1\,\rangle_3\,|\,2\,\rangle_{3'} - |\,2\,\rangle_3\,|\,1\,\rangle_{3'}\right) = -\left(|\,1\,\rangle_3\,|\,2\,\rangle_{3'} - |\,2\,\rangle_3\,|\,1\,\rangle_{3'}\right)$$

so that A is diagonal with the same eigenvalues as in the triplet. The reader can verify that B also has the triplet form so that the antisymmetric product of two triplets is a triplet, as derived earlier.

3.5 Real and complex representations

Consider an irrep $\mathbf{r}$ represented by non-singular matrices $\mathcal{M}^{[\mathbf{r}]}(g)$, for any element g of any finite group. It is obvious that the complex conjugates of these matrices $\overline{\mathcal{M}}^{[\mathbf{r}]}(g)$ also form an irrep, $\mathbf{r}'$, since they satisfy the same closure property. What is the relation of this representation to $\mathbf{r}$? Clearly, the characters of elements in the same classes are related by complex conjugation

$$\chi^{[\mathbf{r}]}(g) = \bar{\chi}^{[\mathbf{r}']}(g). \tag{3.95}$$

One can consider two possibilities. If $\mathbf{r}$ happens to have complex characters, $\mathbf{r}'$ is a *distinct* representation with characters conjugate to those of the first; we call $\mathbf{r}'$

the conjugate representation and denote it by $\bar{r}$. In this case the two matrices $\mathcal{M}$ and $\overline{\mathcal{M}}$ cannot be related by a similarity transformation since they have different characters.

If **r** has real characters, then the characters of **r** and **r**$'$ are the same, and their representation matrices must be related by a similarity transformation

$$\mathcal{M} = \mathcal{S}\,\overline{\mathcal{M}}\,\mathcal{S}^{-1}. \tag{3.96}$$

Using unitarity,

$$\left(\mathcal{M}^{-1}\right)^T = \overline{\mathcal{M}}, \tag{3.97}$$

where the superscript T means transposition, we deduce that

$$\mathcal{S} = \mathcal{M}\,\mathcal{S}\,\mathcal{M}^T. \tag{3.98}$$

Taking the transpose and inverse of this equation, we arrive at

$$\mathcal{S}^T\,\mathcal{S}^{-1}\,\mathcal{M} = \mathcal{M}\,\mathcal{S}^T\,\mathcal{S}^{-1}. \tag{3.99}$$

Since $\mathcal{M}$ represents an irreducible representation, $\mathcal{S}^T\mathcal{S}^{-1}$ must be proportional to the unit matrix, which implies that $\mathcal{S}$ is proportional to its transpose; hence it is either symmetric or antisymmetric

$$\mathcal{S}^T = \pm\mathcal{S}. \tag{3.100}$$

In both cases, **r** is the same (equivalent) to **r**$'$. However when $\mathcal{S}$ is symmetric, the symmetric product of the representation contains a singlet, while in the second case the singlet (invariant) is to be found in the antisymmetric product. Representations with an antisymmetric $\mathcal{S}$ are called *pseudoreal*. In either case the invariant can be written in the form

$$Y^T\bar{\mathcal{S}}\,X, \tag{3.101}$$

where X and Y are column vectors acted upon by $\mathcal{M}^{[\mathbf{r}]}$.

To summarize, representations are of three types: complex with complex characters, real with real characters and with a symmetric quadratic invariant, and pseudoreal with an antisymmetric quadratic invariant.

Complex representations are obviously different from real and pseudoreal representations, while real and pseudoreal representations are harder to distinguish. Fortunately, there is a simple way to find out which is which in terms of characters. We can always express $\bar{\mathcal{S}}$ in the form

$$\bar{\mathcal{S}} = \sum_g \mathcal{M}^T(g)\mathcal{V}\,\mathcal{M}(g), \tag{3.102}$$

in terms of an arbitrary matrix $\mathcal{V}$, with the sum over all group elements. Indeed for any group element g'

$$\mathcal{M}^T(g')\bar{\mathcal{S}}\mathcal{M}(g') = \sum_g \mathcal{M}^T(g')\mathcal{M}^T(g)\mathcal{V}\,\mathcal{M}(g)\mathcal{M}(g'), \tag{3.103}$$

$$= \sum_{g''} \mathcal{M}^T(g'')\mathcal{V}\,\mathcal{M}(g'') = \bar{\mathcal{S}}. \tag{3.104}$$

In terms of matrix elements, we have for the real and pseudoreal representations

$$\bar{\mathcal{S}}_{ij} = \sum_g \mathcal{M}(g)_{ki}\mathcal{V}_{kl}\,\mathcal{M}(g)_{lj} = \pm\sum_g \mathcal{M}(g)_{kj}\mathcal{V}_{kl}\,\mathcal{M}(g)_{li}. \tag{3.105}$$

Hence for any k, l, i and j

$$\sum_g \mathcal{M}(g)_{ki}\mathcal{M}(g)_{lj} = \pm\sum_g \mathcal{M}(g)_{kj}\mathcal{M}(g)_{li}. \tag{3.106}$$

Sum independently over kj and il to obtain

$$\sum_g \mathcal{M}(g)_{ki}\mathcal{M}(g)_{ik} = \pm\sum_g \mathcal{M}(g)_{kk}\mathcal{M}(g)_{ii}, \tag{3.107}$$

that is

$$\sum_g \chi(g^2) = \pm\sum_g \chi(g)\chi(g). \tag{3.108}$$

Since we are dealing with real representations, the right-hand side is just the order of the group. When the representation is complex, $\mathcal{S}$ does not exist and the above expression is zero. Hence our criterion

$$\frac{1}{n}\sum_g \chi(g^2) = \begin{cases} +1 & \text{real} \\ -1 & \text{pseudoreal}\ . \\ 0 & \text{complex} \end{cases} \tag{3.109}$$

Using the character table of the quaternion group, one can show that its doublet representation is pseudoreal, either by using its antisymmetric Kronecker product or by applying this criterion.

3.6 Embeddings

Let $\mathcal{G}$ be a group of order n, with irreps $\mathfrak{R}^{[\alpha]}$, of dimension D_α and characters $\Xi^{[\alpha]}$, represented by matrices $\mathcal{M}^{[\alpha]}(g)$. If $\mathcal{K}$ is an order-k subgroup of $\mathcal{G}$, with irreps $\mathfrak{T}^{[a]}$, of dimensions d_a and characters $\chi^{[a]}$, we wish to study the embeddings of the irreps of $\mathcal{K}$ into those of $\mathcal{G}$. The set of matrices $\mathcal{M}^{[\alpha]}(g)$ form an irrep of $\mathcal{G}$, but specialized to elements k in the subgroup

$$\mathcal{M}^{[\alpha]}(k), \qquad k \in \mathcal{K} \subset \mathcal{G}, \tag{3.110}$$

they form a $(D_\alpha \times D_\alpha)$ *reducible* representation of $\mathcal{K}$. Let $\mathcal{N}^{[a]}(k)$ be the $(d_a \times d_a)$ matrices with characters $\chi^{[a]}$ which represent $\mathcal{K}$. The full reducibility property allows us, after the appropriate similarity transformation of the $(D_\alpha \times D_\alpha)$ matrices, to express them in block-diagonal form, each block being one of the $\mathcal{N}^{[a]}$s

$$\mathcal{M}^{[\alpha]}(k) = \sum_a f^{\alpha}{}_a \mathcal{N}^{[a]}(k), \tag{3.111}$$

or more abstractly

$$\mathfrak{R}^{[\alpha]} = \sum_a f^{\alpha}{}_a \mathfrak{T}^{[a]}. \tag{3.112}$$

The embedding coefficients $f^{\alpha}{}_a$ are positive integers, subject to the dimension constraints,

$$D_\alpha = \sum_a f^{\alpha}{}_a d_a. \tag{3.113}$$

We now show how they are determined from the class structures of $\mathcal{G}$ and of $\mathcal{K}$. The classes C_i of $\mathcal{G}$ break into two categories, those that do not contain any element of $\mathcal{K}$, and those that do. In the latter case, a given class C_i of $\mathcal{G}$ will in general contain elements of $\mathcal{K}$ which live in *different* classes Γ_{i_m} of $\mathcal{K}$.

If the class C_i contains elements of $\mathcal{K}$, we have

$$\Xi_i^{[\alpha]} = \sum_a f^{\alpha}{}_a \chi_{i_m}^{[a]}, \tag{3.114}$$

where $\chi_{i_m}^{[a]}$ is the $\mathcal{K}$ character of *any* of its Γ_{i_m} classes that appear in the C_i class of $\mathcal{G}$. Note that this formula does not depend on m.

We first use the orthogonality of the $\mathcal{G}$ characters to extract

$$\frac{n}{n_i}\delta_{ij} = \sum_\alpha \overline{\Xi}_j^{[\alpha]} \Xi_i^{[\alpha]} = \sum_a \left(\sum_\alpha f^{\alpha}{}_a \overline{\Xi}_j^{[\alpha]} \right) \chi_{i_m}^{[a]}. \tag{3.115}$$

Since this applies to all the classes of $\mathcal{K}$, we then use the orthogonality of its characters, where k_{i_m} is the number of elements in Γ_{i_m}, to find

$$\left(\frac{n}{k}\right) \sum_{i_m} \frac{k_{i_m}}{n_i} \overline{\chi}_{i_m}^{[b]} = \sum_\alpha f^{\alpha}{}_b \overline{\Xi}_i^{[\alpha]}, \tag{3.116}$$

where the sum is over the classes of $\mathcal{K}$ which are contained in the C_i class of $\mathcal{G}$.

If the class $C_\perp$ with character $\Xi_\perp^{[\alpha]}$ has no element in the subgroup, a similar reasoning leads to

$$\sum_\alpha f^\alpha{}_b \, \overline{\Xi}_\perp^{[\alpha]} = 0. \tag{3.117}$$

Sums of characters with positive integer coefficients are called *compound* characters. The length of a compound character

$$\sum_i \frac{n_i}{n} \left(\sum_\alpha f^\alpha{}_b \, \overline{\Xi}_i^{[\alpha]} \right)^2 \tag{3.118}$$

is the number of (simple) characters it contains: a compound character of unit length is a character. These formulæ can be used to find the embedding coefficients, and also determine the $\mathcal{G}$ characters from those of its subgroup $\mathcal{K}$.

Example: $\mathcal{S}_4 \supset \mathcal{A}_4$

Let us find the characters of $\mathcal{S}_4$ from those of $\mathcal{A}_4$, as well as the embedding coefficients. $\mathcal{S}_4$ has five classes. Two of them, C_2 and C_5 contain only odd permutations and do not appear in $\mathcal{A}_4$. The other three split according to

$$C_1 \supset \Gamma_1, \qquad C_3 \supset \Gamma_4, \qquad C_4 \supset \Gamma_2 + \Gamma_3, \tag{3.119}$$

where $\Gamma_{1,2,3,4}$ are the $\mathcal{A}_4$ classes. Hence

$$\sum_\alpha f^\alpha{}_b \, \Xi_2^{[\alpha]} = 0, \qquad \sum_\alpha f^\alpha{}_b \, \Xi_5^{[\alpha]} = 0. \tag{3.120}$$

Applying to the remaining classes yields

$$\sum_\alpha f^\alpha{}_b \, \Xi_1^{[\alpha]} = \frac{24}{12} \chi_1^{[b]}, \text{ for } C_1, \tag{3.121}$$

$$\sum_\alpha f^\alpha{}_b \, \Xi_3^{[\alpha]} = \frac{24}{12} \chi_4^{[b]}, \text{ for } C_3, \tag{3.122}$$

$$\sum_\alpha f^\alpha{}_b \, \Xi_4^{[\alpha]} = \frac{24}{12} \left(\frac{4}{8} \chi_2^{[b]} + \frac{4}{8} \chi_3^{[b]} \right), \text{ for } C_4. \tag{3.123}$$

The next step is to evaluate the compound characters for the four $\mathcal{A}_4$ irreps. Their values are summarized in the following table.

	C_1	$6C_2$	$3C_3$	$8C_4$	$6C_5$	length
1:	2	0	2	2	0	2
$\mathbf{1_1}$:	2	0	2	−1	0	1
$\overline{\mathbf{1}}_1$:	2	0	2	−1	0	1
3:	6	0	−2	0	0	2

Here, we have used $e^{2\pi i/3} + e^{4\pi i/3} = -1$, and indicated their lengths. Since $\mathcal{S}_4$ has a singlet representation with character $\Xi^{[1]}$, it is easy to see that the first compound character contains the singlet once. This yields the first character by subtraction, namely $\Xi^{[1']} = (1, -1, 1, 1, -1)$. This is the one-dimensional representation where the odd permutations are equal to minus one.

The second and third compound characters have unit length: they represent one irrep with character $\Xi^{[2]} = (2, 0, 2, -1, 0)$.

The fourth compound character of length two is made up of two characters. One can check that it is orthogonal to $\Xi^{[1]}$, $\Xi^{[1']}$, and $\Xi^{[2]}$. It must therefore split into two characters each representing a three-dimensional irrep

$$\Xi^{[3_1]} = (3, 1, -1, 0, -1), \qquad \Xi^{[3_2]} = (3, -1, -1, 0, 1). \tag{3.124}$$

The ambiguities in determining their values are fixed by orthogonality with the other characters. This leads us to the following $\mathcal{S}_4$ character table.

$\mathcal{S}_4$	C_1	$6C_2^{[2]}$	$3C_3^{[2]}$	$8C_4^{[3]}$	$6C_5^{[4]}$
$\chi^{[1]}$	1	1	1	1	1
$\chi^{[1']}$	1	−1	1	1	−1
$\chi^{[2]}$	2	0	2	−1	0
$\chi^{[3_1]}$	3	1	−1	0	−1
$\chi^{[3_2]}$	3	−1	−1	0	1

We can read off the embedding coefficients, and deduce the decomposition of the $\mathcal{S}_4$ irreps in terms of those of $\mathcal{A}_4$:

$$\mathcal{S}_4 \supset \mathcal{A}_4 : \qquad \mathbf{1}' = \mathbf{1}; \quad \mathbf{2} = \mathbf{1}_1 + \overline{\mathbf{1}}_1; \quad \mathbf{3}_1 = \mathbf{3}; \qquad \mathbf{3}_2 = \mathbf{3}. \tag{3.125}$$

We see that in going from $\mathcal{A}_4$ to $\mathcal{S}_4$, the $\mathcal{A}_4$ singlet splits into two one-dimensional irreps of $\mathcal{S}_4$. The same happen to the $\mathcal{A}_4$ triplet which splits into two triplets of $\mathcal{S}_4$. On the other hand, the complex $\mathcal{A}_4$ one-dimensional irrep and its conjugate assemble into one real $\mathcal{S}_4$ doublet.

This is a general pattern: some irreps duplicate, and some assemble in going from the subgroup to the group.

There is a systematic way to understand these patterns in terms of the automorphism group of $\mathcal{A}_4$. Some mappings of the group into itself, like conjugation, live inside the group; they are the *inner* automorphisms. Some groups have automorphisms which are not to be found inside the group, called *outer* automorphisms. This is true for $\mathcal{A}_4$ (as we saw earlier), and in fact for any alternating group.

Let $r = (1\,2)$ be the odd permutation of order two which permutes the first two objects, and generates $\mathcal{Z}_2$, the group of outer automorphisms of $\mathcal{A}_4$. It is not in $\mathcal{A}_4$, but maps it into itself by conjugation: $a_i \to r\,a_i r^{-1}$ with $a_i \in \mathcal{A}_4$. This allows the construction of the group $\mathcal{S}_4$, which contains both even and odd permutations. We denote their relation symbolically (as in the ATLAS) by

$$\mathcal{S}_4 = \mathcal{A}_4 \cdot 2, \qquad \mathcal{Z}_2 = \mathcal{S}_4/\mathcal{A}_4, \tag{3.126}$$

with $\mathcal{A}_4$ as commutator subgroup of $\mathcal{S}_4$, and $\mathcal{Z}_2$ is the quotient group.

3.7 $\mathcal{Z}_n$ character table

The cyclic groups of order n are generated by a single element a of order n, with presentation

$$\mathcal{Z}_n : \quad < a \,|\, a^n = e > .$$

All its elements $a, a^2, \ldots, a^{n-1}, e$ commute with one another: all can be represented by (1×1) matrices, that is (complex) numbers. All its n classes are one-dimensional since it is Abelian. The order of the element in each class depends on n. If n is prime, all classes contain an element of order n, but if $n = p \cdot q \cdots$ is a product of integers, some classes have elements of order p, q, etc. For this reason, we do not include the bracketed superscript for the classes. As the presentation requires a to be any one of the nth root of unity,

$$a = 1,\ e^{2i\pi/n}\ e^{2\cdot 2i\pi/n},\ \ldots,\ e^{(n-1)\cdot 2i\pi/n},$$

each choice yields one of n different irreps, labeled by $k = 0, 1, \ldots, n-1$, in agreement with the n one-dimensional conjugacy classes. The first one is the trivial and (very) unfaithful singlet, with all elements equal to one. This information is encoded in the following $\mathcal{Z}_n$ character table.

$\mathcal{Z}_n$	C_1	C_2	C_3	$\cdots$	$\cdots$	C_n
$\chi^{[\mathbf{1}]}$	1	1	1	$\cdots$	$\cdots$	1
$\chi^{[\mathbf{1_1}]}$	1	ϵ	ϵ^2	$\cdots$	$\cdots$	ϵ^{n-1}
$\chi^{[\mathbf{1_2}]}$	1	ϵ^2	ϵ^4	$\cdots$	$\cdots$	ϵ^{2n-2}
$\chi^{[\mathbf{1_3}]}$	1	ϵ^3	ϵ^6	$\cdots$	$\cdots$	ϵ^{3n-3}
.	.	$\cdots$	$\cdots$	$\cdots$	$\cdots$	$\cdots$
$\chi^{[\mathbf{1_{n-1}}]}$	1	ϵ^{n-1}	ϵ^{2n-2}	$\cdots$	$\cdots$	$\epsilon^{(n-1)^2}$

Here, $\epsilon = e^{2i\pi/n}$. Further, since $\epsilon^* = 1/\epsilon$, we see that for even n two representations are real, and $(n-2)/2$ are complex. For n odd $= 2m+1$, only the singlet is real with m complex representations (and their conjugates). In the regular representation a is represented by the $(n \times n)$ matrix

$$a:\quad M(a) = \begin{pmatrix} 0 & 1 & 0 & 0 & \cdots & 0 \\ 0 & 0 & 1 & 0 & \cdots & 0 \\ . & . & . & . & \cdots & . \\ . & . & . & . & \cdots & . \\ 0 & 0 & 0 & 0 & \cdots & 1 \\ 1 & 0 & 0 & 0 & \cdots & 0 \end{pmatrix},$$

which can be brought to diagonal form

$$T\,M(a)\,T^{-1} = \mathrm{Diag}\left(1,\ \epsilon^{(n-1)},\ \epsilon^{(n-2)},\ \cdots,\ \epsilon\right),$$

by

$$T = \frac{1}{\sqrt{n}} \begin{pmatrix} 1 & 1 & 1 & \cdots & 1 \\ 1 & \epsilon & \epsilon^2 & \cdots & \epsilon^{n-1} \\ 1 & \epsilon^2 & \epsilon^4 & \cdots & \epsilon^{2n-2} \\ . & . & . & \cdots & . \\ . & . & . & \cdots & . \\ 1 & \epsilon^{n-1} & \epsilon^{2n-2} & \cdots & \epsilon^{(n-1)^2} \end{pmatrix}.$$

Finally, the Kronecker products of its representations are pretty easy to read off from the character table:

$$\mathbf{1_i} \times \mathbf{1_j} = \mathbf{1_k}, \qquad k = i + j \quad \mathrm{mod}(n).$$

3.8 $\mathcal{D}_n$ character table

A slightly more complicated construction is that of the character tables of the dihedral groups. It is convenient to start from their expression as semi-direct products of cyclic groups

$$\mathcal{D}_n = \mathcal{Z}_n \rtimes \mathcal{Z}_2,$$

with presentation $< a, b \,|\, a^n = b^2 = e \,;\, bab^{-1} = a^{-1} >$, where a and b generate $\mathcal{Z}_n$ and $\mathcal{Z}_2$, respectively. The $2n$ elements of $\mathcal{D}_n$ are then $e, a, a^2, \ldots, a^{n-1}$ and $b, ba, ba^2, \ldots, ba^{n-1}$. The class structure is different as to whether n is even or odd.

For odd n, conjugation under b generates $(n-1)/2$ classes:

$$\left(a, a^{-1}\right), \left(a^2, a^{-2}\right), \ldots, \left(a^{(n-1)/2}, a^{(1-n)/2}\right),$$

and conjugation under a, $aba^{-1} = ba^{-2}$, yields one class

$$\left(b, ba^{-2}, ba^{-4}, \ldots\right)$$

of n elements: $\mathcal{D}_{2m+1}$ has $(m+2)$ classes, and of course as many irreps.

For even n, b-conjugation yields the same $(n-2)/2$ classes of the form (a^k, a^{-k}), plus one additional class of one element $a^{n/2}$. On the other hand, a-conjugation generates two classes, one with elements of the form ba^{even}, the other ba^{odd}: $\mathcal{D}_{2m}$ has $(m+3)$ classes and as many irreps.

To construct their character tables, consider the one-dimensional representations for which a and b are commuting real or complex numbers. It follows that $b = \pm 1$, and $bab^{-1} = a = a^{-1}$, so that we find both $a^2 = a^n = e$.

For odd n, these have only one solution, $a = 1$: $\mathcal{D}_{2m+1}$ has two one-dimensional irreps ($a = 1, b = \pm 1$).

Since $b^2 = e$, the action of $\mathcal{Z}_2$ on $\mathcal{Z}_n$ yields at most two elements of $\mathcal{Z}_n$ (two-dimensional orbits), and we expect $\mathcal{D}_n$ to have two-dimensional irreps. The order of the group formula

$$2n = 1 + 1 + \frac{(n-1)}{2}(2)^2$$

yields $(n-1)/2$ two-dimensional irreps, in which a is represented by a reducible (2×2) matrix. To construct them, go to a basis where a is a diagonal matrix

$$a = \begin{pmatrix} \alpha & 0 \\ 0 & \beta \end{pmatrix},$$

where α and β are nth roots of unity. $\mathcal{Z}_2$ is represented by its regular representation, but since b need not commute with a, it cannot be diagonal, leading to the unitarily equivalent form

$$b = \begin{pmatrix} 0 & 1 \\ 1 & 0 \end{pmatrix}.$$

We see that b-conjugation of a requires $\beta = \alpha^{-1}$, yielding for all two-dimensional representations

$$a = \begin{pmatrix} e^{2\pi ik/n} & 0 \\ 0 & e^{-2\pi ik/n} \end{pmatrix}, \quad \chi_k(a) = 2c_k \equiv 2\cos\left(\frac{2\pi k}{n}\right); \quad b = \begin{pmatrix} 0 & 1 \\ 1 & 0 \end{pmatrix},$$

with $k = 1, 2, \ldots, (n-1)/2$. The character table for $n = 2m + 1$ easily follows.

$\mathcal{D}_{2m+1}$	C_1	$2C_2(a)$	$2C_3(a^2)$	$\cdots$	$2C_{m+1}(a^m)$	$nC_{m+2}(b)$
$\chi^{[\mathbf{1}]}$	1	1	1	$\cdots$	1	1
$\chi^{[\mathbf{1_1}]}$	1	1	1	$\cdots$	1	−1
$\chi^{[\mathbf{2_1}]}$	2	$2c_1$	$2c_2$	$\cdots$	$2c_m$	0
$\chi^{[\mathbf{2_2}]}$	2	$2c_2$	$2c_4$	$\cdots$	$2c_{2m}$	0
.	.	$\cdots$	$\cdots$	$\cdots$	$\cdots$	$\cdots$
$\chi^{[\mathbf{2_m}]}$	2	$2c_m$	$2c_{2m}$	$\cdots$	$2c_{m^2}$	0

The Kronecker products of the two one-dimensional irreps $\mathbf{1}, \mathbf{1_1}$, and the m doublets $\mathbf{2_k}$, $k = 1, 2, \ldots, m$ are rather easy to work out:

$$\mathbf{1_1} \times \mathbf{1_1} = \mathbf{1}, \qquad \mathbf{1_1} \times \mathbf{2_k} = \mathbf{2_k},$$
$$\mathbf{2_k} \times \mathbf{2_k} = \mathbf{1} + \mathbf{1_1} + \mathbf{2_{2k}},$$
$$\mathbf{2_j} \times \mathbf{2_k} = \mathbf{2_{j-k}} + \mathbf{2_{j+k}} \quad j > k,$$

with the identification $\mathbf{2_j} = \mathbf{2_{n-j}}$ for $j = 1, 2, \ldots, (n-1)$.

For even $n = 2m$, the presentation yields two possible solutions for the singlets, $a = \pm 1$: $\mathcal{D}_{2m}$ has four one-dimensional representations ($a = \pm 1, b = \pm 1$). It follows from the order of the group formula

$$2(2m) = 1 + 1 + 1 + 1 + (m-1)(2)^2,$$

that we have $(m-1)$ two-dimensional representations. The slightly more complicated character table reads as follows.

$\mathcal{D}_{2m}$	C_1	$2C_2(a)$	$2C_3(a^2)$	$\cdot\cdot$	$2C_m(a^{m-1})$	$C_{m+1}(a^m)$	$mC_{m+2}(b)$	$mC_{m+3}(ba)$
$\chi^{[\mathbf{1}]}$	1	1	1	$\cdot\cdot$	1	1	1	1
$\chi^{[\mathbf{1_1}]}$	1	1	1	$\cdot\cdot$	1	1	−1	−1
$\chi^{[\mathbf{1_2}]}$	1	−1	1	$\cdot\cdot$	$(-1)^{m-1}$	$(-1)^m$	1	−1
$\chi^{[\mathbf{1_3}]}$	1	−1	1	$\cdot\cdot$	$(-1)^{m-1}$	$(-1)^m$	−1	1
$\chi^{[\mathbf{2_1}]}$	2	$2c_1$	$2c_2$	$\cdot\cdot$	$2c_{m-1}$	$2c_m$	0	0
$\chi^{[\mathbf{2_2}]}$	2	$2c_2$	$2c_4$	$\cdot\cdot$	$2c_{2m-2}$	$2c_{2m}$	0	0
$\cdots$	.	$\cdots$	$\cdots$	$\cdot\cdot$	$\cdots$	$\cdots$	$\cdots$	$\cdots$
$\chi^{[\mathbf{2_{m-1}}]}$	2	$2c_{m-1}$	$2c_{2m-2}$	$\cdot\cdot$	$2c_{(m-1)^2}$	$2c_{m(m-1)}$	0	0

The doublet representations are of the same form as in the odd case. The Kronecker products of the four one-dimensional $\mathbf{1}$ and $\mathbf{1_r}$, $r = 1, 2, 3$ irreps and $(m-1)$ doublets $\mathbf{2_k}$, $k = 1, 2, \ldots, (m-1)$ are tediously worked with the results

$$\begin{aligned}
&\mathbf{1_r} \times \mathbf{1_r} = \mathbf{1},\\
&\mathbf{1_r} \times \mathbf{1_s} = \mathbf{1_t} \quad r, s, t = 1, 2, 3, \quad r \neq s \neq t,\\
&\mathbf{1_1} \times \mathbf{2_k} = \mathbf{2_k},\\
&\mathbf{1_2} \times \mathbf{2_k} = \mathbf{1_3} \times \mathbf{2_k} = \mathbf{2_{|m-k|}}.
\end{aligned}$$

The Kronecker products of two different doublets satisfy more complicated rules:

$$\begin{aligned}
\mathbf{2_j} \times \mathbf{2_k} &= \mathbf{2_{j-k}} + \mathbf{2_{j+k}}, \quad j > k \neq (m-j)\\
\mathbf{2_j} \times \mathbf{2_{m-j}} &= \mathbf{1_2} + \mathbf{1_3} + \mathbf{2_{|m-2j|}}, \quad j \neq 2m.
\end{aligned}$$

The product of the same doublet is simpler

$$\mathbf{2_k} \times \mathbf{2_k} = \mathbf{1} + \mathbf{1_1} + \mathbf{2_{2k}}, \quad k \neq m/2,$$

For m even, the "half-way" doublet satisfies its own Kronecker,

$$\mathbf{2_{m/2}} \times \mathbf{2_{m/2}} = \mathbf{1} + \mathbf{1_1} + \mathbf{1_2} + \mathbf{1_3}.$$

The maximal subgroup of $\mathcal{D}_{2m+1}$ is $\mathcal{Z}_{2m+1}$, which rotates the $(2m+1)$ vertices. $\mathcal{D}_{2m}$ has two maximal subgroups, $\mathcal{Z}_{2m}$ which rotates the $2m$ vertices of the polygons, but also $\mathcal{D}_m$ which is the symmetry of the m-gon that is inscribed by m vertices of the original $2m$ vertices (skipping one). The two ways to choose such an m-gon yield two equivalent embeddings. Finally we note that the commutator subgroup of $\mathcal{D}_{2m}$ is $\mathcal{Z}_m$, that of $\mathcal{D}_{2m+1}$ is $\mathcal{Z}_{2m+1}$, indicating that the dihedral groups are not simple.

3.9 $\mathcal{Q}_{2n}$ character table

The dicyclic groups $\mathcal{Q}_{2n}$ of order $4n$ have the presentation $< a, b \,|\, a^{2n} = e,\ a^n = b^2\ ;\ bab^{-1} = a^{-1} >$. They have four one-dimensional irreps: when the generators commute with one another, the presentation reduces to $a^{2n} = e, a^n = b^2, a = a^{-1}$. Since a is its own inverse, and the $(2n)$th root of unity, it can only be ± 1. In addition, b is a fourth root of unity, with four possible values, $\pm 1, \pm i$. The final constraint $a^n = b^2$ amounts to $(\pm 1)^n = b^2$. This leaves the following solutions:

- n even: four one-dimensional irreps, $a = \pm 1, b = \pm 1$;
- n odd: four one-dimensional irreps: $a = 1, b = \pm 1$ and $a = -1, b = \pm i$.

It is just as easy to construct two-dimensional irreps. To wit, choose a basis where a is diagonal

$$a = \begin{pmatrix} \eta^k & 0 \\ 0 & \eta^l \end{pmatrix}, \qquad b = \begin{pmatrix} r & s \\ t & u \end{pmatrix},$$

where $\eta^{2n} = 1, k, l$ range over $1, 2, \ldots, (2n-1)$, and r, s, t, u are to be determined. The presentation yields four equations

$$(1-\eta^{2k})\,r = 0; \quad (1-\eta^{2l})\,u = 0; \quad (1-\eta^{k+l})\,s = 0; \quad (1-\eta^{k+l})\,t = 0.$$

If r, s, t, u are all different from zero, we are left with $k = l = n$ and a becomes $\pm$ the unit matrix: this case reduces to one-dimensional irreps.

There is only one type of non-trivial solution to these equations $(s, t \neq 0, r = u = 0)$; the case $(r, u \neq 0, s = t = 0)$ yields only the previously obtained one-dimensional irreps.

Setting $(s, t \neq 0, r = u = 0)$ requires $k = -l$, as well as

$$b^2 = a^n \quad \rightarrow \quad st = \eta^{nk} = \eta^{-nk} = (-1)^k,$$

so that the generators become

$$a = \begin{pmatrix} \eta^k & 0 \\ 0 & \eta^{-k} \end{pmatrix}, \qquad b = \begin{pmatrix} 0 & 1 \\ (-1)^k & 0 \end{pmatrix},$$

where we have set $s = 1$ with a harmless similarity transformation. For $k = n$, this representation reduces to two singlets. Also, the representations for $k = (n+1)$, $(n+2), \ldots, (2n-1)$ are the same as those for $k = 1, 2, \ldots, (n-1)$, after a trivial relabeling of the matrix elements. Hence we have $(n-1)$ independent doublet irreps. By checking the dimension of the regular representation,

$$4n = 1 + 1 + 1 + 1 + (n-1)2^2,$$

we conclude that $\mathcal{Q}_{2n}$ contains no more irreps. It follows that we have $(n-1)+4 = (n+3)$ different classes.

The character tables are different depending on whether n is even or odd. For n is even, it is given by the following.

$\mathcal{Q}_{2n}$	C_1	$2C_2(a)$	$2C_3(a^2)$	$\cdot\cdot$	$2C_n(a^{n-1})$	$C_{n+1}(a^n)$	$nC_{n+2}(b)$	$nC_{n+3}(ba)$
$\chi^{[1]}$	1	1	1	$\cdot\cdot$	1	1	1	1
$\chi^{[1_1]}$	1	1	1	$\cdot\cdot$	1	1	-1	-1
$\chi^{[1_2]}$	1	-1	1	$\cdot\cdot$	$(-1)^{n-1}$	$(-1)^n$	1	-1
$\chi^{[1_3]}$	1	-1	1	$\cdot\cdot$	$(-1)^{n-1}$	$(-1)^n$	-1	1
$\chi^{[2_1]}$	2	$2c_{1/2}$	$2c_1$	$\cdot\cdot$	$2c_{(n-1)/2}$	$2c_{n/2}$	0	0
$\chi^{[2_2]}$	2	$2c_1$	$2c_2$	$\cdot\cdot$	$2c_{n-1}$	$2c_n$	0	0
$\cdots$	$\cdot$	$\cdots$	$\cdots$	$\cdot\cdot$	$\cdots$	$\cdots$	$\cdots$	$\cdots$
$\chi^{[2_{n-1}]}$	2	$2c_{(n-1)/2}$	$2c_{n-1}$	$\cdot\cdot$	$2c_{(n-1)^2/2}$	$2c_{n(n-1)/2}$	0	0

It is the same as the character table of $\mathcal{D}_{2n}$, so that the Kronecker product of its irreps is also the same.

When n is odd, the character table is slightly different.

$\mathcal{Q}_{2n}$	C_1	$2C_2(a)$	$2C_3(a^2)$	$\cdots$	$2C_n(a^{n-1})$	$C_{n+1}(a^n)$	$nC_{n+2}(b)$	$nC_{n+3}(ba)$
$\chi^{[1]}$	1	1	1	$\cdots$	1	1	1	1
$\chi^{[1_1]}$	1	1	1	$\cdots$	1	1	-1	-1
$\chi^{[1_2]}$	1	-1	1	$\cdots$	$(-1)^{n-1}$	$(-1)^n$	i	$-i$
$\chi^{[1_3]}$	1	-1	1	$\cdots$	$(-1)^{n-1}$	$(-1)^n$	$-i$	i
$\chi^{[2_1]}$	2	$2c_{1/2}$	$2c_1$	$\cdots$	$2c_{(n-1)/2}$	$2c_{n/2}$	0	0
$\chi^{[2_2]}$	2	$2c_1$	$2c_2$	$\cdots$	$2c_{n-1}$	$2c_n$	0	0
$\cdots$	$\cdot$	$\cdots$	$\cdots$	$\cdots$	$\cdots$	$\cdots$	$\cdots$	$\cdots$
$\chi^{[2_{n-1}]}$	2	$2c_{(n-1)/2}$	$2c_{n-1}$	$\cdots$	$2c_{(n-1)^2/2}$	$2c_{n(n-1)/2}$	0	0

The products of its irreps are given by

$$\mathbf{1_1} \times \mathbf{1_1} = \mathbf{1}, \qquad \mathbf{1_2} \times \mathbf{1_2} = \mathbf{1_3} \times \mathbf{1_3} = \mathbf{1_1},$$
$$\mathbf{1_1} \times \mathbf{1_2} = \mathbf{1_3}, \qquad \mathbf{1_1} \times \mathbf{1_3} = \mathbf{1_2}, \qquad \mathbf{1_2} \times \mathbf{1_3} = \mathbf{1},$$
$$\mathbf{1_1} \times \mathbf{2_k} = \mathbf{2_k}, \qquad \mathbf{1_2} \times \mathbf{2_k} = \mathbf{1_3} \times \mathbf{2_k} = \mathbf{2_{(n-k)}},$$
$$\mathbf{2_k} \times \mathbf{2_k} = \mathbf{1} + \mathbf{1_1} + \mathbf{2_{2k}},$$
$$\mathbf{2_j} \times \mathbf{2_k} = \mathbf{2_{j-k}} + \mathbf{2_{j+k}} \quad (j+k) \neq n, \quad j > k,$$
$$\mathbf{2_j} \times \mathbf{2_{(n-j)}} = \mathbf{1_2} + \mathbf{1_3} + \mathbf{2_{|2j-n|}},$$

with the identification $\mathbf{2_j} = \mathbf{2_{2n-j}}$ for $j = 1, 2, \ldots, (n-1)$.

None of the dicyclic groups is simple, since $\mathcal{Z}_n$ is the commutator subgroup of $\mathcal{Q}_{2n}$.

We are fortunate to be able to build these character tables and representation in such a straightforward manner. In order to do the same for more complicated groups, new machinery and concepts need to be introduced.

3.10 Some semi-direct products

We have seen that the construction of the representations of the dihedral group was facilitated by its semi-direct product structure. In this section, we further elaborate on semi-direct products.

We begin by introducing the *holomorph* of a group as the semi-direct product of a group with its automorphism group (the group which maps its multiplication table to itself). Take the cyclic groups as an example. The holomorph of a cyclic group $\mathcal{H}ol(Z_n)$ depends on n. When $n = p$ is prime, $\mathcal{A}ut\ \mathcal{Z}_p = \mathcal{Z}_{p-1}$, with multiplication defined mod(p). For instance, the elements of $\mathcal{A}ut\ \mathcal{Z}_7$ are the integers $(1, 2, 3, 4, 5, 6)$; multiplication is mod(7), such that $5 \cdot 4 = 20 = 6$, etc. This yields the family of semi-direct products

$$\mathcal{H}ol(Z_p) = \mathcal{Z}_p \rtimes \mathcal{Z}_{p-1}, \tag{3.127}$$

of order $p(p-1)$. We use the convention that the group acted upon appears first. The second should be called the actor group.

When $n = 2, 4$, or $n = p^m, 2p^m$, with $p \neq 2$ is prime, $\mathcal{A}ut\ \mathcal{Z}_n$ is also cyclic, with as many elements as the number of integers less than n with no integers in common (coprime): $\mathcal{A}ut\ \mathcal{Z}_6$ is the cyclic group $\mathcal{Z}_2$, with two elements $1, 5$, coprimes to 6. In general, the number of coprimes of n is given by Euler's totient $\phi(n)$ function. Hence we can build the semi-direct product groups

$$\mathcal{H}ol(\mathcal{Z}_n) = \mathcal{Z}_n \rtimes \mathcal{Z}_{\phi(n)}, \qquad n = 2,\ 4,\ p^m,\ 2p^m\ p \text{ prime} \neq 2. \tag{3.128}$$

Low-lying examples are

$$\mathcal{H}ol(\mathcal{Z}_5) = \mathcal{Z}_5 \rtimes \mathcal{Z}_4 : \quad < a^5 = b^4 = e \,|\, b^{-1}ab = a^3 >, \tag{3.129}$$

$$\mathcal{H}ol(\mathcal{Z}_7) = \mathcal{Z}_7 \rtimes \mathcal{Z}_6 : \quad < a^7 = b^6 = e \,|\, b^{-1}ab = a^5 > . \tag{3.130}$$

In other cases, the automorphism group need not be cyclic: $\mathcal{A}ut\ \mathcal{Z}_8$ is $\mathcal{D}_2 = \mathcal{Z}_2 \times \mathcal{Z}_2$, Klein's Vierergruppe, yielding the order 32 holomorph

$$\begin{aligned}&\mathcal{H}ol(\mathcal{Z}_8) = \mathcal{Z}_8 \rtimes (\mathcal{Z}_2 \times \mathcal{Z}_2) : \\ &\quad < a^8 = b^2 = c^2 = e \,|\, b^{-1}ab = a^{-1}\ ;\ c^{-1}ac = a^5 > .\end{aligned} \tag{3.131}$$

In all cases, $\mathcal{A}ut\ \mathcal{Z}_n$ contains $\phi(n)$ elements, with the multiplication table defined modulo n.

There are many other ways to build semi-direct products of two cyclic groups, as the action of one cyclic group on the other may not be unique. Consider a special case: for any two integers $m < n$, let a, $a^n = e$ generate $\mathcal{Z}_n$, and b, $b^m = e$ generate $\mathcal{Z}_m$, and define the action of $\mathcal{Z}_m$ on $\mathcal{Z}_n$ by conjugation

$$b\,ab^{-1} = a^r.$$

It follows that

$$b^2\,ab^{-2} = b\,a^r b^{-1} = (b\,a\,b^{-1})^r = a^{r \cdot r},$$

so that eventually,

$$b^m\,ab^{-m} = a = a^{r^m},$$

which requires the necessary condition

$$r^m = 1 + l\,n,$$

where l is integer. This enables us to construct the semi-direct product of order nm

$$\mathcal{Z}_n \rtimes_r \mathcal{Z}_m : \quad < a, b \mid a^n = e,\, b^m = e\,;\, bab^{-1} = a^r >, \quad r^m = 1 + ln, \tag{3.132}$$

which may have several solutions. For example, this yields several groups of order 32:

$$\mathcal{Z}_8 \rtimes_5 \mathcal{Z}_4 : \quad < a, b \mid a^8 = e,\, b^4 = e\,;\, bab^{-1} = a^5 >, \quad 5^4 = 625 = 1 \bmod (8),$$

$$\mathcal{Z}_8 \rtimes_{-1} \mathcal{Z}_4 : \quad < a, b \mid a^8 = e,\, b^4 = e\,;\, bab^{-1} = a^{-1} >, \quad (-1)^4 = 1 \bmod (8),$$

$$\mathcal{Z}_8 \rtimes_3 \mathcal{Z}_4 : \quad < a, b \mid a^8 = e,\, b^4 = e\,;\, bab^{-1} = a^3 >, \quad 3^4 = 81 = 1 \bmod (8),$$

showing different semi-direct product constructions, and

$$\mathcal{Z}_{16} \rtimes_7 \mathcal{Z}_2 : \quad < a, b \mid a^{16} = e,\, b^2 = e\,;\, bab^{-1} = a^7 >, \quad 7^2 = 49 = 1 \bmod (16).$$

The infinite family of dihedral groups can be constructed in this way

$$\mathcal{D}_n = \mathcal{Z}_n \rtimes_{-1} \mathcal{Z}_2, \tag{3.133}$$

corresponding to $r = -1$, $r^2 = 1$.

Note that although the dicyclic groups $\mathcal{Q}_{2n}$ are not in general semi-direct products, there are special isomorphisms such as $\mathcal{Q}_6 = \mathcal{Z}_3 \rtimes_3 \mathcal{Z}_4$.

Consider semi-direct products of the form $(\mathcal{Z}_r \times \mathcal{Z}_s) \rtimes \mathcal{Z}_t$, where r, s, t are integers. The action of $\mathcal{Z}_t$ on the $(\mathcal{Z}_r \times \mathcal{Z}_s)$ may be more complicated, as it can act on either of the product groups or both. Let us take as an example the action of $\mathcal{Z}_2$ on $(\mathcal{Z}_4 \times \mathcal{Z}_2)$. Let $a \in \mathcal{Z}_4$, $a^4 = e$, and $b \in \mathcal{Z}_2$, $b^2 = e$ with $ab = ba$ generate the direct product. The action of the first $\mathcal{Z}_2$ on a must send it into an element of order 4 in $\mathcal{Z}_4 \times \mathcal{Z}_2$. There is only one such element a^3b as $(a^3b)^4 = e$. Hence we can define the action as

$$\mathcal{Z}_2 : \qquad a \rightarrow a^3 b, \qquad b \rightarrow b,$$

or it can act on the $\mathcal{Z}_2$ in the direct product as

$$\mathcal{Z}_2 : \qquad a \rightarrow a, \qquad b \rightarrow a^2 b,$$

since $(a^2b)^2 = e$. If the actor group is generated by c, $c^2 = 1$, these correspond respectively to

$$c^{-1}ac = a^3 b, \qquad c^{-1}bc = b,$$

and

$$c^{-1}ac = a, \qquad c^{-1}bc = a^2 b.$$

It follows that there are two semi-direct products involving these groups. The richness of the semi-direct product construction is evidenced by looking at the order sixteen groups, where all but one of the nine non-Abelian groups can be expressed either as a direct or semi-direct product

$$\mathcal{Q}_8, \mathcal{D}_8(= \mathcal{Z}_8 \rtimes_{-1} \mathcal{Z}_2), \mathcal{Q} \times \mathcal{Z}_2, \mathcal{D}_4 \times \mathcal{Z}_2, \mathcal{Z}_8 \rtimes_3 \mathcal{Z}_2,$$

$$\mathcal{Z}_8 \rtimes_5 \mathcal{Z}_2, \mathcal{Z}_4 \rtimes_3 \mathcal{Z}_4, (\mathcal{Z}_4 \times \mathcal{Z}_2) \rtimes \mathcal{Z}_2, (\mathcal{Z}_4 \times \mathcal{Z}_2) \rtimes' \mathcal{Z}_2.$$

The last two groups correspond to the action of the actor group on the first and second group we just discussed.

3.11 Induced representations

This is a method which constructs representations of a group using the representations of its subgroups. It is particularly well-suited for groups which are semi-direct products. Let $\mathcal{H}$ be a subgroup of $\mathcal{G}$ of index N

$$\mathcal{G} \supset \mathcal{H}. \tag{3.134}$$

Suppose we know a d-dimensional representation $\mathbf{r}$ of $\mathcal{H}$, that is for every $h \in \mathcal{H}$,

$$h: \, |\, i\,\rangle \to |\, i'\,\rangle = \mathcal{M}(h)^{[\mathbf{r}]}_{i'j} |\, j\,\rangle, \tag{3.135}$$

acting on the d-dimensional Hilbert space $\mathfrak{H}$, spanned by $|\, i\,\rangle$, $i = 1, 2, \ldots, d$. We want to show how to construct a representation of $\mathcal{G}$, starting from the known representation on $\mathfrak{H}$. It is called the *induced representation* of $\mathcal{G}$ by the $\mathbf{r}$ representation of $\mathcal{H}$.

We begin by considering the coset made up of N points,

$$\mathcal{H} \oplus g_1\mathcal{H} \oplus g_2\mathcal{H} \oplus \cdots \oplus g_{N-1}\mathcal{H}, \tag{3.136}$$

where g_k are elements of $\mathcal{G}$ not in $\mathcal{H}$. Consider the set of elements

$$\mathcal{G} \times \mathfrak{H}: \quad (g, \, |\, i\,\rangle); \qquad g \in \mathcal{G}, \; |\, i\,\rangle \in \mathfrak{H}. \tag{3.137}$$

In order to establish a one-to-one correspondence with the coset, we assume that whenever the $\mathcal{G}$ group element is of the form

$$g = g_k h_a, \tag{3.138}$$

we make the identification

$$\left(g_k h_a, \, |\, i\,\rangle\right) = \left(g_k, \, \mathcal{M}(h_a)^{[\mathbf{r}]}_{ij} |\, j\,\rangle\right), \tag{3.139}$$

so that both g and gh are equivalent in the sense that they differ only by reshuffling $\mathfrak{H}$. In this way, we obtain N copies of $\mathfrak{H}$, the Hilbert space of the $\mathbf{r}$ representation of $\mathcal{H}$, one at each point of the coset, which we take to be

$$(e,\ \mathfrak{H}),\ (g_1,\ \mathfrak{H}),\ (g_2,\ \mathfrak{H}),\ \cdots,\ (g_{N-1},\ \mathfrak{H}). \tag{3.140}$$

The action of $\mathcal{G}$ on this set is simply group multiplication

$$g: \quad (g_k,\ |\,i\,\rangle) \to (gg_k,\ |\,i\,\rangle), \qquad g\ \in \mathcal{G}, \quad k = 1, 2, \cdots, N. \tag{3.141}$$

Suppose that $g = h_a$, the coset decomposition tells us that

$$h_a g_k = g_l h_b, \tag{3.142}$$

where g_l and h_b are uniquely determined. Hence,

$$h_a: \quad \big(g_k,\ |\,i\,\rangle\big) \to \big(h_a g_k,\ |\,i\,\rangle\big) = \big(g_l h_b,\ |\,i\,\rangle\big) = \Big(g_l,\ \mathcal{M}(h_b)^{[\mathbf{r}]}_{ij}|\,j\,\rangle\Big). \tag{3.143}$$

Similarly, when $g = g_l$, the product $g_l g_k$ is itself a group element and can be rewritten uniquely as

$$g_l g_k = g_m h_b, \tag{3.144}$$

for some g_m and h_b. It follows that

$$\begin{aligned} g_l: \quad (g_k,\ |\,i\,\rangle) \to (g_l g_k,\ |\,i\,\rangle) &= (g_m h_b,\ |\,i\,\rangle) \\ &= \Big(g_m,\ \mathcal{M}(h_b)^{[\mathbf{r}]}_{ij}|\,j\,\rangle\Big). \end{aligned} \tag{3.145}$$

We have shown that these N copies of the vector space are linearly mapped into one another under the action of $\mathcal{G}$. The action of $\mathcal{G}$ is thus represented by a $(dN \times dN)$ matrix. It may or may not be irreducible.

In the special case when $\mathcal{H}$ is Abelian, its irreducible representations are one-dimensional, and the induced representation on $\mathcal{G}$ by an irrep of $\mathcal{H}$ is in terms of $(N \times N)$ matrices. This suggest that we think of the induced representation as matrices in block form, $(N \times N)$ matrix, whose elements are themselves $(d \times d)$ matrices.

For the more mathematically minded, we define N mapping (called sections) which single out N vectors in $\mathfrak{H}$:

$$f_l: \quad (g_k,\ |\,i\,\rangle) \longrightarrow \begin{cases} |\,v_k\,\rangle, & l = k, \\ 0, & l \neq k. \end{cases} \tag{3.146}$$

This defines a basis, and we can arrange these N vectors into a column, which is being acted on by the induced representation.

As an example, consider the group of order $3n^2$

$$\Delta(3n^2) = (\mathcal{Z}_n \times \mathcal{Z}_n) \rtimes \mathcal{Z}_3, \qquad (3.147)$$

generated by $a \in \mathcal{Z}_3$, $c \in \mathcal{Z}_n$ and $d \in \mathcal{Z}_n$, with the presentation

$$< a, c, d \,|\, a^3 = c^n = d^n = e, cd = dc, aca^{-1} = c^{-1}d^{-1}, ada^{-1} = c > . \qquad (3.148)$$

Let us construct the representation induced by the one-dimensional representation of $\mathcal{H} = \mathcal{Z}_n \times \mathcal{Z}_n$, for which

$$c = \eta^k; \qquad d = \eta^l, \qquad \eta^n = 1, \qquad (3.149)$$

acting on complex numbers z. The coset contains just three points, and we consider the action of $\mathcal{G}$ on

$$(e,\, z), \qquad (a,\, z), \qquad (a^2,\, z). \qquad (3.150)$$

Then

$$a(e,\, z) = (a,\, z), \qquad a(a,\, z) = (a^2,\, z), \qquad a(a^2,\, z) = (e,\, z), \qquad (3.151)$$

so that a is represented by the (3×3) permutation matrix

$$a = \begin{pmatrix} 0 & 1 & 0 \\ 0 & 0 & 1 \\ 1 & 0 & 0 \end{pmatrix}. \qquad (3.152)$$

Next, we consider the action of c

$$c(e,\, z) = (c,\, z) = (e,\, cz) = (e,\, \eta^k z) = \eta^k(e,\, z). \qquad (3.153)$$

Using

$$ca = ad; \qquad ca^2 = a^2c^{-1}d^{-1}, \qquad (3.154)$$

derived from the presentation, we find

$$c(a,\, z) = (ca,\, z) = (ad,\, z) = (a,\, dz) = (e,\, \eta^l z) = \eta^l(e,\, z), \qquad (3.155)$$

as well as

$$\begin{aligned} c(a^2,\, z) &= (ca^2,\, z) = (a^2c^{-1}d^{-1},\, z) = (a^2,\, c^{-1}d^{-1}z) \\ &= (e,\, \bar{\eta}^k\bar{\eta}^l z) = \eta^{-k-l}(a^2,\, z). \end{aligned} \qquad (3.156)$$

Similarly, the action of d can be derived, using

$$da = c^{-1}d^{-1}; \qquad da^2 = a^2c. \qquad (3.157)$$

The above show that c and d are represented by the diagonal matrices

$$c = \begin{pmatrix} \eta^k & 0 & 0 \\ 0 & \eta^l & 0 \\ 0 & 0 & \eta^{-k-l} \end{pmatrix}, \qquad d = \begin{pmatrix} \eta^l & 0 & 0 \\ 0 & \eta^{-k-l} & 0 \\ 0 & 0 & \eta^k \end{pmatrix}, \qquad (3.158)$$

completing the induced (3×3) representation.

In general, the induced representation is reducible, but when one builds it out of a subgroup that has a low index, one has a better chance of obtaining an irrep from this procedure. This is the case for this example.

Alas, it is not so easy to work out the character tables for more complicated groups. One runs out of tricks, and must rely on numerical methods. Fortunately for us, character tables and other useful group properties have been tabulated in the *Atlas of Simple Groups* for the simple groups, in book form (Oxford University Press) and also on the internet: http://brauer.maths.qmul.ac.uk/Atlas.

A word of caution about the ATLAS's dot notation. By $\mathcal{A}_4 \cdot 2$, they mean the group $\mathcal{S}_4 \supset \mathcal{A}_4$, and derive specific matching rules for their irreps $\mathcal{A}_5 \rightarrow \mathcal{S}_4$: $\mathbf{1} \rightarrow \mathbf{1} + \mathbf{1}_1$, $\mathbf{1}_1 + \bar{\mathbf{1}}_1 \rightarrow \mathbf{2}$, and $\mathbf{3} \rightarrow \mathbf{3}_1 + \mathbf{3}_2$.

On the other hand, the group $2 \cdot \mathcal{A}_4$ is the binary extension or double cover of $\mathcal{A}_4$, to be discussed in detail later, but it does *not* contain $\mathcal{A}_4$ as a subgroup; rather $\mathcal{A}_4$ is the quotient group of the binary tetrahedral group with $\mathcal{Z}_2$.

It is useful to tabulate group properties in a convenient form that generalizes the tables of Thomas and Wood. In Appendix 1, we gather in one place important information (presentation, classes, character table, Kronecker products, embeddings, representations) for groups which appear ubiquitously in physical applications.

3.12 Invariants

In physics, invariance manifests itself when the Hamiltonian (or *action* in field theory) can be expressed in terms of group invariants. Typically, the Hamiltonian is a function of the dynamical variables which transform as some representation of the group. Mathematically, this amounts to constructing invariants out of polynomials built out of group representations. Invariants appear whenever the (multiple) Kronecker products of representations contains $\mathbf{1}$, the *singlet* irrep.

For instance, the $\mathcal{A}_4$ Kronecker products show two quadratic invariants in the product of two irreps

$$[\mathbf{3} \times \mathbf{3}]_{\text{singlet}}, \qquad \mathbf{1}_1 \times \bar{\mathbf{1}}_1. \qquad (3.159)$$

There are more ways to form cubic invariants. We can form cubic invariants in four ways

$$\left[(\mathbf{3}\times\mathbf{3})_{\mathbf{3}}^{\rm sym}\times\mathbf{3}\right]_{\rm singlet},\qquad \left[(\mathbf{3}\times\mathbf{3})_{\mathbf{3}}^{\rm antisym}\times\mathbf{3}\right]_{\rm singlet},$$

$$\left[(\mathbf{3}\times\mathbf{3})_{\mathbf{1}_1}^{\rm sym}\times\bar{\mathbf{1}}_1\right]_{\rm singlet},\qquad \left[(\mathbf{3}\times\mathbf{3})_{\bar{\mathbf{1}}_1}^{\rm sym}\times\mathbf{1}_1\right]_{\rm singlet},$$

where the representation subscript denotes the projection of the product on that representation: $\mathbf{3}$ appears in both the symmetric and antisymmetric product of two triplets. The second cubic invariant requires at least two triplets.

The same reasoning leads to four quartic invariants built out of triplets alone:

$$\left[(\mathbf{3}\times\mathbf{3})_{\mathbf{3}}^{\rm sym}\times(\mathbf{3}\times\mathbf{3})_{\mathbf{3}}^{\rm sym}\right]_{\rm singlet},\qquad \left[(\mathbf{3}\times\mathbf{3})_{\mathbf{3}}^{\rm antisym}\times(\mathbf{3}\times\mathbf{3})_{\mathbf{3}}^{\rm sym}\right]_{\rm singlet},$$

$$\left[(\mathbf{3}\times\mathbf{3})_{\mathbf{3}}^{\rm antisym}\times(\mathbf{3}\times\mathbf{3})_{\mathbf{3}}^{\rm antisym}\right]_{\rm singlet},\qquad \left[(\mathbf{3}\times\mathbf{3})_{\mathbf{1}_1}^{\rm sym}\times(\mathbf{3}\times\mathbf{3})_{\bar{\mathbf{1}}_1}^{\rm sym}\right]_{\rm singlet},$$

but they are not all independent. This procedure, while very useful to identify invariants in the product of representations, produces too many redundancies. How do we find out how many of those are independent? Furthermore, their actual construction requires detailed knowledge encoded in the Clebsch–Gordan decompositions.

The identification and construction of finite group invariants is clearly a formidable task, but it is tractable. The determination of invariants constructed out of several irreducible representations, can be somewhat alleviated by the fact that all irreps of most groups can be generated by the Kronecker products of its fundamental irrep. There are exceptions to this rule: in some cases, these products generate different irreps always in the same combination, making it impossible to distinguish them by this construction, but in practice a good first step is to find all invariants made out of the fundamental irrep.

Consider a finite group $\mathcal{G}$ of order N, with n irreducible representations $\mathbf{r_a}$, with $\mathbf{r}_1 = \mathbf{1}$ the singlet. We express the k-fold symmetric Kronecker product of any irrep as a sum of irreps as

$$\underbrace{(\mathbf{r_a}\times\mathbf{r_a}\times\cdots\times\mathbf{r_a})_s}_{k}=\sum_b \mathcal{N}^{[k]}(\mathbf{r_b};\mathbf{r_a})\,\mathbf{r_b}. \tag{3.160}$$

The integer coefficient $\mathcal{N}^{[k]}(\mathbf{r_b};\mathbf{r_a})$ denotes the number of $\mathbf{r_b}$ irreps in the k-fold product of $\mathbf{r_a}$ irreps. Consider Molien's remarkable generating function

$$M(\mathbf{r_b};\mathbf{r_a};\lambda)=\frac{1}{N}\sum_{i=1}^{n} n_i\frac{\overline{\chi}_i^{[\mathbf{r_b}]}}{\det\left(1-\lambda A_i^{[\mathbf{r_a}]}\right)}, \tag{3.161}$$

where i labels the class C_i of $\mathcal{G}$, with n_i elements. $A_i^{[\mathbf{r_a}]}$ is any group element in C_i expressed in the $\mathbf{r_a}$ representation, and $\chi_i^{[\mathbf{r_b}]}$ are the characters of the $\mathbf{r_b}$ representation. Its power series

$$M(\mathbf{r_b};\, \mathbf{r_a};\, \lambda) = \sum_{k=0}^{\infty} \mathcal{N}^{[k]}(\mathbf{r_b};\, \mathbf{r_a})\, \lambda^k, \tag{3.162}$$

yields the desired coefficients. When we set $\mathbf{r_b}$ to be the singlet irrep, the Molien function yields the number of possible invariants constructed out of one irreducible representation

$$M(\mathbf{1};\, \mathbf{r};\, \lambda) = \frac{1}{N} \sum_{i=1}^{n} \frac{n_i}{\det\left(1 - \lambda A_i^{[\mathbf{r}]}\right)} = \sum_{k=0}^{\infty} \mathcal{N}^{[k]}\, \lambda^k, \tag{3.163}$$

where $\mathcal{N}^{[k]}$ denotes the number of invariants of order k. What makes the Molien function particularly useful is that it can always be written in the form

$$M(\mathbf{1};\, \mathbf{r};\, \lambda) = \frac{(1 + \sum d_k\, \lambda^k)}{(1 - \lambda^{a_1})^{n_1}(1 - \lambda^{a_2})^{n_2} \ldots}, \tag{3.164}$$

where the numerator is a *finite* polynomial in λ, and the d_k, a_k and n_k are *positive* integers. The numerator yields d_k invariants of order k, while expansion of each factor in the denominator generates n_k invariants of order a_k, etc. This infinite number of invariants can be expressed as products of a finite set of *basic* invariants, but the Molien expansion does not single them out. However, the subset of powers in the denominators a_k which are smaller than those in the numerator determine invariants of lesser order, and those must form the bulk of the basic invariants, but whenever they are comparable, the distinction becomes blurred. To make matters worse, invariants satisfy non-linear relations among themselves called *syzygies*.

In some cases, it can even happen that the invariants of order a_1, a_2, etc., from the denominator of the Molien function, do not satisfy syzygies among themselves. They are called *free invariants*. The powers in the numerator then refer to *constrained* invariants which satisfy syzygies with the free invariants. Unfortunately, this neat distinction among invariants is of limited validity.

We apply this approach to the tetrahedral group $\mathcal{A}_4$, with four irreps, $\mathbf{1}$, $\mathbf{1}_1$, $\bar{\mathbf{1}}_1$, and $\mathbf{3}$. Using its character table, it is an easy matter to compute the Molien function for each. For the singlet irrep, we have of course

$$M(\mathbf{1};\, \mathbf{1};\, \lambda) = \frac{1}{1 - \lambda}, \tag{3.165}$$

corresponding to the one-dimensional trivial invariant. Applied to $\mathbf{1}_1$, the other one-dimensional representations, we find

$$M(\mathbf{1};\,\mathbf{1}_1;\,\lambda) = M(\mathbf{1};\,\bar{\mathbf{1}}_1;\,\lambda) = \frac{1}{1-\lambda^3}, \tag{3.166}$$

yielding one invariant of cubic order with the $\mathbf{1}_1$. Calling its one component z, it means that z^3 is invariant, as one can easily check.

For the triplet, a straightforward computation of the Molien function yields

$$M(\mathbf{1};\,\mathbf{3};\,\lambda) = \frac{1+\lambda^6}{(1-\lambda^2)(1-\lambda^3)(1-\lambda^4)}. \tag{3.167}$$

We infer three free invariants, of order 2, 3, and 4, and one sixth-order constrained invariant. There is no easy way to construct these invariants, so we simply state their expression in terms of the three real coordinates x_i, $i = 1, 2, 3$ spanning the triplet irrep. The quadratic invariant is the length of the vector, as in $SO(3)$

$$\langle 2\rangle = \left(x_1^2 + x_2^2 + x_3^2\right). \tag{3.168}$$

The free cubic and quartic invariants are found to be

$$\langle 3\rangle = x_1x_2x_3, \qquad \langle 4\rangle = x_1^4 + x_2^4 + x_3^4, \tag{3.169}$$

and the constrained sixth-order invariant is

$$\langle 6\rangle = \left(x_1^2 - x_2^2\right)\left(x_2^2 - x_3^2\right)\left(x_3^2 - x_1^2\right). \tag{3.170}$$

Its square is related by one syzygy, to sums of polynomials in the free invariants:

$$4\langle 6\rangle^2 - 2\langle 4\rangle^3 + 108\langle 3\rangle^4 + \langle 2\rangle^6 + 36\langle 4\rangle\langle 3\rangle^2\langle 2\rangle - 20\langle 3\rangle^2\langle 2\rangle^3 + 5\langle 4\rangle^2\langle 2\rangle^2 - 4\langle 4\rangle\langle 2\rangle^4 = 0, \tag{3.171}$$

which gives an idea of the level of complication.

3.13 Coverings

We have noted the close similarity between the dihedral and dicyclic groups, indeed their presentations are almost the same

$$\mathcal{D}_n : \; < a, b \,|\, a^n = b^2 = e \,;\, bab^{-1} = a^{-1} >,$$
$$\mathcal{Q}_{2n} : \; < a, b \,|\, a^{2n} = e \,;\, a^n = b^2 \,;\, bab^{-1} = a^{-1} > .$$

In both, a^n and b^2 are equal to one another, but in the dicyclic group, they are of order two. Hence $\mathcal{Q}_{2n}$ contains a $\mathcal{Z}_2$ subgroup generated by b^2, called the *Schur multiplier*. In symbols (following the ATLAS), we write

$$\mathcal{Q}_{2n} = 2 \cdot \mathcal{D}_n,$$

and we say that $\mathcal{Q}_{2n}$ is the double cover of $\mathcal{D}_n$.

This is a special case of a general construction. Let G be a finite group generated by a set of elements a_i (letters), with presentation defined by a bunch of relations $w_\alpha(a_1, a_2, \ldots) = e$ (words). We define a new group G' where the words are no longer equal to the unit element, and commute with the letters, that is, $w_\alpha(a_1, a_2, \ldots) \neq e$, but such that they commute with the generators of G, that is, $w_\alpha\, a_i = a_i\, w_\alpha$. The w_α generate a subgroup C of G', which commutes with G. G' is called the *covering group* of G by a central subgroup C with $G = G'/C$. It is clearly not simple, and the Schur multiplier is the intersection of C with its commutator subgroup. In the dicyclic group, there is only one w of order two. Dicyclic groups have spinor representations while the dihedral groups does not. For physicists, this is a familiar situation: the spinor representations are obtained by going from $SO(3)$ to $SU(2)$ where $SU(2)$ is the cover of $SO(3)$.

4
Hilbert spaces

Physics happens in Hilbert spaces. Physicists like to describe Hilbert spaces in terms of their basis, in such a way that any vector can be represented as a linear combination of basis vectors with complex coefficients. The number of basis vectors may be finite or infinite, with the basis described in terms of discrete or continuous variables defined over various intervals. All can be constructed out of products of Hilbert spaces whose basis vectors are labeled by only one variable, continuous or discrete. In the continuous case, the range of the variable may be unbounded or bounded. We start with the simpler finite cases.

4.1 Finite Hilbert spaces

Since discrete groups are irreducibly represented by unitary matrices in finite Hilbert spaces, their study requires special attention. A D-dimensional Hilbert space has a basis made up of D kets $\{\,|\,1\,\rangle, |\,2\,\rangle, \ldots, |\,D-1\,\rangle, |\,D\,\rangle\,\}$. Its dual is made up of bras $\langle\, j\,|,\ j = 1, 2, \ldots, D$. They satisfy completeness and orthonormality, the hallmark of Hilbert spaces, that is

$$\sum_{j=1}^{D} |\,j\,\rangle\langle\,j\,| = 1, \qquad \langle\,j\,|\,k\,\rangle = \delta_{jk}. \tag{4.1}$$

In this space, one can build D^2 linear operators, of the form $|\,j\,\rangle\langle\,k\,|$, which generates $|\,j\,\rangle$ from $|\,k\,\rangle$. By completeness, one of them

$$\sum_{j=1}^{D} |\,j\,\rangle\langle\,j\,|, \tag{4.2}$$

reduces to the identity transformation, leaving $(D^2 - 1)$ non-trivial linear operators. By taking appropriate linear combinations, all can be made into hermitian operators:

$$\big(|\,k\,\rangle\langle\,j\,| + |\,j\,\rangle\langle\,k\,|\big), \quad i\big(|\,k\,\rangle\langle\,j\,| - |\,j\,\rangle\langle\,k\,|\big), \quad \text{for} \quad j \neq k, \tag{4.3}$$

and the $(D-1)$ diagonal combinations which we can write in the physics way as

$$\big(|\,1\,\rangle\langle\,1\,| - |\,2\,\rangle\langle\,2\,|\big), \big(|\,1\,\rangle\langle\,1\,| + |\,2\,\rangle\langle\,2\,| - 2\,|\,3\,\rangle\langle\,3\,|\big),$$

$$\ldots\ldots$$

$$\big(|\,1\,\rangle\langle\,1\,| + |\,2\,\rangle\langle\,2\,| + \cdots + |\,D-1\,\rangle\langle\,D-1\,| - (D-1)|\,D\,\rangle\langle\,D\,|\big).$$

It might be inferred that a D-dimensional Hilbert space can sustain at least (D^2-1) operators which can be used to generate an infinite number of linear transformations. We shall see later that in some special cases, there can be even more symmetry generators.

4.2 Fermi oscillators

A favorite way to construct finite Hilbert spaces is by means of n operators b^i and their hermitian conjugates $b^\dagger_i$, $i = 1, 2, \ldots, n$. They describe fermionic harmonic oscillators, which satisfy canonical *anti*commutation relations

$$\{b^i,\, b^\dagger_j\} = \delta^i{}_j, \qquad \{b^i,\, b^j\} = \{b^\dagger_i,\, b^\dagger_j\} = 0, \tag{4.4}$$

where the anticommutator is defined as

$$\{A,\, B\} \;\equiv\; A\,B + B\,A. \tag{4.5}$$

We construct a Hilbert space from the vacuum state $|\,\Omega\,\rangle$, and its dual $\langle\,\Omega\,|$, subject to the n annihilation conditions

$$b^i|\,\Omega\,\rangle = 0, \qquad \langle\,\Omega\,|b^\dagger_i = 0. \tag{4.6}$$

and normalization

$$\langle\,\Omega\,|\,\Omega\,\rangle = 1. \tag{4.7}$$

Using the anticommutation relations, we see that the n states

$$|\,i\,\rangle = b^\dagger_i|\,\Omega\,\rangle, \tag{4.8}$$

are normalized and orthogonal to the vacuum state

$$\langle\,i\,|\,j\,\rangle = \delta^i{}_j, \qquad \langle\,i\,|\,\Omega\,\rangle = 0. \tag{4.9}$$

The rest of the Hilbert space is spanned by the states

$$|\,[i_1\,i_2\,\cdots\,i_k]\,\rangle = b^\dagger_{i_1} b^\dagger_{i_2}\cdots b^\dagger_{i_k}|\,\Omega\,\rangle, \tag{4.10}$$

where k is limited to n since the square of any $b^\dagger_i$ is zero. In addition, they are antisymmetric under any $i_j \leftrightarrow i_l$ interchange, and thus in one-to-one correspondence with totally antisymmetric tensors (called *forms* by some). There are $\binom{n}{k}$ such states which generate a Hilbert space with dimension 2^n, given by the binomial formula

$$\sum_{k=0}^{n} \binom{n}{k} = 2^n. \tag{4.11}$$

This very rich finite-dimensional Hilbert space sustains many algebraic structures, in particular continuous and finite groups. We postpone the discussion of the continuous groups, and focus on the finite groups.

One such group is the finite Dirac group (see Lomont [15]). Arrange the fermionic harmonic oscillators into hermitian combinations

$$\gamma_j = b^j + b^\dagger_j, \qquad \gamma_{n+j} = i(b^j - b^\dagger_j), \tag{4.12}$$

which we write in a unified way as γ_a, $a = 1, 2, \ldots, 2n$. These obey the *Dirac* or *Clifford algebra*

$$\{\gamma_a, \gamma_b\} = 2\delta_{ab}. \tag{4.13}$$

Now consider elements of the form [15]

$$\pm\gamma_1^{l_1}\gamma_2^{l_2}\cdots\gamma_{2n}^{l_{2n}}, \tag{4.14}$$

where $l_k = 0, 1$, since $(\gamma_1)^2 = (\gamma_2)^2 = \cdots = 1$. There are 2^{2n+1} elements of this kind, including 1 and -1. Since the γ_a are their own inverses, all elements have inverses as well; for example, $(\gamma_1\gamma_4)^{-1} = \gamma_4\gamma_1 = -\gamma_1\gamma_4$. Closed under multiplication, they form a finite group of order 2^{2n+1}, called the Dirac group. For $n = 2$, its 32 elements are just

$$\pm 1, \qquad \pm\gamma_a, \qquad \pm\gamma_a\gamma_b, \qquad \pm\gamma_a\gamma_b\gamma_c, \qquad \pm\gamma_1\gamma_2\gamma_3\gamma_4, \tag{4.15}$$

with $a < b < c$. Since for $a \neq b$,

$$\gamma_a\gamma_b\gamma_a^{-1} = -\gamma_b, \tag{4.16}$$

we deduce that all group elements, except for 1 and -1, split into classes each containing the element and its negative: $(\gamma_a, -\gamma_a)$, $(\gamma_a\gamma_b, -\gamma_a\gamma_b)$, $a \neq b$, etc. The Dirac group has thus $2^n + 1$ classes, and as many irreducible representations.

Its only normal subgroup, $\mathcal{Z}_2 = (\mathbf{1}, -\mathbf{1})$, is also its commutator (derived) subgroup, so that it has 2^{2n} one-dimensional irreps. The dimension d of the extra irrep is deduced from the order formula

$$2^{2n+1} = 2^{2n} + d^2, \tag{4.17}$$

so that $d = 2^n$. In this representation, the γ_a are simply the well-known $(2^n \times 2^n)$ Dirac matrices.

Since its elements split into two types, those which square to 1, and those which square to -1, the reality of the representation can easily be determined by computing

$$\frac{1}{2^{2n+1}} \sum_{g} \chi(g^2) = \frac{1}{2^{2n+1}} \left[n_+ \chi(1) + n_- \chi(-1) \right], \tag{4.18}$$

where $n_\pm$ is the number of elements which square to ± 1, and $\chi(\pm 1)$ are the characters. We find

$$n_+ = 2^{2n} + i^{n(n-1)}\, 2^n, \qquad n_- = 2^{2n+1} - n_+ = 2^2 - i^{n(n-1)}\, 2^n, \tag{4.19}$$

so that they add up to the order. Hence

$$\frac{1}{2^{2n+1}} \sum_{g} \chi(g^2) = i^{n(n-1)}. \tag{4.20}$$

The representation is real whenever n is a multiple of four; otherwise it is pseudoreal.

4.3 Infinite Hilbert spaces

A Hilbert space is best described by its basis vectors, called *kets*. Consider first a Hilbert space with an infinite-dimensional basis, which we choose to label with the positive integers, $|\, n \,\rangle$, where $n = 0, 1, 2, 3, \ldots, \infty$. Any vector (ket) can be written as a linear combination

$$|\, \Psi \,\rangle = \sum_{n=0}^{\infty} z_n \,|\, n \,\rangle, \tag{4.21}$$

where the z_n are arbitrary complex numbers. In order to associate numbers to these kets, we need to introduce the *dual* vector space, which has a mirror basis made of *bras*, $\langle\, n \,|$, with $n = 0, 1, 2, 3, \ldots$. An arbitrary bra can be written as a linear combination of bras with arbitrary complex coefficients

$$\langle\, \Xi \,| = \sum_{n=0}^{\infty} w_n \,\langle\, n \,|, \tag{4.22}$$

where w_n are complex numbers. In particular, the bra associated with the state Ψ is given by

$$\langle\, \Psi \,| = \sum_{n=0}^{\infty} \bar{z}_n \,\langle\, n \,|, \tag{4.23}$$

where the bar denotes complex conjugation. The vector space and its dual are needed to map bras and kets into complex numbers, which "Di rac" calls bra kets:

$$\langle \Psi \mid \mid \Phi \rangle \rightarrow \langle \Psi \mid \Phi \rangle = \text{complex number.} \tag{4.24}$$

It is particularly convenient to choose an *orthonormal basis* in the sense that

$$\langle n \mid m \rangle = \delta_{m\,n}. \tag{4.25}$$

Such a basis also satisfies the *completeness relation*

$$\sum_{n=0}^{\infty} \mid n \rangle \langle n \mid = 1. \tag{4.26}$$

A vector space over the complex numbers with its dual space defined so as to have a complete orthonormal basis is called a Hilbert space. Then

$$\langle \Psi \mid \Psi \rangle = \sum_{n=0}^{\infty} |z_n|^2, \tag{4.27}$$

is the norm of the state vector represented by Ψ; it is positive definite. In this space dwell an infinite number of operators which can be written in the form

$$C = \sum_{n,m} c^n{}_m \mid m \rangle \langle n \mid, \tag{4.28}$$

where the $c^n{}_m$ are complex numbers. We define the hermitian operator

$$C^\dagger \equiv \sum_{n,m} \bar{c}^m{}_n \mid n \rangle \langle m \mid, \tag{4.29}$$

in such a way that the action of C on a ket is the same as that of $C^\dagger$ on its corresponding bra. That is

$$\begin{aligned} \langle \Psi \mid C \mid \Phi \rangle &= \langle \Psi \mid \Big(C \mid \Phi \rangle \Big) = \langle \Psi \mid \Phi' \rangle, \\ &= \Big(\langle \Psi \mid C \Big) \mid \Phi \rangle = \langle \Psi' \mid \Phi \rangle, \end{aligned}$$

where

$$\mid \Phi' \rangle = C \mid \Phi \rangle, \qquad \langle \Psi' \mid = \langle \Psi \mid C. \tag{4.30}$$

Then,

$$\mid \Psi' \rangle = C^\dagger \mid \Psi \rangle. \tag{4.31}$$

It follows that if C is a hermitian operator,

$$C = C^\dagger \quad \rightarrow \quad c^n{}_m = \bar{c}^m{}_n : \tag{4.32}$$

any hermitian operator is a matrix which is equal to itself when its elements are transposed *and* conjugated. We can view any operator as an infinite-dimensional matrix which acts on kets from the left to produce other kets, and on bras from the right to produce other bras. Of special interest are the operations that preserve the norm of the states. Consider the transformation

$$|\,\Psi\,\rangle \;\rightarrow\; |\,\Psi'\,\rangle = \mathcal{U}\,|\,\Psi\,\rangle, \tag{4.33}$$

such that

$$\langle\,\Psi'\,|\,\Psi'\,\rangle = \langle\,\Psi\,|\,\Psi\,\rangle. \tag{4.34}$$

If this is true for *any* state Ψ, it follows that

$$\mathcal{U}^{\dagger}\mathcal{U} = 1, \rightarrow\; \mathcal{U}^{-1} = \mathcal{U}^{\dagger}. \tag{4.35}$$

The operator $\mathcal{U}$ is said to be *unitary*. Now let us write it as

$$\mathcal{U} = e^{M}, \tag{4.36}$$

where M is another operator. The condition means that

$$M^{\dagger} = -M, \tag{4.37}$$

and M is an antihermitian operator, which can always be written as i times a hermitian operator.

While there is an infinite number of operators in an infinite-dimensional Hilbert space, some particular ones have proven to be of great mathematical and physical interest, such as

$$a \;\equiv\; \sum_{m=0}^{\infty} \sqrt{m+1}\,|\,m\,\rangle\langle\,m+1\,|, \tag{4.38}$$

which changes each basis vector to its lower sibling and annihilates the zero state

$$a\,|\,n\,\rangle = \sqrt{n}\,|\,n-1\,\rangle, \qquad a\,|\,0\,\rangle = 0. \tag{4.39}$$

With its hermitian conjugate,

$$a^{\dagger} \;\equiv\; \sum_{m=0}^{\infty} \sqrt{m+1}\,|\,m+1\,\rangle\langle\,m\,|, \tag{4.40}$$

it forms the simple commutation relation

$$[\,a, a^{\dagger}\,] \;\equiv\; a\,a^{\dagger} - a^{\dagger}\,a = \sum_{n=0}^{\infty} |\,n\,\rangle\langle\,n\,| = 1. \tag{4.41}$$

It follows that any ket is viewed as the action of the creation operator $a^\dagger$ on the "vacuum" state $|\,0\,\rangle$:

$$|\,n\,\rangle = \frac{(a^\dagger)^n}{\sqrt{n!}}\,|\,0\,\rangle. \tag{4.42}$$

It is well-known that we can change basis to a continuous variable. Namely, introduce the hermitian combinations

$$X \equiv \frac{1}{\sqrt{2}}(a^\dagger + a), \qquad P \equiv \frac{i}{\sqrt{2}}(a^\dagger - a), \tag{4.43}$$

so that

$$[\,X, P\,] = i, \tag{4.44}$$

is the Heisenberg algebra. With the proper units, X can be interpreted as a position operator whose eigenvalue denotes a coordinate which ranges from $-\infty$ to $+\infty$. We define the eigenket $|\,x\,\rangle$ through

$$X\,|\,x\,\rangle = x\,|\,x\,\rangle. \tag{4.45}$$

They form a new basis which satisfies the normalization

$$\langle\,x\,|\,x'\,\rangle = \delta(x - x'), \tag{4.46}$$

and completeness relation

$$\int_{-\infty}^{+\infty} dx\,|\,x\,\rangle\langle\,x\,| = 1. \tag{4.47}$$

The relation between the two bases is defined through

$$|\,x\,\rangle = \sum_{n=0}^{\infty} K_n(x)\,|\,n\,\rangle, \tag{4.48}$$

where

$$K_n(x) = \langle\,n\,|\,x\rangle, \tag{4.49}$$

is a function in the continuous variable x. It can be written in the form

$$K_n(x) = H_n(x)\,e^{-x^2/2}, \tag{4.50}$$

where H_n is the Hermite polynomial of degree n. Completeness and orthogonality of our basis translates in the same way for the H_n polynomials

$$\int_{-\infty}^{+\infty} dx\, e^{-x^2}\, H_n(x) H_m(x) = \delta_{nm}, \tag{4.51}$$

$$\sum_{n=0}^{+\infty} H_n(x)\, H_n(y) = e^{-(x^2+y^2)/2}\, \delta(x-y). \tag{4.52}$$

We conclude this section with two remarks and a caveat.

First remark. The physical relevance of operators in Hilbert space depends on their stability under time evolution. In Quantum Mechanics, the state of a physical system at any time is represented by a state vector $|\,\Psi\,\rangle_t$. At a later time, this vector has evolved into another one $|\,\Psi\,\rangle_{t'}$. Time evolution is regarded as an infinite number of infinitesimal steps from t to $t+\delta t$. The Hamiltonian is the operator which takes the state ket into its infinitesimal future

$$|\,\Psi\,\rangle_{t+\delta t} \;\approx\; (1 + iH\,\delta t)|\,\Psi\,\rangle_t. \tag{4.53}$$

The Hamiltonian H is a hermitian operator which generates time translations. Its exponentiation generates a finite time evolution

$$e^{iH(t-t')}\,|\,\Psi\,\rangle_{t'} = |\,\Psi\,\rangle_t, \tag{4.54}$$

so that the operator

$$\mathcal{U}(t-t') \;\equiv\; e^{iH(t-t')}, \tag{4.55}$$

is unitary

$$\mathcal{U}^\dagger\,\mathcal{U} = \mathcal{U}\,\mathcal{U}^\dagger = 1. \tag{4.56}$$

Second remark. In Quantum Mechanics, any Hilbert space operator can be viewed as generating *some* transformation; the Hamiltonian which generates time translation, is often used to define a physical system. Let A and B be two operators, and consider their commutator

$$[\,A, B\,].$$

It has two different meanings: the infinitesimal change in the operator A due to a transformation generated by B, and that of B due to A. For instance if $A = H$, our commutator means either the change in B under time evolution, or the change in H under the transformation generated by B. Thus if the commutator is zero it means that H is invariant under the B transformation *and* that B is invariant under time evolution: it represents a constant of the motion. For example if H commutes with the angular momentum operators (which generate space rotations), it means that the energy (eigenvalue of H) is invariant under rotations, *and* that angular momentum is a constant of the motion (generated by the Hamiltonian).

Caveat. We described a Hilbert space with a simple basis labeled by an unbounded continuous variable. A product of such Hilbert spaces yields a space described by many such coordinates, which can then serve as a label for a point in space. But what if the coordinates are not *globally* defined, as happens when there are topological obstructions? Then one has to use coordinate patches for different regions, with complicated matching conditions. What if there are "holes" in the sense of singular points? How does one treat the innards of black holes? The Universe is, we think, represented by a state in Hilbert space, and the determination of its Hilbert space is at the heart of physics. Such knowledge for a particular physical system is a crucial step towards the understanding of its dynamics. With the success of the Standard Model of particle physics, one may even speculate that the dynamics may well be uniquely determined by symmetries. Our purpose at hand, however, is not to tackle such deep questions but to learn how to represent transformation groups in Hilbert spaces.

5

$SU(2)$

Continuous symmetry operations, e.g. rotation about some axis, can be viewed as analytic functions of its parameters, in this case a continuous angle, defined modulo 2π. All can be understood, except for topological obstructions, as stemming from the repeated application of one operator which generates (in this case) infinitesimal rotations about that axis. Generators of rotations about different axes do not in general commute; rather they form a *Lie algebra*, which closes under commutation. This chapter is devoted to the study of the most elementary but non-trivial Lie algebra. In the process we will introduce many concepts that will be of use in the study of arbitrary Lie algebras. This algebra ubiquitously lies at the root of many areas of physics, and mathematics.

5.1 Introduction

The smallest non-trivial Lie algebra is realized in a two-dimensional Hilbert space, in which roam three linear operators:

$$T^+ \equiv |\,1\,\rangle\langle\,2\,|; \qquad T^- \equiv |\,2\,\rangle\langle\,1\,|; \qquad T^3 \equiv \tfrac{1}{2}\Big(|\,1\,\rangle\langle\,1\,| - |\,2\,\rangle\langle\,2\,|\Big). \tag{5.1}$$

They satisfy an algebra under commutation

$$[\,T^+, T^-\,] = 2\,T^3; \qquad [\,T^3, T^\pm\,] = \pm\,T^\pm. \tag{5.2}$$

Define as before hermitian operators

$$T^1 \equiv \frac{1}{2}\Big(T^- + T^+\Big); \qquad T^2 \equiv \frac{i}{2}\Big(T^- - T^+\Big). \tag{5.3}$$

In terms of which

$$[\,T^A, T^B\,] = i\epsilon^{ABC}\,T^C, \tag{5.4}$$

$A, B, C = 1, 2, 3$, and where ϵ^{ABC} is the Levi–Civita symbol. An algebra closed under commutation is called a Lie algebra provided it also satisfies the Jacobi identity, in this case

$$[T^A, [T^B, T^C]] + \text{cyclic permutations} = 0. \tag{5.5}$$

It is trivially satisfied by our construction.

This is our first Lie algebra. It is so ubiquitous that it has many names: mathematicians call it A_1, B_1, C_1, and physicists call it $SU(2)$, $SO(3)$, $Sp(2)$. The meaning of these names will become evident later. Any polynomial function of the elements of a Lie algebra that commutes with the Lie algebra is called a *Casimir operator*. In this case there is only one; it is quadratic and given by

$$C_2 \equiv (T^1)^2 + (T^2)^2 + (T^3)^2; \qquad [C_2, T^A] = 0. \tag{5.6}$$

Our explicit construction represents this algebra in a two-dimensional Hilbert space, our first representation. We say that the Lie algebra $SU(2)$ has a two-dimensional irrep, written as **2**. We note that the generators satisfy the additional relation $(T^{\pm})^2 = 0$, which is specific to the representation, and not to the algebra.

It is not too difficult to work out all the irreducible unitary representations of this algebra. There are many ways of doing it, but we will use a technique which will generalize naturally to more complicated algebras.

First some remarks. (1) In Quantum Mechanics, states are labeled by the eigenvalues of a maximal number of commuting operators. Our algebra has two such operators, the Casimir operator C_2, and any one of its three elements T^A, $A = 1, 2, 3$. Let us take C_2 and T^3. The algebra will be represented on eigenstates of these two operators. (2) The Casimir operator is positive definite, which means that the spectrum of any of the T^A is bounded, for a finite value of C_2. Roughly speaking, we expect the largest and lowest values of any of the T^A to be like $\pm\sqrt{C_2}$.

So we look for states $|\ c\,;\ m\ \rangle$ which satisfy

$$C_2 |\ c\,;\ m\ \rangle = c\,|\ c\,;\ m\ \rangle, \qquad T^3 |\ c\,;\ m\ \rangle = m\,|\ c\,;\ m\ \rangle, \tag{5.7}$$

where c and m are real since these operators are hermitian. We can use the remaining operators in the algebra to create new states. It is easily seen from the commutation relations that T^+ raises the value of T^3 by one unit, and does not change the value of C_2. Hence the state $T^+ |\ c\,;\ m\ \rangle$ must be proportional to $|\ c\,;\ m+1\ \rangle$ (since there are no other labels), so we set

$$T^+ |\ c\,;\ m\ \rangle = d_m^{(+)}\,|\ c\,;\ m+1\ \rangle, \tag{5.8}$$

where $d_m^{(+)}$ is to be determined later. We can generate an infinite set of states with T^3 eigenvalues $m, m+1, m+2, \ldots$, by repeated application of T^+, but, as we have remarked, the spectrum of T^3 is bounded, and this sequence must stop: there is a state where T^3 assumes its maximum value, j:

$$T^+ \mid c;\, j\,\rangle = 0; \qquad T^3 \mid c;\, j\,\rangle = j \mid c;\, j\,\rangle. \tag{5.9}$$

This state is called the *highest weight state* of the representation. We can easily work out the value of the Casimir operator on the highest weight state, using the commutation relations

$$\begin{aligned} C_2 \mid c;\, j\,\rangle &= \left((T^3)^2 + (T^+T^- + T^-T^+)/2\right) \mid c;\, j\,\rangle, \\ &= \left((T^3)^2 + [\,T^+, T^-\,]/2\right) \mid c;\, j\,\rangle, \\ &= \left((T^3)^2 + T^3\right) \mid c;\, j\,\rangle = j\,(j+1) \mid c;\, j\,\rangle, \end{aligned}$$

so that $c = j(j+1)$. The maximum eigenvalue of T^3 determines the Casimir operator. We thus relabel the kets by j instead of c: $\mid c\,;\, m\,\rangle \rightarrow \mid j\,;\, m\,\rangle$.

Similarly, we can generate states of lower T^3 values by repeated application of T^-, but this process must terminate as well, yielding a *lowest weight state* that is annihilated by T^-

$$T^- \mid j\,;\, k\,\rangle = 0; \qquad T^3 \mid j\,;\, k\,\rangle = k \mid j\,;\, k\,\rangle, \tag{5.10}$$

where k is in the sequence $j, j-1, j-2, \ldots$. However, this state has the same value of C_2, and thus

$$\begin{aligned} C_2 \mid j\,;\, k\,\rangle &= \left((T^3)^2 + T^-T^+/2\right) \mid j\,;\, k\,\rangle, \\ &= \left((T^3)^2 + [\,T^-, T^+\,]/2\right) \mid j\,;\, k\,\rangle, \\ &= \left((T^3)^2 - T^3\right) \mid j\,;\, k\,\rangle = k\,(k-1) \mid j\,;\, k\,\rangle, \end{aligned}$$

so that

$$k(k-1) = j(j+1). \tag{5.11}$$

This quadratic equation has two solutions, $k = j + 1$, which is impossible since k is always less than j, and this leaves the only possibility, $k = -j$.

With little work we have found that any irrep of this algebra contains $(2j+1)$, so that $2j$ must be an integer, with T_3 eigenvalues ranging from j to $-j$ in integer steps. Physicists call j the total angular momentum, while mathematicians call $2j$ the *Dynkin label* of the representation. The Lie algebra of $SU(2)$ has therefore an infinite number of irreps, one for each integer value of $2j$. The value $j = 0$ corresponds to the trivial *singlet* representation, $j = 1/2$ is the famous *spinor* representation, $j = 1$ describes the three states of a vector, etc.

The action of T^- produces a state of lower T^3 value by one unit. Since there is no other label, it must be that it is proportional to $\mid j\,;\, m\,\rangle$:

$$T^- \mid j\,;\, m\,\rangle = d_m^{(-)} \mid j\,;\, m-1\,\rangle. \tag{5.12}$$

Using

$$[\,T^+, T^-\,] \mid j\,;\, m\,\rangle = 2T^3 \mid j\,;\, m\,\rangle, \tag{5.13}$$

we find that

$$d_{m-1}^{(+)}\, d_m^{(-)} - d_m^{(+)} d_{m+1}^{(-)} = 2m, \tag{5.14}$$

subject to $d_j^{(+)} = d_{-j}^{(-)} = 0$. Since $\mid j\,;\, m\,\rangle$ and $\mid j\,;\, m'\,\rangle$ are eigenstates of a hermitian operator, their inner product must vanish for $m \neq m'$. Furthermore we can determine the constants $d_m^{(\pm)}$ so as to make them have unit norm. A little computation shows that if

$$d_m^{(+)} = \sqrt{(j-m)(j+m+1)}, \qquad d_m^{(-)} = \sqrt{(j+m)(j-m+1)}, \tag{5.15}$$

the states form an orthonormal set

$$\langle\, j\,; m \mid j\,; m'\,\rangle = \delta_{m,m'}. \tag{5.16}$$

In addition, completeness is satisfied *over the states of the representation*, that is

$$\sum_{m=-j}^{m=+j} \mid j\,;\, m\,\rangle\langle\, j\,;\, m \mid = 1. \tag{5.17}$$

It is useful to represent the Lie algebra in this representation as $(2j+1) \times (2j+1)$ matrices

$$\left(T^A\right)_{pq}, \tag{5.18}$$

where p, q run from $1, 2, \ldots, 2j+1$. We can choose to represent T^3 as a diagonal matrix with entries $j,\, j-1,\, j-2, \ldots, -j$, in which case $T^+(T^-)$ are upper(lower) triangular matrix (homework). The representation matrices are always traceless, since each is the commutator of two matrices, which is always traceless because of the cyclic property of the trace.

Two representations with "celebrity" status, are worthy of special mention. (1) The smallest one, with $j = 1/2$, in which the generators are represented in terms of the three (2×2) Pauli spin matrices

$$T^A = \frac{\sigma^A}{2}, \tag{5.19}$$

which satisfy

$$\sigma^A \sigma^B = \delta^{AB} + i\,\epsilon^{ABC}\,\sigma^C; \qquad \sigma^{A*} = -\sigma^2 \sigma^A \sigma^2, \tag{5.20}$$

where star denotes charge conjugation.

(2) The second celebrity representation is that which has as many states as elements in the Lie algebra. It is called the *adjoint* representation, with $2j = 2$, where the generators can be represented by (3×3) hermitian matrices

$$T^1 = \begin{pmatrix} 0 & 0 & 0 \\ 0 & 0 & i \\ 0 & -i & 0 \end{pmatrix}, \quad T^2 = \begin{pmatrix} 0 & 0 & i \\ 0 & 0 & 0 \\ -i & 0 & 0 \end{pmatrix}, \quad T^3 = \begin{pmatrix} 0 & i & 0 \\ -i & 0 & 0 \\ 0 & 0 & 0 \end{pmatrix}. \tag{5.21}$$

We can associate an infinite *lattice* with a Lie algebra, as an array of points, each representing a state in the totality of its irreps. In the $SU(2)$ case the lattice is one-dimensional, with evenly spaced points. Each point corresponds to a particular value of T^3, and represents a state that can appear in an infinite number of representations. If its axis is taken to be $2T^3$, they sit at the integer points:

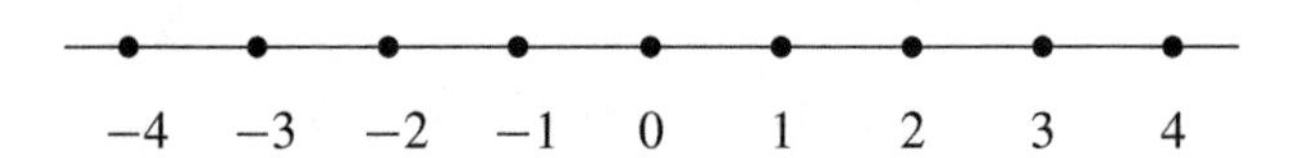

We can draw some lessons from the way we have constructed these representations. (1) The $SU(2)$ irreps can be identified by their highest weight state; (2) all other states of the irrep are generated by repeated application of the lowering operator T^-. $SU(2)$ has only one label and one lowering operator, but for bigger Lie algebras, the highest weight state and its descendants will have several labels and more lowering operators. $SU(2)$ has one Casimir operator and one label; Lie algebras will have in general as many labels as Casimir operators. Their number is called the *rank* of the algebra, and it is equal to the number of commuting elements of the Lie algebra. The $SU(2)$ Lie algebra has rank one (hence the subscript in its mathematical names).

5.2 Some representations

There is another way to generate the irreducible representations, dear to the heart of physicists. It is based on the fact that all representations can be generated by taking direct products of the smallest representations, in this case the doublet **2**.

Suppose we clone our Lie algebra, producing two identical copies, which we denote by $T^A_{(1)}$ and $T^A_{(2)}$, each satisfying the same Lie algebra, but commuting with one another.

$$[\, T^A_{(a)},\, T^B_{(a)}\,] = i\,\epsilon^{ABC}\, T^C_{(a)}, \quad a = 1, 2 \qquad [\, T^A_{(1)},\, T^B_{(2)}\,] = 0. \tag{5.22}$$

Each acts on its own Hilbert space, spanned by $|\,\cdots\rangle_{(1)}$ and $|\,\cdots\rangle_{(2)}$. The sum of the generators generates the same algebra

$$\left[\, T^A_{(1)} + T^A_{(2)},\; T^B_{(1)} + T^B_{(2)}\,\right] = i\,\epsilon^{ABC}\,\left(T^C_{(1)} + T^C_{(2)}\right), \tag{5.23}$$

acting on the direct product states

$$|\,\ldots\rangle_{(1)}|\,\ldots\rangle_{(2)}, \tag{5.24}$$

which span a new Hilbert space, constructed out of the original two. Since their sums satisfies the same commutation relations, one should be able to represent their action in terms of the representations we have previously derived. When $SU(2)$ is identified with three-dimensional rotations, physicists call this construction, the addition of angular momentum.

Consider the (Kronecker) direct product of two arbitrary representations $\mathbf{2j+1}$ and $\mathbf{2k+1}$. The highest weight direct product state, which we label only with the T^3 eigenvalue, since we know the range of its eigenvalues:

$$|\,j\,\rangle_{(1)}\,|\,k\,\rangle_{(2)}, \tag{5.25}$$

satisfies

$$T^3|\,j\,\rangle_{(1)}\,|\,k\,\rangle_{(2)} = \left(T^3_{(1)} + T^3_{(2)}\right)|\,j\,\rangle_{(1)}\,|\,k\,\rangle_{(2)} = (j+k)\,|\,j\,\rangle_{(1)}\,|\,k\,\rangle_{(2)}, \tag{5.26}$$

and

$$T^+|\,j\,\rangle_{(1)}\,|\,k\,\rangle_{(2)} = \left(T^+_{(1)} + T^+_{(2)}\right)|\,j\,\rangle_{(1)}\,|\,k\,\rangle_{(2)} = 0. \tag{5.27}$$

It must be the highest weight state of the $\mathbf{2(j+k)+1}$ representation. In order to generate the rest of its states, we apply to it the sum of the two lowering operators. The action of T^- generates the linear combination (neglecting the numerical coefficient in front of each ket)

$$|\,j-1\,\rangle_{(1)}\,|\,k\,\rangle_{(2)} + |\,j\,\rangle_{(1)}\,|\,k-1\,\rangle_{(2)}. \tag{5.28}$$

We form its orthogonal combination

$$|\,j-1\,\rangle_{(1)}\,|\,k\,\rangle_{(2)} - |\,j\,\rangle_{(1)}\,|\,k-1\,\rangle_{(2)}, \tag{5.29}$$

on which $T^3 = j + k - 1$, and check (homework) that it is annihilated by T^+: it is the highest weight of another representation, with highest weight $\mathbf{2(j+k-1)+1}$. We repeat the process, and find new states which are linear combinations of the three states

$$|\,j-2\,\rangle_{(1)}\,|\,k\,\rangle_{(2)}, \qquad |\,j-1\,\rangle_{(1)}\,|\,k-1\,\rangle_{(2)}, \qquad |\,j\,\rangle_{(1)}\,|\,k-2\,\rangle_{(2)}. \tag{5.30}$$

Two linear combinations are lower weight states of the two representations $\mathbf{2(j+k)+1}$ and $\mathbf{2(j+k-1)+1}$; the third is the highest weight state of a new $\mathbf{2(j+k-2)+1}$ representation. The pattern becomes clear, we keep generating at each level one new representation until we run out of states because we hit the lower bound of the two original representations. It is easy to see (homework) that we stop at the $|j-k|$ representation, and thus

$$[\mathbf{2j+1}] \times [\mathbf{2k+1}] = [\mathbf{2(j+k)+1}] + [\mathbf{2(j+k-1)+1}] + \cdots + [\mathbf{2(j-k)+1}], \tag{5.31}$$

taking here $j \geq k$. Hence by taking the product of the lowest representation we generate all the representations of the algebra. This is a very economical way to proceed as long as one does not have to deal with large representations which, thankfully is usually the case in physics.

To see how this works in detail, we give two examples. First take the direct product of two spinor representations. Start from the direct product highest weight state

$$|\uparrow\uparrow\rangle \equiv |\tfrac{1}{2},\tfrac{1}{2}\rangle_{(1)}|\tfrac{1}{2},\tfrac{1}{2}\rangle_{(2)}, \tag{5.32}$$

and act on it with the sum of the lowering operators

$$T^- \equiv T^-_{(1)} + T^-_{(2)}, \tag{5.33}$$

to get the linear combination

$$|\downarrow\uparrow\rangle + |\uparrow\downarrow\rangle \equiv |\tfrac{1}{2},-\tfrac{1}{2}\rangle_{(1)}|\tfrac{1}{2},\tfrac{1}{2}\rangle_{(2)} + |\tfrac{1}{2},\tfrac{1}{2}\rangle_{(1)}|\tfrac{1}{2},-\tfrac{1}{2}\rangle_{(2)}, \tag{5.34}$$

and do it again to obtain

$$|\downarrow\downarrow\rangle \equiv |\tfrac{1}{2},-\tfrac{1}{2}\rangle_{(1)}|\tfrac{1}{2},-\tfrac{1}{2}\rangle_{(2)}. \tag{5.35}$$

Further action of the lowering operator on this state yields zero. Hence these three states form the three-dimensional representation of the algebra. Note that the direct product Hilbert space has four states, the fourth being the orthogonal linear combination

$$|\tfrac{1}{2},-\tfrac{1}{2}\rangle_{(1)}|\tfrac{1}{2},\tfrac{1}{2}\rangle_{(2)} - |\tfrac{1}{2},\tfrac{1}{2}\rangle_{(1)}|\tfrac{1}{2},-\tfrac{1}{2}\rangle_{(2)}. \tag{5.36}$$

It is easy to see that it is killed by either lowering or raising operators; it is therefore a singlet state. We have just proven that two spin one-half form themselves symmetrically into a spin one combination or antisymmetrically into a spin zero state. We indicate this as

$$\mathbf{2} \times \mathbf{2} = \mathbf{3} + \mathbf{1}, \tag{5.37}$$

indicating here the representation by its dimensionality.

For a second example, consider the Kronecker product of spin one-half with spin one representations. In our language, it is $\mathbf{2} \times \mathbf{3}$ which, according to what we have just said,

$$\mathbf{2} \times \mathbf{3} = \mathbf{4} + \mathbf{2}, \tag{5.38}$$

and contains contains two representations, spin $3/2$ and $1/2$. To construct these states explicitly, starting from the highest weight states, we need the action of the lowering operators on the $j = 1/2$ and $j = 1$ representations

$$T^- |\, \tfrac{1}{2}, \tfrac{1}{2} \,\rangle = |\, \tfrac{1}{2}, -\tfrac{1}{2} \,\rangle; \quad T^- |\, 1, 1 \,\rangle = \sqrt{2} \,|\, 1, 0 \,\rangle; \quad T^- |\, 1, 0 \,\rangle = \sqrt{2} \,|\, 1, -1 \,\rangle. \tag{5.39}$$

We first build the states of the spin $3/2$ representation, starting from the highest weight of the constituent representations:

$$|\, \tfrac{3}{2}, \tfrac{3}{2} \,\rangle = |\, \tfrac{1}{2}, \tfrac{1}{2} \,\rangle \,|\, 1, 1 \,\rangle. \tag{5.40}$$

This state is uniquely determined, as we can see from its value of T^3. We generate all the remaining states of this representation by use of the sum of the lowering operators. Using eq. (5.39), we have

$$\begin{aligned} |\, \tfrac{3}{2}, \tfrac{1}{2} \,\rangle &= \frac{1}{\sqrt{3}} T^- \,|\, \tfrac{3}{2}, \tfrac{3}{2} \,\rangle, \\ &= \sqrt{\tfrac{1}{3}} \,|\, \tfrac{1}{2}, -\tfrac{1}{2} \,\rangle \,|\, 1, 1 \,\rangle + \sqrt{\tfrac{2}{3}} \,|\, \tfrac{1}{2}, \tfrac{1}{2} \,\rangle |\, 1, 0 \,\rangle. \end{aligned} \tag{5.41}$$

Then we continue the process

$$\begin{aligned} |\, \tfrac{3}{2}, -\tfrac{1}{2} \,\rangle &= \tfrac{1}{2} T^- \,|\, \tfrac{3}{2}, \tfrac{1}{2} \,\rangle, \\ &= \frac{1}{2\sqrt{3}} \Big(2\sqrt{2} \,|\, \tfrac{1}{2}, -\tfrac{1}{2} \,\rangle \,|\, 1, 0 \,\rangle + 2 \,|\, \tfrac{1}{2}, \tfrac{1}{2} \,\rangle \,|\, 1, -1 \,\rangle \Big), \\ &= \sqrt{\tfrac{2}{3}} \,|\, \tfrac{1}{2}, -\tfrac{1}{2} \,\rangle \,|\, 1, 0 \,\rangle + \sqrt{\tfrac{1}{3}} \,|\, \tfrac{1}{2}, \tfrac{1}{2} \,\rangle \,|\, 1, -1 \,\rangle. \end{aligned} \tag{5.42}$$

Finally, we lower one more time to get the lowest weight state

$$\begin{aligned} |\, \tfrac{3}{2}, -\tfrac{3}{2} \,\rangle &= \frac{1}{\sqrt{3}} T^- \,|\, \tfrac{3}{2}, -\tfrac{1}{2} \,\rangle, \\ &= |\, \tfrac{1}{2}, -\tfrac{1}{2} \,\rangle \,|\, 1, -1 \,\rangle. \end{aligned}$$

The remaining four states are then obtained as the combinations that are orthogonal to (5.41). In this case there is only one

$$\sqrt{\tfrac{2}{3}} \,|\, \tfrac{1}{2}, -\tfrac{1}{2} \,\rangle \,|\, 1, 1 \,\rangle - \sqrt{\tfrac{1}{3}} \,|\, \tfrac{1}{2}, \tfrac{1}{2} \,\rangle, \tag{5.43}$$

and since it is annihilated by T^+, it is the highest weight state of the spin one-half representation, that is

$$|\,\tfrac{1}{2},\tfrac{1}{2}\,\rangle = \sqrt{\tfrac{2}{3}}\,|\,\tfrac{1}{2},-\tfrac{1}{2}\,\rangle\,|\,1,1\,\rangle - \sqrt{\tfrac{1}{3}}\,|\,\tfrac{1}{2},\tfrac{1}{2}\,\rangle. \tag{5.44}$$

The coefficients which appear in these decompositions are called the *Clebsch–Gordan* coefficients, familiarly know as "Clebsches." They are so important to physical applications, that they have been tabulated. See for example the Particle Data Group (http://pdg.gov/).

There are countless ways to represent the $SU(2)$ algebra. Our favorite is in terms of two fermionic harmonic oscillators. In this case, the Hilbert space contains just four states

$$|\,0\,\rangle, \qquad b_1^\dagger|\,0\,\rangle, \qquad b_2^\dagger|\,0\,\rangle, \qquad b_1^\dagger b_2^\dagger|\,0\,\rangle. \tag{5.45}$$

The three $SU(2)$ generators are simply

$$T_i^{\;j} = b_i^\dagger b^j - \frac{1}{2}\delta_i^{\;j}\, b_k^\dagger b^k, \qquad i,j = 1,2. \tag{5.46}$$

It is easily seen that the four states break up into two singlets and one doublet. This construction generalizes to an arbitrary number of harmonic oscillators.

5.3 From Lie algebras to Lie groups

What kind of group does this algebra generate? We have seen that a general unitary operator can be written as the exponential of i times a hermitian matrix. The generators of the $SU(2)$ Lie algebra are hermitian, so that any operator of the form

$$\mathcal{U}(\vec{\theta}) = e^{i\,\theta_A T^A}, \tag{5.47}$$

where the θ_A are three real parameters, is a unitary operator. These operators satisfy the group axioms. Each has an inverse, with $\vec{\theta} \to -\vec{\theta}$. One, with $\vec{\theta} = 0$, is the unit element. Less obvious is their closure under multiplication, that is

$$\mathcal{U}(\vec{\theta})\mathcal{U}(\vec{\eta}) = \mathcal{U}(\vec{\phi}(\vec{\theta},\vec{\eta})), \tag{5.48}$$

as their associativity

$$\mathcal{U}(\vec{\theta})\big(\mathcal{U}(\vec{\eta})\mathcal{U}(\vec{\chi})\big) = \big(\mathcal{U}(\vec{\theta})\mathcal{U}(\vec{\eta})\big)\mathcal{U}(\vec{\chi}). \tag{5.49}$$

The proof of closure is based on the Baker–Hausdorf theorem which expresses the product of two exponentials as an exponential, with exponent in the Lie algebra. We do not prove it here, as the reader can convince him/herself by expanding both sides to second order in the parameters. We can easily verify it for $SU(2)$ in the $j = 1/2$ representation. A little mathematical exercise shows that

$$\mathcal{U}(\vec{\theta}) = \exp\left(i\,\theta^A\,\frac{\sigma^A}{2}\right) = \cos(\frac{\theta}{2}) + i\,\hat{\theta}^A\,\sigma^A\,\sin(\frac{\theta}{2}), \tag{5.50}$$

where

$$\hat{\theta}^A = \frac{\theta^A}{\theta}; \qquad \theta = \sqrt{\theta^A\theta^A}. \tag{5.51}$$

In this form both closure and associativity are manifest: $\mathcal{U}(\vec{\theta})$ satisfies the group axioms. It is a continuous (Lie) group because the $\vec{\theta}$ are continuous parameters.

It describes a group transformation that is periodic only after a 4π rotation. A 2π transformation results in a reversal in sign of the vector on which it acts. Viewed as an operator in a two-dimensional Hilbert space, this transformation leaves the quadratic form,

$$|z_1|^2 + |z_2|^2, \tag{5.52}$$

invariant. This is the reason why the group of unitary transformation in two dimensions associated with the Lie algebra $SU(2)$ is called $SU(2)$. The ever-careful mathematicians distinguish the algebra from the group by using lower-case letters for the algebra ($su(2)$), and upper-case letters for the group ($SU(2)$). The U signifies that the transformations are unitary, and the S that these transformations have unit determinant. Their general form is

$$\begin{pmatrix} \alpha & \beta \\ -\overline{\beta} & \overline{\alpha} \end{pmatrix}, \tag{5.53}$$

subject to

$$|\alpha|^2 + |\beta|^2 = 1. \tag{5.54}$$

Any group element can be represented as a point on the surface of a four-sphere S_3, the *group manifold* of $SU(2)$. A succession of transformations will trace a continuous curve on S_3. In particular any closed curve can be shrunk to a point since one cannot lasso a basketball, even in four dimensions! We say that the $SU(2)$ group manifold is *simply connected.*

Not all group manifolds are simply connected. To see this, let us look at the exponential mapping of the same algebra, but in the $j = 1$ representation, where the $SU(2)$ generators are represented by matrices

$$\left(T^A\right)_{BC} = -i\epsilon^{ABC}. \tag{5.55}$$

The group elements are expressed as (3×3) rotation matrices

$$\mathcal{R}_{BC}(\vec{\theta}) = \exp\left(\theta^A\epsilon_{ABC}\right), \tag{5.56}$$

which generates rotations in space, acting on the three components of a vector

$$\vec{\mathbf{v}} \to \vec{\mathbf{v}}' = \mathcal{R}(\vec{\theta})\, \vec{\mathbf{v}}. \tag{5.57}$$

This element generates the group of rotations in real three-dimensional space, called $SO(3)$. It is the group of orthogonal transformations of unit determinant that leave the length of a three-dimensional vector invariant

$$\vec{\mathbf{v}} \cdot \vec{\mathbf{v}} = \vec{\mathbf{v}}' \cdot \vec{\mathbf{v}}' \to \mathcal{R}(\vec{\theta})^T \mathcal{R}(\vec{\theta}) = 1, \tag{5.58}$$

where T means matrix transpose. By expanding the exponential, we find (homework) that

$$\mathcal{R}_{BC}(\vec{\theta}) = \delta_{BC} \cos\theta + \epsilon_{BCA} \hat{\theta}_A \sin\theta + \hat{\theta}_B \hat{\theta}_C (1 - \cos\theta), \tag{5.59}$$

which describes a rotation by an angle θ about the axis $\hat{\theta}_A$. The group element is then given by the unit vector $\hat{\theta}_A$ and the angle θ. We can use this expression to determine the $SO(3)$ group manifold. It is a three-dimensional space with the angles θ_A labeling its orthogonal axes. The group element is the same for θ and $\theta + 2\pi$, so that we limit its range to

$$-\pi \le \theta \le \pi, \tag{5.60}$$

and allow negative values for the angles. The $SO(3)$ group manifold is the inside and the surface of a sphere of radius π, but since $\theta = \pi$ and $\theta = -\pi$ lead to the same group element, antipodal points on the surface of the sphere are identified with one another. Alternatively, a point in the manifold is a vector with components

$$\xi_1 = \frac{\theta}{\pi} \sin\eta \cos\phi, \qquad \xi_2 = \frac{\theta}{\pi} \sin\eta \sin\phi, \qquad \xi_3 = \frac{\theta}{\pi} \cos\eta, \tag{5.61}$$

subject to the constraint

$$(\xi_1)^2 + (\xi_2)^2 + (\xi_3)^2 = \left(\frac{\theta}{\pi}\right)^2, \tag{5.62}$$

with $|\theta| \le \pi$, with antipodal identification on the sphere $|\theta| = \pi$. This describes the $SO(3)$ *group manifold*.

To investigate its topological properties, consider a rotation that starts at the origin and does not make it to the surface of the sphere and comes back, forming a closed curve on the manifold. That curve can be shrunk to a point without obstruction. Now consider another path which starts at the origin and makes its way to the surface of the sphere, and then reappears at the antipodal point before rejoining the origin. This closed curve cannot be deformed to a point: $SO(3)$ is multiply connected. Next, consider a curve which describes a path that pierces the sphere at two points. We leave it to the reader to show that this path can also be deformed to a point. It means that $SO(3)$ is *doubly connected*.

It can be shown that our algebra generates yet another group, $Sp(2)$, the group of quaternionic transformations, a notion we will examine in detail when we consider symplectic algebras.

The same Lie algebra can generate groups with different group manifolds with different topologies. When a Lie algebra generates several groups, the one whose manifold is simply connected is called the *universal covering group*: $SU(2)$ is the universal covering group of $SO(3)$.

5.4 $SU(2) \to SU(1,1)$

The $SU(2)$ Lie algebra has yet another interesting property: one can reverse the sign of two of its elements without changing the algebra. For example

$$T^{1,2} \to -T^{1,2}; \qquad T^3 \to T^3. \tag{5.63}$$

This operation is called an *involutive automorphism* of the algebra. It is an operation which forms a finite group of two elements. Its eigenvalues are therefore ± 1, and it splits the Lie algebra into even and odd subsets.

One can use this to obtain a closely related algebra by multiplying all odd elements by i. This generates another Lie algebra with three elements L^A, with $L^{1,2} \equiv iT^{1,2}$, $L^3 \equiv T^3$. Note that $L^{1,2}$ are no longer hermitian. The new algebra is

$$[L^1\,,\, L^2\,] = -i\, L^3, \qquad [L^2\,,\, L^3\,] = i\, L^1; \qquad [L^3\,,\, L^1\,] = i\, L^2. \tag{5.64}$$

It is still a Lie algebra because this replacement does not affect the Jacobi identity. The minus sign in the first commutator has crucial consequences because the Casimir operator,

$$Q = (L^1)^2 + (L^2)^2 - (L^3)^2, \qquad [\, Q,\, L^i\,] = 0, \tag{5.65}$$

is no longer positive definite. This means that the spectrum of L^3 need not be bounded: the natural realm of this Lie algebra is an infinite-dimensional Hilbert space. What a difference a sign makes! This algebra is called a *non-compact* or *real form* of the $SU(2)$ algebra. As you can see from the Casimir operator, the invariant quadratic form is not positive definite, but looks like the length of a vector in three-dimensional Minkowski space with two spatial and one time directions, hence this algebra is called $SO(2,1)$ (also $SU(1,1)$) by physicists.

One can go further and absorb the i on the right-hand side of the commutators, and generate the algebra in terms of the real matrices (hence the name real form?)

$$\begin{pmatrix} 1 & 0 \\ 0 & -1 \end{pmatrix}, \begin{pmatrix} 0 & 1 \\ 1 & 0 \end{pmatrix}, \begin{pmatrix} 0 & -1 \\ 1 & 0 \end{pmatrix}. \tag{5.66}$$

Its representation theory is far more complicated than that of its compact counterpart. Its simplest unitary representations can be found in the infinite Hilbert space generated by one bosonic harmonic oscillator. To see this, consider the operators

$$L^+ \equiv \frac{1}{2\sqrt{2}}\, a^\dagger a^\dagger; \qquad L^- \equiv \frac{1}{2\sqrt{2}}\, a\, a = (L^+)^\dagger. \tag{5.67}$$

Compute their commutator

$$[L^+,\, L^-] = -\frac{1}{4}\,(1 + 2a^\dagger a) \equiv -L^3. \tag{5.68}$$

The new operator itself has nice commutation relations with $L^\pm$

$$[L^3,\, L^\pm] = \pm L^\pm, \tag{5.69}$$

completing the non-compact $SO(2, 1)$ Lie algebra, as we can see by defining the hermitian generators through

$$L^+ \equiv \frac{1}{\sqrt{2}}\left(L^1 + iL^2\right), \qquad L^- \equiv \frac{1}{\sqrt{2}}\left(L^1 - iL^2\right). \tag{5.70}$$

In this representation the Casimir operator reduces to a c-number

$$Q = \frac{3}{16}. \tag{5.71}$$

The action of these operators is particularly simple. Starting from the vacuum state, $|\,0\,\rangle$, we note that

$$L^-\,|\,0\,\rangle = 0, \tag{5.72}$$

while

$$L^+\,|\,0\rangle = \tfrac{1}{2}\,|\,2\,\rangle, \tag{5.73}$$

and then by repeated application of the raising operator L^-, we generate an infinite tower of states

$$(L^+)^n\,|\,0\,\rangle = \frac{\sqrt{(2n)!}}{2\sqrt{2}}\,|\,2n\,\rangle. \tag{5.74}$$

Each state is an eigenstate of L^3,

$$L^3\,|\,2n\,\rangle = \frac{(1+4n)}{4}\,|\,2n\,\rangle. \tag{5.75}$$

Hence the infinite tower of states $|\,0\,\rangle, |\,2\,\rangle, \ldots, |\,2n\,\rangle, \ldots$ forms a representation of the algebra. Furthermore since the operators are hermitian, this representation is unitary and manifestly irreducible. The L^3 spectrum is bounded from below, starting at $1/4$ and increasing forever in integer steps.

There is another tower of states, $|\,1\,\rangle, |\,3\,\rangle, \ldots, |\,2n+1\,\rangle, \ldots$ which are also related to one another with the action of the generators of our Lie algebra. In this case, the L^3 spectrum starts at $3/4$. The harmonic oscillator space splits into two representations of $SO(2, 1)$ of even and occupation number states. They are called *singleton* representations since they involve only one harmonic oscillator. The one-dimensional harmonic oscillator has no particular symmetry of its own (except for parity $x \rightarrow -x$), so where does this algebra fit in? The clue is that its Hamiltonian

$$H = a^\dagger a + \tfrac{1}{2} = 2L^3, \tag{5.76}$$

is itself a member of the algebra. The algebra relates states of different energy; such an algebra is called a *spectrum generating algebra*; each representation contains all of its states of a given parity.

There are of course many other representations of this algebra. We can generate as many finite-dimensional representations as we want, by starting from the $(2j+1)$-dimensional hermitian representation matrices of the compact algebra and multiplying any two generators by i. Of course none is unitary.

As we have seen, in the $j = 1/2$ representation, the three generators are

$$\frac{1}{2}\begin{pmatrix} 0 & 1 \\ -1 & 0 \end{pmatrix}; \quad \frac{1}{2}\begin{pmatrix} 1 & 0 \\ 0 & -1 \end{pmatrix}; \quad \frac{i}{2}\begin{pmatrix} 0 & 1 \\ 1 & 0 \end{pmatrix}. \tag{5.77}$$

The corresponding group elements generated by these (2×2) matrices are of the form

$$e^M; \qquad M = \tfrac{1}{2}\begin{pmatrix} i\theta_3 & -\theta_1 + i\theta_2 \\ -\theta_1 - i\theta_2 & -i\theta_3 \end{pmatrix}. \tag{5.78}$$

They are not unitary, with matrices of the form

$$h = \begin{pmatrix} \alpha^* & \beta^* \\ \beta & \alpha \end{pmatrix}, \tag{5.79}$$

of unit determinant

$$|\,\alpha\,|^2 - |\,\beta\,|^2 = 1. \tag{5.80}$$

These matrices generate a group called $SU(1, 1)$, since it leaves the above quadratic form (5.80) invariant. This representation is *not* unitary. The Casimir operator expressed in terms of hermitian operators is not positive definite, which means that unitary representations of this non-compact algebra have to be *infinite-dimensional*.

To illustrate the complexity of the unitary representations of non-compact groups, we list without proof the many different types of unitary irreducible representations (irreps) of $SO(2, 1) = SU(1, 1)$. We express the Casimir operator as $Q = -j(j-1)$. One finds, after much work, five types of unitary irreps. All are infinite-dimensional, the states within each irrep being labeled by l_3, the eigenvalues of L^3:

Continuous class

- Non-exceptional interval: $j = -\frac{1}{2} - is, \quad 0 \leq s < \infty$.
  (i) Integral case, $l_3 = 0, \pm 1, \pm 2, \ldots$.
  (ii) Half-integral case: $l_3 = 0, \pm\frac{1}{2}, \pm\frac{3}{2}, \ldots$.
  They are realized on square-integrable functions on the unit circle $f(\varphi)$, with inner product

$$(f, g) = \frac{1}{2\pi} \int_0^{2\pi} d\varphi \, f^*(\varphi) \, g(\varphi),$$

$$\mathcal{U}(h) \, f(\varphi) = | \, \alpha^* - \beta \, e^{i\varphi} \, |^{-1-2is} \, f(\Phi),$$

where

$$e^{i\Phi} = \frac{\alpha \, e^{i\varphi} - \beta^*}{\alpha^* - \beta \, e^{-i\varphi}}.$$

- Exceptional interval: $j = -\frac{1}{2} - \sigma, \quad 0 < \sigma < \frac{1}{2}; \qquad l_3 = 0, \pm 1, \pm 2, \ldots$.
  They describe functions on the circle such that

$$\mathcal{U}(h) \, f(\varphi) = | \, \alpha^* - \beta \, e^{i\varphi} \, |^{-1-2\sigma} \, f(\Phi),$$

in a space with inner product

$$(f, g)_\sigma = \frac{\Gamma(\sigma + 1/2)}{(2\pi)^{3/2} 2^\sigma \Gamma(\sigma)} \int_0^{2\pi} d\varphi_1 \int_0^{2\pi} d\varphi_2 \Big[1 - \cos(\varphi_1 - \varphi_2) \Big]^{\sigma - 1/2} f^*(\varphi_1) g(\varphi_2).$$

Discrete class

- $j = \frac{1}{2}, 1, \frac{3}{2}, \ldots; \qquad l_3 = j, j+1, j+2, \ldots$.
- $j = \frac{1}{2}, 1, \frac{3}{2}, \ldots; \qquad l_3 = -j, -j-1, -j-2, \ldots$.
- $j = \frac{1}{4}, \frac{3}{4}; \qquad l_3 = j, j+1, j+2, \ldots$ (singleton representations).

  They are realized on functions of a complex variable, $f(z)$, defined on the open circle $| \, z \, | < 1$,

$$\mathcal{U}(h) \, f(z) = (\alpha^* + i\beta z)^{2j} \, f\left(\frac{\alpha z - i\beta^*}{\alpha^* + i\beta z} \right)$$

with inner product

$$(f, g)_j = \frac{-2j - 1}{\pi} \oint d^2 z \left[1 - |z|^2 \right]^{-2j-2} f^*(z) \, g(z).$$

In a sense the singletons are the simplest unitary irreps of this algebra, but they are actually part of a representation of a *super algebra* which is an algebra closed under *both* commutators and anticommutators.

Indeed the lowest states of the two singleton representations, $| \, 0 \, \rangle$ and $| \, 1 \, \rangle$, can be related to one another by the application of the creation operator $a^\dagger$. Hence we can extend the Lie algebra to include the operators a and $a^\dagger$ that relate the two singleton representations. However, while their commutator is not in the Lie algebra, their anticommutator

$$\{a,\, a^\dagger\} \equiv a\,a^\dagger + a^\dagger\, a = 1 + 2a^\dagger\, a = 4L_3, \tag{5.81}$$

is. We further check that

$$[\,L^+, a\,] = -\frac{1}{\sqrt{2}}\,a^\dagger, \qquad [\,L_3, a\,] = -\tfrac{1}{2}\,a. \tag{5.82}$$

Thus, by extending the Lie algebra operation to include both commutators and anti-commutators, we obtain a *super-Lie algebra*. Its singleton representation includes both singleton representations of its Lie sub-algebra, and is the Hilbert space generated by one harmonic oscillator. See how much one can do with one set of creation and annihilation operators! Could it be that *any* Hilbert space can be thought of as a representation of some algebraic structure?

5.5 Selected $SU(2)$ applications

The smallest Lie algebra we have just studied and its related groups have many applications in physics (for particle physics, see the books by Cahn [1] and Georgi [6]). In the following we have chosen a few examples which illustrate its importance.

5.5.1 The isotropic harmonic oscillator

The quantum-mechanical isotropic harmonic oscillator is treated in many books, and we simply state its main features. It is a physical system with time evolution generated by the Hamiltonian

$$H = \frac{\vec{p}\cdot\vec{p}}{2m} + \tfrac{1}{2}\,k\,\vec{x}\cdot\vec{x}, \tag{5.83}$$

where $\vec{x}$ and $\vec{p}$ are the three-dimensional position and momentum vectors. It is solved by rewriting the Hamiltonian in terms of the three creation and annihilation operators $A_i^\dagger$ and A_i,

$$H = \hbar\omega(\vec{A}^\dagger \cdot \vec{A} + \tfrac{3}{2}), \tag{5.84}$$

with

$$[\,A_i,\, A_j^\dagger\,] = \delta_{ij}, \quad i, j = 1, 2, 3. \tag{5.85}$$

Its eigenstates are

$$H\,|\,n_1, n_2, n_3\,\rangle = \hbar\omega(n_1 + n_2 + n_3 + \tfrac{3}{2})|\,n_1, n_2, n_3\,\rangle, \tag{5.86}$$

with

$$| n_1, n_2, n_3 \rangle = \prod_i \frac{(A_i^\dagger)^{n_i}}{\sqrt{n_i!}} | 0 \rangle. \tag{5.87}$$

We want to analyze this spectrum using group theory. Its lowest state is the vacuum state $| 0 \rangle$, annihilated by all three A_i. At the next energy level, one finds *three* states $| 1, 0, 0 \rangle$, $| 0, 1, 0 \rangle$ and $| 0, 0, 1 \rangle$, all with the same energy. Define three transition operators among these

$$P_{12}| 1, 0, 0 \rangle = | 0, 1, 0 \rangle \; ; \; P_{13}| 1, 0, 0 \rangle = | 0, 0, 1 \rangle \; ; \; P_{23}| 0, 1, 0 \rangle = | 0, 0, 1 \rangle .$$

Since these states have the same energy, that means

$$[H, \; P_{ij}] = 0 \tag{5.88}$$

acting on any of these three kets: the degeneracy of the first excited state tells us that there exist operators which commute with the Hamiltonian, at least when acting on the first excited states. These operators generate symmetries of the Hamiltonian. The action of T_{ij} is to destroy $A_i^\dagger$ and replace it with $A_j^\dagger$, so they can be written as hermitian operators of the form

$$P_{ij} = i(A_i^\dagger \, A_j - A_j^\dagger \, A_i), \tag{5.89}$$

which satisfy the $SU(2)$ Lie algebra, and generate rotations in three dimensions. They are of course the angular momentum operators

$$L_1 = P_{23} = x_2 \, p_3 - x_3 \, p_2, \text{ etc.} \tag{5.90}$$

They commute with the Hamiltonian and reflects its invariance under space rotations. The commutator

$$[L_i, \; A_j^\dagger] = -i\epsilon_{ijk} \, A_k^\dagger, \tag{5.91}$$

denotes how the three creation operators transform under rotation: they transform into one another like the components of a vector, but they are tensor operators. Hence they span the **3** representation of $SU(2)$. Hence the Nth level transforms as the N-fold *symmetric* product of the vector

$$\left(\mathbf{3} \times \mathbf{3} \times \cdots \times \mathbf{3} \times \mathbf{3}\right)_{sym} .$$

This enables us to decompose any excited level in terms of this product. As an example take the second excited level. Since

$$\mathbf{3} \times \mathbf{3} = \mathbf{5} + \mathbf{1},$$

it breaks up into the quadrupole representation and a singlet. Specifically the quadrupole is the second rank symmetric traceless tensor,

$$\left[A_i^\dagger A_j^\dagger + A_i^\dagger A_j^\dagger - \frac{1}{6}\delta_{ij}(\vec{A}^\dagger \cdot \vec{A}^\dagger)\right]|\,0\,\rangle,$$

while the singlet is just the dot product

$$(\vec{A}^\dagger \cdot \vec{A}^\dagger)|\,0\,\rangle.$$

We note that in this spectrum, only integer spin representations appear. Our analysis shows that we can think of all the states as symmetric tensors, with the traces removed by the δ_{ij} tensor. It is called an invariant tensor because it is mapped into itself by an $SO(3)$ transformation.

What about the rest of the representations, such as antisymmetric tensors and half-odd integer representations? Can we generate them with creation and annihilation operators? Of course: we can do **anything** with ladder operators! We have already seen how to generate *all* representations by taking Kronecker products of the two-component spinor representation. This suggests that we think of the two spinor states as being created by *two* creation operators, $a_\alpha^\dagger$, $\alpha = 1, 2$. The operators, simply given in terms of the Pauli matrices

$$T^A = \frac{1}{2} a_\alpha^\dagger \, (\sigma^A)_{\alpha\beta} \, a_\beta, \tag{5.92}$$

satisfy the $SU(2)$ commutation relations as long as

$$[\, a_\alpha, \, a_\beta^\dagger \,] = \delta_{\alpha\beta}. \tag{5.93}$$

These operators (first constructed by J. Schwinger) are not to be confused with those of the isotropic harmonic oscillator problem. They are $SU(2)$ spinors, while those for the isotropic harmonic oscillator are $SU(2)$ vectors. The Fock space generated by arbitrary products of the spinor creation operators generate the $SU(2)$ lattice, while that generated by the vector creation operators generate only a subset of representations.

5.5.2 The Bohr atom

Bohr's model of the hydrogen atom is a cornerstone of Quantum Mechanics. It is described by the Hamiltonian of a particle of momentum $\vec{p}$, charge e and mass m, bound by an attractive Coulomb potential

$$H = \frac{\vec{p} \cdot \vec{p}}{2m} - \frac{e^2}{r}; \qquad r = \sqrt{\vec{x} \cdot \vec{x}}. \tag{5.94}$$

Its spectrum of negative energy bound states, familiar to physics students, is given by

$$E_n = -\frac{\mathcal{R}_y}{n^2}, \tag{5.95}$$

where $\mathcal{R}_y$ is Rydberg's constant and n is the principal quantum number $n = 1, 2, \ldots$. Each level is populated by states of angular momentum $l = 0, 1, 2, \ldots, n-1$. The Bohr atom is manifestly rotationally invariant, meaning that the angular momentum operators commute with the Bohr Hamiltonian

$$L_i = \epsilon_{ijk}\, x_j\, p_k, \qquad [\, H,\ L_i\,] = 0. \tag{5.96}$$

As we have seen in the previous example, this invariance implies that levels of the same energy will come in sets of representations of the rotation algebra. Indeed this happens, but the first excited level is made up of *two* $SO(3)$ representations, the ones with $l = 1$ and $l = 0$. This implies that there is more symmetry in the problem than meets the eye. In particular it hints at the existence of three operators that map each of the $l = 1$ states into the $l = 0$ state. We infer the existence of an operator which commutes with the Hamiltonian, and is a vector operator under $SO(3)$.

Indeed, the existence of such a vector had been known for a long time (Laplace, Runge and Lenz) in the context of Kepler's problem, which uses the same potential. Its classical form is

$$A_i^{\text{classical}} = \epsilon_{ijk}\, p_j\, L_k - m\, e^2\, \frac{x_i}{r}. \tag{5.97}$$

For use in quantum theory, we symmetrize on the momenta and angular momenta, to get its hermitian form

$$\begin{aligned} A_i &= \tfrac{1}{2}\, \epsilon_{ijk}\, (p_j\, L_k + L_k\, p_j) - m\, e^2\, \frac{x_i}{r}, \\ &= \epsilon_{ijk}\, p_j\, L_k - m\, e^2\, \frac{x_i}{r} - i\, p_i. \end{aligned}$$

It is not difficult to see that $\vec{A}$ is orthogonal to $\vec{L}$

$$\vec{L} \cdot \vec{A} = \vec{A} \cdot \vec{L} = 0. \tag{5.98}$$

Also it is manifestly a vector operator, which can be verified by explicit computation of its commutator with L_i

$$[\, L_i,\ A_j\,] = i\, \epsilon_{ijk}\, A_k. \tag{5.99}$$

Finally, a careful computation yields

$$[\, H,\ A_j\,] = 0, \tag{5.100}$$

which means that this vector is a constant of the motion, *and* that the Bohr Hamiltonian is invariant under the transformations it generates. This extra invariance is not at all manifest; it is a *hidden symmetry*.

It provides for a quick and elegant solution of the Bohr atom. A few months after the creation of Heisenberg's paper, Pauli was the first to solve the quantum mechanical Bohr atom. He used group theoretical methods, which we present below.

We begin by computing the commutator of A_i with itself. The result is rather simple

$$[\,A_i,\,A_j\,] = i\,\epsilon_{ijk}\,L_k\,(-2m\,H). \tag{5.101}$$

Since the Bohr Hamiltonian describes a bound state system, we expect the eigenvalues of H to be negative. Also, since H commutes with both $\vec{L}$ and $\vec{A}$, we can define without ambiguity the new operator

$$\hat{A}_i \equiv \frac{A_i}{\sqrt{-2m\,H}}, \tag{5.102}$$

which satisfies the simpler commutation relation

$$[\,\hat{A}_i,\,\hat{A}_j\,] = i\,\epsilon_{ijk}\,L_k, \tag{5.103}$$

closing the algebra. We note that the linear combinations

$$X_i^{(+)} \equiv \tfrac{1}{2}(L_i + \hat{A}_i), \qquad X_i^{(-)} \equiv \tfrac{1}{2}(L_i - \hat{A}_i), \tag{5.104}$$

not only commute with one another

$$[\,X_i^{(+)},\,X_j^{(-)}\,] = 0, \tag{5.105}$$

but also form *two* independent $SU(2)$ Lie algebras

$$[\,X_i^{(+)},\,X_j^{(+)}\,] = i\,\epsilon_{ijk}\,X_k^{(+)}; \qquad [\,X_i^{(-)},\,X_j^{(-)}\,] = i\,\epsilon_{ijk}\,X_k^{(-)}. \tag{5.106}$$

The invariance algebra of the Bohr atom is the direct product of two $SU(2)$ algebras, $SU(2) \times SU(2)$. The two Casimir operators of these algebras must be constants of the motion, but as we see from eq. (5.98), they are the same since

$$C_2^{(+)} = \frac{1}{4}\left(L_i + \hat{A}_i\right)\left(L_i + \hat{A}_i\right) = \frac{1}{4}\left(L_i - \hat{A}_i\right)\left(L_i - \hat{A}_i\right) = C_2^{(-)}, \tag{5.107}$$

that is

$$C_2^{(+)} = j_1\,(j_1 + 1) = C_2^{(-)} = j_2\,(j_2 + 1) \equiv j\,(j + 1), \tag{5.108}$$

where $j = j_1 = j_2$.

The next step is to express the Hamiltonian in terms of the Casimir operators by calculating the length of the LRL vector. This is a tricky calculation because we are dealing with operators, so we give some of the details. We split the result into direct and cross terms. The direct terms easily yield

$$A_i\,A_i\Big|_{\text{direct}} = \epsilon_{ijk}\,\epsilon_{imn}\,p_j\,p_m\,L_k\,L_n - \frac{2m\,e^2}{r}\,\epsilon_{ijk}x_i\,p_j\,L_k + m^2\,e^4$$
$$= (p_j\,p_j)\,L_i\,L_i - p_j\,p_j - \frac{2m\,e^2}{r}\,L_i\,L_i + m^2e^4.$$

The mass-independent cross terms are

$$-i\epsilon_{ijk}(p_j\,L_k\,p_i + p_i\,p_j\,L_k) = 2p_i\,p_i, \tag{5.109}$$

using

$$[\,L_k,\;p_j\,] = i\epsilon_{kji}p_i.$$

The first of the mass-dependent cross terms yields

$$-m\,e^2\epsilon_{ijk}\left(p_j\,L_k\,\frac{x_i}{r} + \frac{x_i}{r}\,p_j, L_k\right) = -\frac{2m\,e^2}{r}\,L_k\,L_k - 2im\,e^2\,p_i\,\frac{x_i}{r},$$

while the second gives

$$im\,e^2\left(p_i\,\frac{x_i}{r} + \frac{x_i}{r}\,p_i\right).$$

Their sum is

$$-\frac{2m\,e^2}{r}\,L_k\,L_k - \frac{3m\,e^2}{r} - im\,e^2\left[\,p_i\,x_i,\;\frac{1}{r}\right].$$

Evaluation of the commutator yields the final result

$$A_i\,A_i = \left(\vec{p}\cdot\vec{p} - \frac{2m\,e^2}{r}\right)\left(L_i\,L_i + 1\right) + m^2\,e^4\,. \tag{5.110}$$

We divide both sides by $-2mH$, and solve for H, to get

$$H = -\frac{m\,e^4/2}{\vec{L}\cdot\vec{L} + \vec{\hat{A}}\cdot\vec{\hat{A}} + 1} = -\frac{m\,e^4/2}{4\,C_2^{(+)} + 1}. \tag{5.111}$$

The relation between the Hamiltonian and the Casimir operator is explicit; in terms of j,

$$H = -\frac{m\,e^4}{2(2j+1)^2}. \tag{5.112}$$

This clearly shows that j is *not* the angular momentum but is related to the principal quantum number

$$n = 2\,j + 1.$$

The angular momentum is the sum of the two independent Lie algebras

$$L_i = X_i^{(+)} + X_i^{(-)}, \tag{5.113}$$

so that finding the degeneracy of the levels with $n = 2j + 1$ amounts to listing the states in the direct product of $(\mathbf{2j+1}) \times (\mathbf{2j+1})$, the classical problem as the addition of two angular momenta with the same j, yielding states with angular momentum $l = 2j = n - 1, 2j - 1 = n - 2, \ldots, 1, 0$, which is the well-known result.

By recognizing the symmetries of Bohr's hydrogen, and using his knowledge of the representations of $SU(2)$, Pauli was the first (*Zs. f. Phys.* **36**, 336 (1926)) to derive the spectrum of its Hamiltonian. Such was the power of Pauli and of group theory!

5.5.3 Isotopic spin

Nuclear forces appear to act equally on protons and neutrons, the constituents of nuclei. This led W. Heisenberg to think of them as the two components of an *iso* doublet of a new $SU(2)$ which he called *isotopic spin*. It cannot be an exact concept since protons are charged and neutrons are not. Isotopic spin invariance applies only to nuclear forces, and it appears as a *broken symmetry* when the weak and electromagnetic forces are taken into account. The nuclear potential was soon understood as being caused by new elementary particles, the pions. They appear in three charge states, π^+, π^0 and π^-, suggesting that they could be assembled into an *isotriplet*.

Fermi–Yang model

In this model (*Phys. Rev.* **76**, 1739 (1949)), Fermi and Yang suggest that pions, the particles behind nuclear forces, might be nucleon–antinucleon bound states. Think of nucleons and antinucleons as constructed from Fermi oscillator states, that is

$$|\,p\,\rangle = b_1^\dagger\,|\,0\,\rangle, \qquad |\,n\,\rangle = b_2^\dagger\,|\,0\,\rangle, \tag{5.114}$$

for the nucleons, and

$$|\,\bar{p}\,\rangle = \bar{b}_1^\dagger\,|\,0\,\rangle, \qquad |\,\bar{n}\,\rangle = \bar{b}_2^\dagger\,|\,0\,\rangle, \tag{5.115}$$

for their antiparticles. The oscillators satisfy the usual anticommutation relations

$$\left\{b_i,\, b_j^\dagger\right\} = \delta_{ij}, \qquad \left\{\bar{b}_i,\, \bar{b}_j^\dagger\right\} = \delta_{ij}, \tag{5.116}$$

all other anticommutators vanishing. We construct the $SU(2) \times U(1)$ generators,

$$I_j = \frac{1}{2}\, b^\dagger\, \sigma_j\, b - \frac{1}{2}\, \bar{b}^\dagger\, \sigma_j^*\, \bar{b}, \quad j = 1, 2, 3, \quad I_0 = \frac{1}{2}\left(b^\dagger\, b - \bar{b}^\dagger\, \bar{b}\right), \tag{5.117}$$

where $*$ denotes conjugation, and the matrix indices have been suppressed. Fermi and Yang noted that the pions have the same quantum numbers as nucleon–antinucleon pairs, leading to

$$|\,\pi^+\,\rangle = b_1^\dagger\, \bar{b}_2^\dagger\, |\,0\,\rangle, \quad |\,\pi^0\,\rangle = \frac{1}{2}\left(b_1^\dagger\, \bar{b}_1^\dagger - b_2^\dagger\, \bar{b}_2^\dagger\right), |\,0\,\rangle, \quad |\,\pi^-\,\rangle = b_2^\dagger\, \bar{b}_1^\dagger\, |\,0\,\rangle, \tag{5.118}$$

that is pions look like nucleon–antinucleons bound together; while erroneous as stated, their idea survives today, since pions contain quark–antiquark pairs. This is an early example of the use of group theory to the taxonomy of elementary particles. Soon we will generalize it to include strange particles.

The electric charge for the pion isotriplet is simply equal to the third component of isospin, while for the nucleons, it is the linear combination

$$Q = I_3 + I_0, \tag{5.119}$$

where I_0 is equal to $1/2$ for nucleons, $-1/2$ for antinucleons.

Wigner supermultiplet model

Wigner (*Phys. Rev.* **51**, 106 (1939), and independently Stückelberg (*Helv. Phys. Acta* **11**, 225 (1938)), extended isospin symmetry by combining it with spin. This leads to

$$SU(4) \supset SU(2)_I \times SU(2)_{spin}, \tag{5.120}$$

with the nucleons transforming both as spinors and isospinors,

$$|\,N\,\rangle \sim \mathbf{4} = (\mathbf{2}, \mathbf{2}_{spin}\,), \qquad |\,\overline{N}\,\rangle \sim \bar{\mathbf{4}} = (\mathbf{2}, \mathbf{2}_{spin}\,),$$

where the second labels the spin. The pions belong to the fifteen-dimensional adjoint representation built from

$$\mathbf{4} \times \bar{\mathbf{4}} = \mathbf{15} + \mathbf{1}, \tag{5.121}$$

with the result

$$\mathbf{15} = (\,\mathbf{3}, \mathbf{3}_{spin}\,) + (\,\mathbf{1}, \mathbf{3}_{spin}\,) + (\,\mathbf{3}, \mathbf{3}_{spin}\,). \tag{5.122}$$

It contains additional spin one states which are found in experiments: one isovector $(\,\mathbf{3}, \mathbf{3}_{spin}\,)$, the ρ^+, ρ^0, ρ^-, vector mesons, and one isoscalar vector meson, ω. More massive than the pions, they generate corrections to the nuclear potential which provide the short-range repulsion (because they are vectors) as required by nuclear data. They can also be viewed *à la* Fermi–Yang, as nucleon–antinucleon pairs. Today we know them to be quark–antiquark pairs.

Epilogue

We cannot close this brief survey of the uses of $SU(2)$ in physics without mentioning its use as the *gauge group* which underlies the modern theory of β decay. This discussion is relegated to the section where we discuss the groups of the Standard Model.

6

$SU(3)$

While many of the techniques introduced in the previous chapter carry over to the study of more general Lie algebras (see for instance Gilmore [7]), there are non-trivial generalizations which are best discussed in the context of the more complicated $SU(3)$ algebra.

6.1 $SU(3)$ algebra

Consider a Hilbert space and its dual spanned by three kets and three bras, which form an orthonormal and complete basis,

$$\langle a \,|\, b \rangle = \delta_{ab}, \quad a, b = 1, 2, 3; \qquad \sum_{a=1}^{3} |\, a \rangle\langle a \,| = 1. \tag{6.1}$$

It is the natural habitat of nine operators

$$X^a{}_b \equiv |\, a \rangle \, \langle b \,|, \quad a, b = 1, 2, 3; \tag{6.2}$$

one combination is the identity operator by completeness, while the others obey the commutation relations

$$[\, X^a{}_b,\; X^c{}_d \,] = \delta^c_b \, X^a{}_d - \delta^a_d \, X^c{}_b. \tag{6.3}$$

Together with $X^a{}_a = 0$, these commutation relations define the abstract description of the eight-dimensional Lie algebra known as $SU(3)$ to physicists and A_2 to mathematicians. Combinations of the form

$$(X \cdot X)^a{}_b \equiv X^a{}_n \, X^n{}_b, \quad (X \cdot X \cdot X)^a{}_b \equiv X^a{}_p \, X^p{}_q \, X^q{}_b, \ldots \tag{6.4}$$

transform the same way as the generators,

$$[\, X^a{}_b,\; (X \cdots X \cdots X)^c{}_d \,] = \delta^c_b \, (X \cdots X \cdots X)^a{}_d - \delta^a_d \, (X \cdots X \cdots X)^c{}_b. \tag{6.5}$$

Hence their traces over a, b are invariants

$$[\,X^a{}_b,\ (X\cdot X)^n{}_n\,] = [\,X^a{}_b,\ (X\cdot X\cdot X)^n{}_n\,] = \cdots = 0, \tag{6.6}$$

leading to an apparently infinite number of invariants. In fact there are only two independent invariants. To see this, construct the *traceless* antisymmetric tensor

$$A^{ac}_{bd} = X^a{}_b\, X^c{}_d - X^c{}_b\, X^a{}_d - X^a{}_d\, X^c{}_b + X^c{}_d X^a{}_b - \text{trace terms}, \tag{6.7}$$

with the trace terms determined by demanding

$$A^{an}_{bn} = A^{mn}_{mn} = 0. \tag{6.8}$$

Then by use of the Levi–Civita tensor on three indices, we write

$$C^n{}_m = \epsilon_{mac}\epsilon^{nbd} A^{ac}_{bd}, \qquad A^{ac}_{bd} = \epsilon^{mac}\epsilon_{nbd}\, C^n{}_m. \tag{6.9}$$

By taking traces we see that C^n_m vanishes and thus so does A^{ac}_{bd}. It is then a simpler matter by multiplying with X^n_m to deduce that

$$(X\cdot X\cdot X)^a{}_b = \alpha\,\delta^a_b + \beta\, X^a{}_b + \gamma\,(X\cdot X)^a{}_b, \tag{6.10}$$

where α, β, γ are expressible in terms of traces. Hence all higher-order combinations can be reduced and $SU(3)$ has only two Casimir invariants, one quadratic and one cubic

$$C_2 = (X\cdot X)^a{}_a, \qquad C_3 = (X\cdot X\cdot X)^a{}_a. \tag{6.11}$$

Proceeding as for $SU(2)$, we single out two "diagonal" operators,

$$\begin{aligned} T^3 &\equiv \tfrac{1}{2}\Big(|\,1\,\rangle\langle\,1\,| - |\,2\,\rangle\langle\,2\,|\Big),\\ T^8 &\equiv \tfrac{1}{2\sqrt{3}}\Big(|\,1\,\rangle\langle\,1\,| + |\,2\,\rangle\langle\,2\,| - 2\,|\,3\,\rangle\langle\,3\,|\Big), \end{aligned} \tag{6.12}$$

normalized in the same way so that

$$\mathrm{Tr}\,(T^3\,T^3) = \mathrm{Tr}\,(T^8\,T^8) = \tfrac{1}{2}; \qquad \mathrm{Tr}\,(T^3\,T^8) = 0. \tag{6.13}$$

These two operators clearly commute with one another, and their commutators with the remaining six operators are given by

$$[\,T^3,\ X^1{}_2\,] = X^1{}_2;\quad [\,T^3,\ X^1{}_3\,] = \tfrac{1}{2}\,X^1{}_3;\quad [\,T^3,\ X^2{}_3\,] = -\tfrac{1}{2}\,X^2{}_3; \tag{6.14}$$

$$[\,T^8,\ X^1{}_2\,] = 0;\quad [\,T^8,\ X^1{}_3\,] = \tfrac{\sqrt{3}}{2}\,X^1{}_3;\quad [\,T^8,\ X^2{}_3\,] = \tfrac{\sqrt{3}}{2}\,X^2{}_3, \tag{6.15}$$

together with their hermitian conjugates, while

$$[\,X^1{}_2,\ X^{1\dagger}_{\ 2}\,] = 2\,T^3, \tag{6.16}$$

and finally

$$[X^1{}_3, X^1{}_3^\dagger] = T^3 + \sqrt{3}\,T^8, \qquad [X^2{}_3, X^2{}_3^\dagger] = -T^3 + \sqrt{3}\,T^8. \tag{6.17}$$

This allows for a simple generalization of Pauli's spin matrices to the (3×3) hermitian *Gell-Mann matrices*

$$\lambda_1 = \begin{pmatrix} 0 & 1 & 0 \\ 1 & 0 & 0 \\ 0 & 0 & 0 \end{pmatrix}, \quad \lambda_2 = \begin{pmatrix} 0 & -i & 0 \\ i & 0 & 0 \\ 0 & 0 & 0 \end{pmatrix}, \quad \lambda_3 = \begin{pmatrix} 1 & 0 & 0 \\ 0 & -1 & 0 \\ 0 & 0 & 0 \end{pmatrix}$$

$$\lambda_4 = \begin{pmatrix} 0 & 0 & 1 \\ 0 & 0 & 0 \\ 1 & 0 & 0 \end{pmatrix}, \quad \lambda_5 = \begin{pmatrix} 0 & 0 & -i \\ 0 & 0 & 0 \\ i & 0 & 0 \end{pmatrix}, \quad \lambda_6 = \begin{pmatrix} 0 & 0 & 0 \\ 0 & 0 & 1 \\ 0 & 1 & 0 \end{pmatrix}$$

$$\lambda_7 = \begin{pmatrix} 0 & 0 & 0 \\ 0 & 0 & -i \\ 0 & i & 0 \end{pmatrix}, \quad \lambda_8 = \frac{1}{\sqrt{3}} \begin{pmatrix} 1 & 0 & 0 \\ 0 & 1 & 0 \\ 0 & 0 & -2 \end{pmatrix}.$$

Now we study this algebra in a language that generalizes to arbitrary Lie algebras. Think of the two diagonal operators as generating a vector in a two-dimensional space, and introduce the notation H^i, $i = 1, 2$ with

$$H^1 \equiv T^3; \qquad H^2 \equiv T^8. \tag{6.18}$$

Associate the remaining six "off-diagonal" elements with two-dimensional vectors

$$X^1{}_2 \equiv E_{\boldsymbol{\beta}^{(1)}}; \qquad X^1{}_3 \equiv E_{\boldsymbol{\beta}^{(2)}}; \qquad X^2{}_3 \equiv E_{\boldsymbol{\beta}^{(3)}}; \tag{6.19}$$

where the components of the two-dimensional vectors $\boldsymbol{\beta}^{(a)}$, $a = 1, 2, 3$, are defined by the commutators

$$[H^i, E_{\boldsymbol{\beta}^{(a)}}] = \beta^{(a)\,i}\, E_{\boldsymbol{\beta}^{(a)}}. \tag{6.20}$$

It is easy to see that

$$\boldsymbol{\beta}^{(1)} = \{1, 0\}, \qquad \boldsymbol{\beta}^{(2)} = \{\tfrac{1}{2}, \tfrac{\sqrt{3}}{2}\}, \qquad \boldsymbol{\beta}^{(3)} = \{-\tfrac{1}{2}, \tfrac{\sqrt{3}}{2}\}, \tag{6.21}$$

which expresses the vectors in the *H-basis*. The $\boldsymbol{\beta}^{(a)}$ are called the *roots* of the Lie algebra. Since

$$E_{\boldsymbol{\beta}^{(a)}}{}^\dagger = E_{-\boldsymbol{\beta}^{(a)}}, \tag{6.22}$$

their negatives are also root vectors. The three roots, living in a two-dimensional space, are not independent: they satisfy

$$\boldsymbol{\beta}^{(2)} = \boldsymbol{\beta}^{(1)} + \boldsymbol{\beta}^{(3)}. \tag{6.23}$$

The commutator of the elements associated with a root with their hermitian conjugates yield linear combinations of the H elements, allowing us to write

$$[\, E_{\boldsymbol{\beta}^{(a)}},\, E_{-\boldsymbol{\beta}^{(a)}}\,] = \beta^{(a)}{}_{i}\, H^{i}, \tag{6.24}$$

thus defining the root components with *lower* indices. Evaluation of the commutators show them to be simply related to the components with upper indices

$$\beta^{(a)}{}_{i} \equiv g_{ij}\, \beta^{(a)\, j}, \tag{6.25}$$

thus introducing the metric and its inverse

$$g_{ij} = 2\,\delta_{ij}, \qquad g^{ij} = \frac{1}{2}\,\delta^{ij}. \tag{6.26}$$

Distinguishing upper and lower indices may seem like overkill at this stage, but it is warranted for general Lie algebras. This enables us to define a scalar product

$$(\,\boldsymbol{\beta}^{(a)},\, \boldsymbol{\beta}^{(b)}\,) \equiv \beta^{(a)}{}_{i}\, \beta^{(b)\, i} = g^{ij}\beta^{(a)}{}_{i}\beta^{(b)}{}_{j}. \tag{6.27}$$

The orthogonality of the vectors along H^1 and H^2 with respect to this scalar product is the same as the vanishing of the trace of (T^3T^8).

The remaining non-zero commutators of this algebra are easily found to be

$$[\, X^1{}_2,\, X^3{}_1\,] = -X^3{}_2; \qquad [\, X^1{}_2,\, X^2{}_3\,] = X^1{}_3; \qquad [\, X^1{}_3,\, X^3{}_2\,] = X^1{}_2, \tag{6.28}$$

and their hermitian conjugates. The commutator of any two root elements is either proportional to another root element, or zero. The criterion is that the root associated with the commutator of two elements is the sum of the two roots associated with each generator in the commutator, that is

$$[\, E_{\boldsymbol{\beta}},\, E_{\boldsymbol{\beta}'}\,] \;\sim\; E_{\boldsymbol{\beta}+\boldsymbol{\beta}'}. \tag{6.29}$$

This implies that the commutator

$$[\, H^i,\, [\, E_{\boldsymbol{\beta}^{(a)}},\, E_{-\boldsymbol{\beta}^{(a)}}\,]\,],$$

must vanish, which is easily verified using the Jacobi identity.

Next, consider a zero commutator, such as

$$[\, X^1{}_2,\, X^1{}_3\,] = [\, E(\boldsymbol{\beta}^{(1)}),\, E(\boldsymbol{\beta}^{(2)})\,] = 0. \tag{6.30}$$

The sum of these two roots is

$$\boldsymbol{\beta}^{(1)} + \boldsymbol{\beta}^{(2)} = \{\tfrac{3}{2},\, \tfrac{\sqrt{3}}{2}\}, \tag{6.31}$$

which is **not** a root vector: the rule seems to be that the commutator

$$[\, E_{\boldsymbol{\beta}},\, E_{\boldsymbol{\beta}'}\,]$$

vanishes whenever the vector $(\boldsymbol{\beta} + \boldsymbol{\beta}')$ is not a root.

One can plot the eight elements of the Lie algebra by their root vectors in the $H^1 - H^2$ plane, including the two zero-length roots associated with H^1 and H^2, yielding the following $SU(3)$ root diagram.

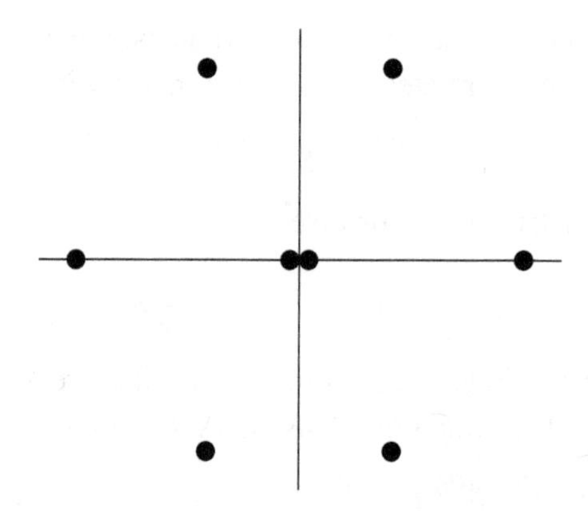

In this simple example, we have introduced some important general concepts: to each element of the Lie algebra corresponds a root vector defined in a space endowed with a metric and scalar product. The dimension of that space is equal to the number of commuting generators (*rank*) of the algebra, two for $SU(3)$. The roots are not linearly independent, since their number is larger than the dimension of the space they span.

While it is obviously desirable to describe Lie algebras in basis-independent terms, some bases are more useful than others, depending on the application.

For $SU(3)$, bases are defined by the choice of two independent roots; below we give several choices, and show in detail how they are related.

6.2 α-Basis

We begin by splitting the six non-zero roots into positive and negative ones. The actual definition of which is positive and which is negative is arbitrary. Here we define a positive root as one with a positive H^2 component, and a positive H^1 component if its H^2 component is zero.

Hence all three roots $\boldsymbol{\beta}^{(1)}$, $\boldsymbol{\beta}^{(2)}$, $\boldsymbol{\beta}^{(3)}$ are positive. We take the dependent positive root to be the one expressed as a linear combination of the other two with *positive* coefficients. This leads to the unique choice

$$\boldsymbol{\beta}^{(2)} = \boldsymbol{\beta}^{(1)} + \boldsymbol{\beta}^{(3)}. \tag{6.32}$$

The remaining two positive roots, $\boldsymbol{\beta}^{(1)}$ and $\boldsymbol{\beta}^{(3)}$are called the *simple roots* of the Lie algebra. We define

$$\boldsymbol{\alpha}_1 \equiv \boldsymbol{\beta}^{(1)}, \qquad \boldsymbol{\alpha}_2 \equiv \boldsymbol{\beta}^{(3)}, \tag{6.33}$$

and express all roots in terms of their components along these two simple roots. This defines the α-basis; unlike the H-basis, it is not orthogonal. All properties of the algebra can be expressed in terms of the simple roots. Explicitly,

$$(\boldsymbol{\alpha}_1, \boldsymbol{\alpha}_1) = (\boldsymbol{\alpha}_2, \boldsymbol{\alpha}_2) = 2, \qquad (\boldsymbol{\alpha}_1, \boldsymbol{\alpha}_2) = -1. \tag{6.34}$$

The two simple roots of $SU(3)$ have equal length and lie at 120° from one another. This information can be summarized in several ways. First, through the *Cartan matrix* with elements

$$A_{ij} \equiv 2\frac{(\boldsymbol{\alpha}_i, \boldsymbol{\alpha}_j)}{(\boldsymbol{\alpha}_j, \boldsymbol{\alpha}_j)}, \tag{6.35}$$

which for $SU(3)$ is

$$\begin{pmatrix} 2 & -1 \\ -1 & 2 \end{pmatrix}.$$

Second, through a two-dimensional diagram: draw a point for each simple root. Here the two points are the same indicating the roots have equal length. Connect the points by a line when the roots are at $2\pi/3$ from one another. This is the $SU(3)$ *Dynkin diagram.*

This dumbbell diagram contains all the relevant information about the algebra, if you know how to read it, of course.

6.3 ω-Basis

A closely related basis is the ω-basis, spanned by two vectors $\boldsymbol{\omega}_1$ and $\boldsymbol{\omega}_2$. They are defined such that the coefficients of the simple roots in the α-basis expressed in the ω-basis, are the rows of the Cartan matrix, that is

$$\begin{aligned} \boldsymbol{\alpha}_1 &\equiv 2\boldsymbol{\omega}_1 - \boldsymbol{\omega}_2, \\ \boldsymbol{\alpha}_2 &\equiv -\boldsymbol{\omega}_1 + 2\boldsymbol{\omega}_2. \end{aligned}$$

Their H-basis components are

$$\boldsymbol{\omega}_1 = \{\tfrac{1}{2}, \tfrac{1}{2\sqrt{3}}\}, \qquad \boldsymbol{\omega}_2 = \{0, \tfrac{1}{\sqrt{3}}\}, \tag{6.36}$$

showing that they have equal length, and make a $\pi/3$ angle, as shown below.

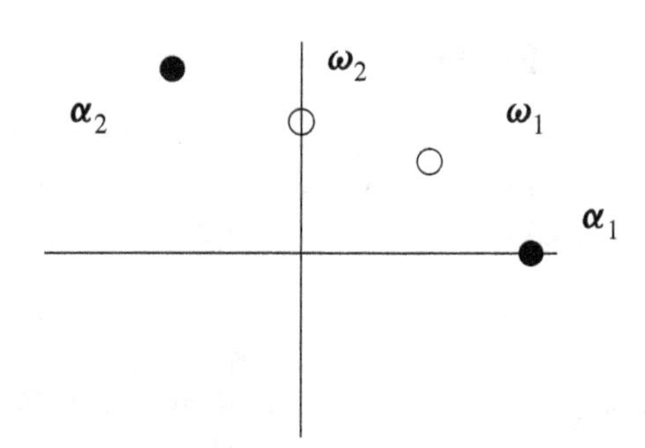

We can represent the elements of the Lie algebra in terms of their components in these two bases: the commuting elements H^1, H^2 have zero components in all bases. In the H-basis, the components of the six roots contain ugly square roots and the like; in the α- and ω-bases, they become beautiful integers.

Roots:	$\beta^{(1)}$	$\beta^{(3)}$	$\beta^{(2)}$	$-\beta^{(1)}$	$-\beta^{(3)}$	$-\beta^{(2)}$
α-basis:	(1,0)	(0,1)	(1,1)	(-1,0)	(0,-1)	(-1,-1)
ω-basis:	(2,-1)	(-1,2)	(1,1)	(-2,1)	(1,-2)	(-1,-1)

Together with the two (0, 0) states, they form the defining adjoint representation of the $SU(3)$ Lie algebra.

6.4 α'-Basis

Since we have not justified our procedure nor claimed uniqueness in constructing these bases, the reader's confidence in the method may be frayed, lest we build yet another set of bases.

Suppose that we had defined positive roots as those with a positive H^1 component, so that all three $\boldsymbol{\beta}^{(1)}$, $\boldsymbol{\beta}^{(2)}$, $-\boldsymbol{\beta}^{(3)}$ are positive roots. We then follow the same procedure, using the same criterion to determine the dependent root: the root that can be expressed as a linear combination of the other two with *positive* coefficients is the dependent one, in this case

$$\boldsymbol{\beta}^{(1)} = \boldsymbol{\beta}^{(2)} + [-\boldsymbol{\beta}^{(3)}]. \tag{6.37}$$

We then take $\boldsymbol{\beta}^{(2)}$ and $-\boldsymbol{\beta}^{(3)}$ to be the simple roots, and define

$$\boldsymbol{\alpha}_1^{'} \equiv \boldsymbol{\beta}^{(2)}, \qquad \boldsymbol{\alpha}_2^{'} \equiv -\boldsymbol{\beta}^{(3)}. \tag{6.38}$$

Call this new basis the α'-basis. The Cartan matrix is the same as before since

$$(\boldsymbol{\alpha}_1', \boldsymbol{\alpha}_1') = (\boldsymbol{\alpha}_2', \boldsymbol{\alpha}_2') = 2, \qquad (\boldsymbol{\alpha}_1', \boldsymbol{\alpha}_2') = -1, \tag{6.39}$$

as is the Dynkin diagram. The new ω'-basis is introduced in the same way, through

$$\boldsymbol{\alpha}_1' \equiv 2\,\boldsymbol{\omega}_1' - \boldsymbol{\omega}_2', \qquad \boldsymbol{\alpha}_2' \equiv -\boldsymbol{\omega}_1' + 2\,\boldsymbol{\omega}_2'. \tag{6.40}$$

These have the H-basis components

$$\boldsymbol{\omega}_1' = \left\{\tfrac{1}{2}, \tfrac{1}{2\sqrt{3}}\right\}, \qquad \boldsymbol{\omega}_2' = \left\{\tfrac{1}{2}, -\tfrac{1}{2\sqrt{3}}\right\}. \tag{6.41}$$

The different primed basis vectors are displayed as follows.

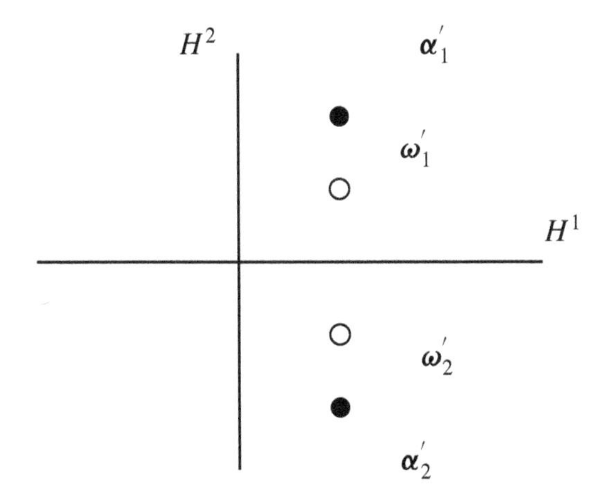

These diagrams look the same except for an overall rotation: $\boldsymbol{\omega}_1 = \boldsymbol{\omega}_1'$, but $\boldsymbol{\omega}_2'$ can be obtained as the reflection of $\boldsymbol{\omega}_2$ about the direction $\boldsymbol{\omega}_1$. Such a reflection is called a *Weyl reflection*. In both bases, the sliver contained inside and on the two $\boldsymbol{\omega}$ vectors is called the *Weyl chamber*.

These bases are widely used for building representations of the algebra. Their construction for the $SU(3)$ Lie algebra follows the same procedures as for $SU(2)$. First we label the states by the eigenvalues of the maximal set of commuting generators, two for $SU(3)$. We set aside for the moment the more traditional labels attached to the Casimir operators, since we saw that for $SU(2)$, the range of eigenvalues determined the Casimir. Here we have two operators, and thus two ranges, which suggests two Casimir operators. It is in fact true that there are as many Casimir operators as the rank, the number of commuting generators. The first is of course quadratic, but the second happens to be cubic, which is not obvious at all at this stage.

6.5 The triplet representation

First we determine the representation of $SU(3)$ in the three-dimensional Hilbert space. Different bases correspond to different labeling schemes. We can label its three states by two numbers, the eigenvalues of $\vec{H}$,

$$H^i \,|\, h^1, h^2 \,\rangle = h^i \,|\, h^1, h^2 \,\rangle,$$

yielding the three states

$$|\, \tfrac{1}{2}, \tfrac{1}{2\sqrt{3}} \,\rangle, \qquad |\, 0, -\tfrac{1}{\sqrt{3}} \,\rangle, \qquad |\, -\tfrac{1}{2}, \tfrac{1}{2\sqrt{3}} \,\rangle.$$

Alternatively, we can express them in the α- and ω-bases as well, yielding for the same three states

$$\alpha\text{-basis:} \quad |\, \tfrac{2}{3}, \tfrac{1}{3} \,\rangle, \qquad |\, -\tfrac{1}{3}, -\tfrac{2}{3} \,\rangle, \qquad |\, -\tfrac{1}{3}, \tfrac{1}{3} \,\rangle,$$

$$\omega\text{-basis:} \quad |\, 1, 0 \,\rangle, \qquad |\, 0, -1 \,\rangle, \qquad |\, -1, 1 \,\rangle.$$

Note that the state labels are rational numbers in the α-basis, and integers in the ω-basis. This feature generalizes to all Lie algebras.

To build representations, we seek the lowering operators. $SU(3)$ has three lowering operators, but one is linearly dependent since its root vector is a linear combination of the simple roots (that is, is the commutator of the other two), so we can express all lowering operators in terms of only two (same as the rank), each associated with a simple root. The application of a lowering operator associated with the simple root $\boldsymbol{\alpha}_1$, say, will lower the state labels according to the value of that root:

$$E_{-\alpha_1} \,|\, h^1, h^2 \,\rangle \sim |\, h^1 - 1, h^2 \,\rangle,$$
$$E_{-\alpha_2} \,|\, h^1, h^2 \,\rangle \sim |\, h^1 + \tfrac{1}{2}, h^2 - \tfrac{\sqrt{3}}{2} \,\rangle.$$

We note that since the weights associated with the lowering (and raising) operators have *integer* components in the α- and ω-bases their application to the states will shift their labels by integers.

Therein lies the great advantage of the ω-basis: starting with a highest weight state with *integer* labels in the ω-basis, we generate only states with (positive and negative) integer labels. Furthermore, the two ω-basis raising or lowering operators are associated with the vectors whose components are the rows of the Cartan matrix, $(2, -1)$ and $(-1, 2)$, and their negatives, respectively. In the eight-dimensional adjoint representation, application of the raising operators to the state $|\, 1, 1 \,\rangle$ gives zero as they would generate either $(3, 0)$ or $(0, 3)$, neither of which are in the representation. This enables us to infer that the highest weight states (annihilated by the raising operators) are those with either positive integers or zero

ω-basis labels. As another example, the application of the lowering operators on the highest weight state with ω-basis labels $(1, 0)$, produces three integer-labeled states

$$|\,1,0\,\rangle \longrightarrow |\,-1,1\,\rangle \longrightarrow |\,0,-1\,\rangle,$$

which form the triplet representation. The third state is a lowest weight state, as its labels have no positive integers. Further application of the lowering operators gives zero. Similarly, we could have started with the state $|\,0,1\,\rangle$ and generated the antitriplet representation. This illustrates the convenience of this basis.

The $SU(3)$ group lattice is an infinite two-dimensional lattice. Its points are labeled by two integers which indicate the projection along $\boldsymbol{\omega}_1$ and $\boldsymbol{\omega}_2$. Thus the point $(5, -2)$ lies five units along $\boldsymbol{\omega}_1$ and two units along $-\boldsymbol{\omega}_2$. Each point corresponds to a possible state. The points with one zero entry lie at the edge of the Weyl chamber. The points inside the Weyl chamber are labeled by two positive integers; those at the edge have one positive integer and one zero entry; all serve to label irreducible representations. The fundamental domain of the $SU(3)$ lattice is just an equilateral triangle.

Scholium. *Conjugation*

We can use this decomposition of the $SU(3)$ Lie algebra in terms of its $SO(3)$ subalgebra to gain an insight in the operation of charge conjugation. As we have seen, the $SU(2)$ and $SU(3)$ algebras operate on complex vectors, and so it is natural to see what conjugation does to them. Consider first the Pauli matrices, and define

$$\tilde{\sigma}_A \equiv -\sigma_A^*. \tag{6.42}$$

It is easy to see that these matrices satisfy the same algebra as the Pauli matrices, and therefore represent the same $SU(2)$ algebra. However, the signs of the charges are reversed and, rather than containing the charges $(+1/2, -1/2)$, they yield the charges $(-1/2, +1/2)$, but these are the same up to a flip. A specific way to see this is to recall that

$$\sigma^{A*} = -\sigma^2\,\sigma^A\,\sigma^2,$$

so that the states of the tilde representation are obtained from those of the untilded one by the action of σ^2, which flips the two states. It follows that these two representations are equivalent. In this case even though there is conjugation, it does not yield any new representation. Wigner calls this case *pseudoreal*.

For $SU(3)$ though, charge conjugation yields entirely new representations. To see this, do the same for the Gell-Mann matrices and define

$$\tilde{\boldsymbol{\lambda}}_A \equiv -\boldsymbol{\lambda}_A^*. \tag{6.43}$$

Since they are hermitian, conjugation is the same as transposition, and this operation does not change the $\boldsymbol{\lambda}^2$, $\boldsymbol{\lambda}^5$ and $\boldsymbol{\lambda}^7$ which generate the $SO(3)$ sub-algebra. The remaining five matrices change sign under this operation. They transform as the quadrupole, and the commutation of any two of them yields an $SO(3)$ generator, so that the tilde matrices satisfy the same algebra. Schematically,

$$[\,\tilde{L},\,\tilde{L}\,] \sim \tilde{L}; \qquad [\,\tilde{L},\,\tilde{Q}\,] \sim \tilde{Q}; \qquad [\,\tilde{Q},\,\tilde{Q}\,] \sim \tilde{L}. \tag{6.44}$$

The tilde matrices therefore satisfy the $SU(3)$ algebra. They act on a three-dimensional representation, but it is different from the one we have already constructed! To see this it suffices to look at the charges of its states, since they are the negatives of the previous ones; in the ω-basis, they are

$$|\,-1,0\,\rangle, \qquad |\,0,1\,\rangle, \qquad |\,1,-1\,\rangle. \tag{6.45}$$

Clearly the highest weight state is $|\,0,1\,\rangle$. This entirely new representation is the conjugate of the triplet representation. It is denoted as $\overline{\mathbf{3}}$. Its Dynkin label is (0 1).

In general, if *any* Lie algebra is satisfied by hermitian matrices T^A, the same Lie algebra is satisfied by minus their transpose. This follows easily by taking the transpose of the defining equation. It follows that these matrices live either in a new Hilbert space, or in the same Hilbert space (as for $SU(2)$).

6.6 The Chevalley basis

In our previous discussion, the choice for H^1 and H^2 was motivated by physics. Mathematicians prefer to construct the algebra by generalizing the $SU(2)$ structure. For $SU(2)$ and $SU(3)$ we found as many commuting generators as simple roots, a feature that generalizes to all Lie algebras. To describe $SU(3)$, we define two copies of $SU(2)$ generators, h_r, e_r and e_{-r}, $r = 1,2$ with commutation relations

$$[\,h_r,\,e_r\,] = 2e_r, \qquad [\,h_r,\,e_{-r}\,] = -2e_{-r}, \qquad [\,e_r,\,e_{-r}\,] = h_r. \tag{6.46}$$

In addition, we have

$$[\,h_1,\,h_2\,] = 0. \tag{6.47}$$

In matrix form, these correspond to

$$h_1 \sim \begin{pmatrix} 1 & 0 & 0 \\ 0 & -1 & 0 \\ 0 & 0 & 0 \end{pmatrix}, \qquad h_2 \sim \begin{pmatrix} 0 & 0 & 0 \\ 0 & 1 & 0 \\ 0 & 0 & -1 \end{pmatrix}, \tag{6.48}$$

to be compared with the Gell-Mann matrices $\boldsymbol{\lambda}_3$ and $\boldsymbol{\lambda}_8$ for H^1 and H^2. Hence it follows that

$$h_1 = 2H^1, \qquad h_2 = \sqrt{3}H^2 - H^1. \tag{6.49}$$

The e_i are identified with the elements of the simple roots

$$e_1 = E_{\alpha_1}, \qquad e_2 = E_{\alpha_2}. \tag{6.50}$$

We will see later that for some Lie algebras, a normalization factor will be required. In the α basis, this yields

$$[\,h_1,\, e_2\,] = -e_2, \qquad [\,h_2,\, e_1\,] = -e_1. \tag{6.51}$$

Happily, we note that these commutators can be neatly written in terms of the Cartan matrix

$$[\,h_r,\, e_s\,] = A_{sr}\, e_s. \tag{6.52}$$

This feature generalizes to all Lie algebras. As we have seen the algebra contains more elements; in this notation, they are obtained as commutators of the simple root elements. For $SU(3)$, the only non-zero element is $[\,e_1,\, e_2\,]$ together with its complex conjugate. In terms of matrices, we have

$$e_1 = \begin{pmatrix} 0 & 1 & 0 \\ 0 & 0 & 0 \\ 0 & 0 & 0 \end{pmatrix}, \qquad e_2 = \begin{pmatrix} 0 & 0 & 0 \\ 0 & 0 & 1 \\ 0 & 0 & 0 \end{pmatrix}, \tag{6.53}$$

while the third root is identified with

$$e_3 \equiv [\,e_1,\, e_2\,] = \begin{pmatrix} 0 & 0 & 1 \\ 0 & 0 & 0 \\ 0 & 0 & 0 \end{pmatrix}. \tag{6.54}$$

There are no other elements because all double commutators vanish

$$[\,e_1,\, [\,e_1,\, e_2\,]] = [\,e_2,\, [\,e_1,\, e_2\,]] = 0. \tag{6.55}$$

The *Serre presentation*, to be discussed later, offers a systematic way to recast these vanishings in terms of the Cartan matrix.

Note that, in this basis, all elements are represented by matrices with integer entries. This is an elementary example of the *Chevalley basis*, in which the algebra is represented by matrices with integer entries, and where all structure functions are integers. It seems appropriate to denote the matrix with unit entry in the ij position as e_{ij}. Thus we have

$$e_1 = e_{12}, \quad e_2 = e_{23}, \quad e_{-1} = e_{21}, \quad e_{-2} = e_{32}$$

as well as

$$h_1 = e_{11} - e_{22}, \qquad h_2 = e_{22} - e_{33}.$$

This shorthand notation is very useful.

6.7 $SU(3)$ in physics

This group manifests itself in many different areas of particle physics, simply because Nature seems to like the number three! There are three families of elementary fermions; there are three colors of quarks, and there are three quarks with mass significantly lower than the inverse diameter of a proton (with the help of Planck's constant $\hbar$ and the speed of light c, mass and length are contragredient).

6.7.1 The isotropic harmonic oscillator redux

Let us take another look at the isotropic harmonic oscillator in the light of $SU(3)$. We found its invariance group by examining the three-fold degeneracy of its first level, with transitions between these states represented by operators which replaced $a_i^\dagger$ with another $a_j^\dagger$, and we suggested their form to be

$$L_{ij} \equiv i(A_i^\dagger A_j - A_j^\dagger A_i).$$

However, these transitions could have been generated just as well by the other hermitian combinations

$$(A_i^\dagger A_j + A_j^\dagger A_i),$$

and nobody asked about them! There are six such operators, and under $SO(3)$ they transform as the five-dimensional "quadrupole" representation represented by the symmetric traceless tensor

$$Q_{ij} = \frac{1}{2}(A_i^\dagger A_j + A_j^\dagger A_i) - \frac{1}{3}\delta_{ij}\vec{A}^\dagger \cdot \vec{A}, \tag{6.56}$$

which satisfies

$$[L_{ij}, Q_{mn}] = i\delta_{jm} Q_{in} + i\delta_{jn} Q_{im} - i\delta_{im} Q_{jn} - i\delta_{in} Q_{jm}, \tag{6.57}$$

while the trace is left invariant

$$Q_0 = \vec{A}^\dagger \cdot \vec{A}, \qquad [L_{ij}, Q_0] = 0. \tag{6.58}$$

Furthermore, their commutators

$$[Q_{ij}, Q_{mn}] = -\frac{i}{4}\Big(\delta_{im} L_{jn} + \delta_{jm} L_{in} + \delta_{in} L_{jm} + \delta_{jn} L_{im}\Big), \tag{6.59}$$

and

$$[Q_0, Q_{mn}] = 0, \tag{6.60}$$

close into the $SU(3)$ Lie algebra with its eight generators expressed in terms of its $SO(3)$ sub-algebra. We say that its adjoint representation splits in terms of $SO(3)$ representations as

$$\mathbf{8} = \mathbf{3} + \mathbf{5} + \mathbf{1}.$$

It is easy to verify that all these generators commute with the Hamiltonian of the isotropic harmonic oscillator

$$[\,H,\; Q_{mn}\,] = 0. \tag{6.61}$$

We were too quick in our previous pass at identifying its invariance group; we see that it is the much larger $SU(3)$. It follows that all excited levels of this system form representations of $SU(3)$: since there are six states at the next level, there must be a six-dimensional representation of $SU(3)$, and so on.

6.7.2 The Elliott model

Think of the nucleus as made up of N nucleons moving independently in a harmonic oscillator potential. Each nucleus is located at $x_i^{(a)}$ with momentum $p_i^{(a)}$, where $i = 1, 2, 3$, and $a = 1, 2, \ldots N$. We convert these in terms of creation and annihilation operators, and write the Hamiltonian as

$$H_0 = \sum_{a=1}^{N} \sum_{i=1}^{3} a_i^{\dagger\,(a)}\, a_i^{(a)} + \frac{3N}{2}. \tag{6.62}$$

It commutes with the nine generators

$$\sum_{a=1}^{N} a_i^{\dagger\,(a)}\, a_j^{(a)}, \tag{6.63}$$

which form the $SU(3) \times U(1)$ algebra. At this stage the energy levels are d-fold degenerate, where d is the dimension of any $SU(3)$ representation. This degeneracy is broken when inter-nucleon interactions are taken into account. They are described by one two-body potential which we approximate as

$$V(\vec{x}^{(a)} - \vec{x}^{(b)}) \;\approx\; m^2(\vec{x}^{(a)} - \vec{x}^{(b)})^2 + \lambda\,(\vec{x}^{(a)} - \vec{x}^{(b)})^4, \tag{6.64}$$

with a harmonic potential and an anharmonic term. The Hamiltonian now becomes

$$H = H_0 + \frac{c}{2} \sum_{i,j=1}^{3} Q_{ij}\, Q_{ij}, \tag{6.65}$$

where c is some constant and

$$Q_{ij} = \sum_{a=1}^{N} Q_{ij}^{(a)}, \tag{6.66}$$

where

$$Q_{ij}^{(a)} = \frac{1}{2}(x_i^{(a)} x_i^{(a)} + p_i^{(a)} p_i^{(a)}), \tag{6.67}$$

is the quadrupole moment of the ath nucleon. This is the Elliott model (J. P. Elliott, *Proc. R. Soc. Lond.* A**245**, 128 (1958)). Physically, the long-range interaction depends on the shape: when one nucleon bulges in one direction and acquires a large quadrupole moment, it interacts most with other nucleons which also have a large quadrupole moment in the same direction. It also breaks the $SU(3) \times U(1)$ symmetry of the harmonic approximation.

As we have seen, these are $SU(3)$ generators, so that this perturbation just splits the degeneracy among members of each $SU(3)$ irrep, without generating mixing between different representations. The combination

$$C_2 = \sum_{i=1}^{3} L_i L_i + \frac{1}{2} \sum_{i,j=1}^{3} Q_{ij} Q_{ij} \tag{6.68}$$

is nothing but the quadratic $SU(3)$ Casimir operator. This allows us to rewrite the Hamiltonian as

$$H = H_0 + \frac{c}{2}\left(C_2 - L(L+1)\right), \tag{6.69}$$

where L is the angular momentum. For a given $SU(3)$ irrep, C_2 depends on two integers, and L occurs in a range of values $L = 0, 1, 2, \ldots, L_{\text{max}}$, where L_{max} is the value of the magnetic quantum number for the highest weight state, which has the largest deviation from a pure sphere. The d-fold degeneracy is thus split into $L_{\text{max}} + 1$ levels.

Following Dyson [4], consider the magnesium nucleus Mg^{24}, described by eight nucleons in six possible orbits ($L = 0$ and $L = 2$ unfilled shells) outside a closed shell. In addition each nucleon has two possible spin values ($\pm 1/2$), and two charges (zero and one), resulting in $6 \times 4 = 24$ possibilities. Hence the number of configurations for the eight nucleons is simply

$$\frac{24!}{8!\,14!},$$

more than a mere million. Diagonalizing such a large Hamiltonian has led nuclear physicists either to madness, or to treat the nuclear levels statistically and develop the theory of random matrices. This simple-minded model brings tremendous simplification to the problem. Comparing the measured energy levels with those calculated from the Elliott model, assuming only *one* $SU(3)$ irrep, we find that 70%–90% of the energy levels are correctly reproduced!

6.7.3 The Sakata model

With the advent of strange particles, the taxonomy of "elementary particles" had to be extended. Sakata (*Prog. Theor. Phys.* **16**, 686 (1956)) suggested that the isospin doublet be augmented by one new spin one-half neutral nucleon which he dubs Λ and its antiparticle. Such a particle had been discovered through its decay into a proton and a π^+, leaving tracks bending in opposite ways, hence its name.

In group-theoretical language, this amounts to generating the $SU(3) \times U(1)$ algebra through

$$T_{ij} = b_i^\dagger b_j - \bar{b}_j^\dagger \bar{b}_i, \quad i, j = 1, 2, 3, \tag{6.70}$$

by including a third Fermi oscillator. The nucleons and antinucleons now form a $SU(3)$ triplet and antitriplet, respectively. The electric charge resides in *both* $SU(3)$ and $U(1)$, as

$$Q = I_3 + \frac{1}{\sqrt{3}} I_8 + \frac{1}{3} N, \tag{6.71}$$

where N is the nucleon number and $I_8 = (1, 1, -2)/(2\sqrt{3})$ is the second diagonal $SU(3)$ generator for the triplet.

Pions are found to reside in the adjoint octet representation

$$\mathbf{3} \times \bar{\mathbf{3}} = \mathbf{8} + \mathbf{1}. \tag{6.72}$$

In the decomposition

$$SU(3) \supset SU(2) \times U(1), \tag{6.73}$$

it contains

$$\mathbf{8} = \mathbf{3}_0 + \mathbf{2}_1 + \mathbf{2}_{-1} + \mathbf{1}_0, \tag{6.74}$$

where the subscript denotes the $U(1)$ hypercharge $Y = 2\sqrt{3} I_8$. The isodoublets are the strange kaons, (K^0, K^+) and $(K^-, \overline{K}^0)$, and the extra singlet is the η meson. They are the pseudoscalar mesons. In picture form, this is represented as follows.

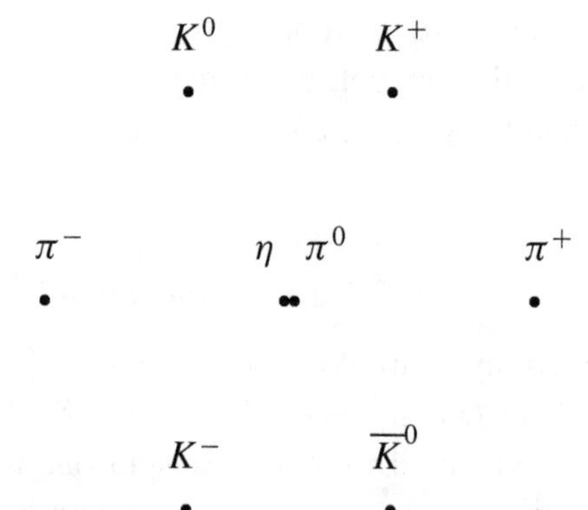

For the octet, the electric charge reduces to the Gell-Mann–Nishijima formula

$$Q = I_3 + \frac{Y}{2}. \tag{6.75}$$

The Sakata model correctly described the pseudoscalar mesons, but did not account for the plethora of baryons, the proton, the heavy unstable spin one-half particles, and spin $3/2$ resonances that were being discovered in experiments, almost on a monthly basis at the time. Hence it serves as an early use of group theory to systematize the taxonomy of elementary particles.

6.7.4 The Eightfold Way

It was Gell-Mann and Y. Ne'eman who independently arrived at the correct assignment for the baryons. In a nutshell, they assigned the proton and neutron to the following $SU(3)$ octet representation.

n p

Σ^- Λ Σ^0 Σ^+

Ξ^- Ξ^0

All eight have spin one-half, but differ by their masses and their "strangeness" S: the nucleons are not strange ($S = 0$), the Λ and Σ are a little strange ($S = -1$), and the Ξ more so ($S = -2$). All are stable with respect to strong decays (i.e. decays with a characteristic decay time around 10^{24} seconds), but take much longer to decay eventually (weakly or electromagnetically) to the neutron.

The neutral Λ has a mass of 1115 MeV; The three Σ baryons center in mass around 1190 MeV. The charge decays either as $\Sigma^+ \to p + \pi^0$ or $\Sigma^+ \to n + \pi^+$ while Σ^0 decays by photon emission: $\Sigma^0 \to \Lambda + \gamma$. The heavier "cascade" baryons Ξ^0 (1314 MeV) and Ξ^- (1321 MeV) decay mostly into $\Lambda + \pi$.

To account for these mass differences, the assumption was made that the Hamiltonian which governs the strong interactions is made up of two pieces: the largest, H_0, is invariant under $SU(3)$ and its contribution is to give the same mass to all particles within a multiplet, since H_0 commutes with the $SU(3)$ generators. The smaller second piece generates the mass differences. Since the strong interactions preserve both isospin and strangeness, this second piece must commute with $SU(2)_I \times U(1)_Y$, but not with the rest of the $SU(3)$ generators. The first place to look for an operator with these requirements is the octet representation which contains an isospin singlet with $Y = 0$.

In equations,

$$H = H_0 + a\, T^3_{\ 3}, \tag{6.76}$$

where the a is some constant, and $T^c_{\ d}$ is an operator which satisfies

$$[\, X^a_{\ b},\ T^c_{\ d}\,] = \delta^c_b\, T^a_{\ d} - \delta^a_d\, T^c_{\ b}. \tag{6.77}$$

Other choices would have been tensor operators with multiple upper and lower indices all aligned along the three direction, but this is the simplest.

Although $T^3_{\ 3}$ is an infinite-dimensional operator, it only shuffles states within the same representation, and it suffices to determine its matrix elements within one representation. As part of the strong Hamiltonian, the generator of time translations, particles belonging to different $SU(3)$ representations will not mix.

We now show that $T^3_{\ 3}$ in a given representation space can be written as a function of the $SU(3)$ generators. Consider any $SU(3)$ irreducible representation, labeled by the kets $|\, i\,\rangle$, $i = 1, 2 \ldots, d$. We have completeness on the states of the same representation,

$$1 = \sum_i |\, i\,\rangle\langle\, i\,|, \tag{6.78}$$

which allows us to write

$$T^3_{\ 3} = \sum_{i,j} |\, i\,\rangle\langle\, i\,|\, T^3_{\ 3}|\, j\,\rangle\langle\, j\,|. \tag{6.79}$$

But, we have seen that $|\, i\, \rangle$ can be written as a polynomial in the (lowering) generators (X^-) acting on a state of highest weight $|\, \mathrm{hw}\, \rangle$. Hence

$$T^3_{\ 3} = \sum_{i,j} \langle\, i\, |\, T^3_{\ 3} |\, j\, \rangle\, f_i(X^-) |\, \mathrm{hw}\, \rangle \langle\, \mathrm{hw}\, |\, f_j(X^+). \tag{6.80}$$

Now comes Okubo's trick: introduce $\mathcal{P}$, the projection operator on the state of highest weight

$$\mathcal{P} |\, \mathrm{hw}\, \rangle = |\, \mathrm{hw}\, \rangle, \qquad \mathcal{P}\, f_i(X^-) |\, \mathrm{hw}\, \rangle = 0. \tag{6.81}$$

Then

$$T^3_{\ 3} = \sum_{i,j} \langle\, i\, |\, T^3_{\ 3}\, |\, j\, \rangle f_i(X^-) \mathcal{P} |\, \mathrm{hw}\, \rangle \langle\, \mathrm{hw}\, |\, f_j(X^+), \tag{6.82}$$

$$= \sum_{i,j} \langle\, i\, |\, T^3_{\ 3} |\, j\, \rangle f_i(X^-) \mathcal{P} \sum_k |\, k\, \rangle \langle\, k\, |\, f_j(X^+), \tag{6.83}$$

$$= \sum_{i,j} \langle\, i\, |\, T^3_{\ 3} |\, j\, \rangle f_i(X^-) \mathcal{P} f_j(X^+). \tag{6.84}$$

Since the matrix elements $\langle\, i\, | T^3_{\ 3} |\, j\, \rangle$ are just numbers and $\mathcal{P}$ can itself be written in terms of the generators in the representation, we see that when restricted to a representation, the operator $T^3_{\ 3}$ is a function of the generators.

Hence the most general form is

$$T^3_{\ 3} = a_0 \delta^3_{\ 3} + a_1\, X^3_{\ 3} + a_2\, (X \cdot X)^3_{\ 3} + a_3\, (X \cdot X \cdot X)^3_{\ 3} + \cdots\,, \tag{6.85}$$

which contains an infinite number of parameters, but use of eq. (6.10) reduces it to

$$T^3_{\ 3} = a_0 \delta^3_{\ 3} + a_1\, X^3_{\ 3} + a_2\, (X \cdot X)^3_{\ 3}, \tag{6.86}$$

which involves only three unknown representation-dependent constants. It is now a simple matter to deduce that

$$X^3_{\ 3} = Y, \qquad (X \cdot X)^3_{\ 3} = \text{constant} + \frac{3}{2} Y + \frac{1}{4} Y^2 - I(I+1), \tag{6.87}$$

leading to the final result

$$< H >= a + bY + c \left(\frac{1}{4} Y^2 - I(I+1) \right), \tag{6.88}$$

where a, b, c only depend on the representation.

The application of this formula to the octet and decuplet representations is wildly successful. This shift in the Hamiltonian is interpreted (in lowest order perturbation theory), as a mass shift for the baryons and a squared mass shift for the mesons. In the octet with four isospin multiplets, we expect one mass formula, given by

$$M_N + M_\Xi = \frac{1}{2}(3\, M_\Lambda + M_\Sigma\,). \tag{6.89}$$

Applied to the neutral members of the isospin multiplets, it gives $2254.4 = 2269.7$, accurate to less than 1%!

The same mass formula works well for the mass squared of the octet pseudoscalar mesons

$$m_K^2 + m_{\overline{K}}^2 = \frac{1}{2}(3\, m_\eta^2 + m_\pi^2\,), \tag{6.90}$$

with $(497)^2 = (478)^2$, which is not as good as for the baryons.

In the eightfold way, the spin $3/2$ baryon resonances fall into the decuplet representation, the symmetric tensor with three upper triplet indices, (its conjugate has three lower indices), that is

$$(\mathbf{3} \times \mathbf{3} \times \mathbf{3})_{\text{sym}} = \mathbf{10}. \tag{6.91}$$

The particle assignment is given as follows.

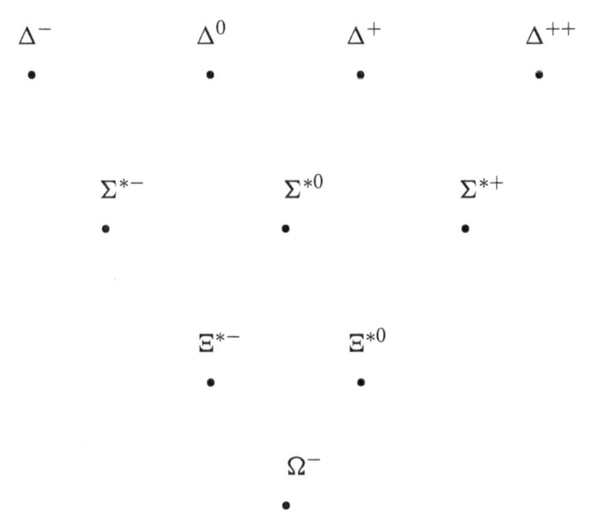

The isospin $3/2$ multiplet contains resonances found in $\pi - N$ scattering, the three other isospin multiplets contain strange resonances. We see that in this multiplet isospin and strangeness are related $I = 1 + Y/2$, so that we have *two* mass relations

$$M_{\Omega^-} - M_{\Xi^*} = M_{\Xi^*} - M_{\Sigma^*} = M_{\Sigma^*} - M_\Delta. \tag{6.92}$$

With (in MeV) $M_\Delta = 1230$, $M_{\Sigma^*} = 1385$, $M_{\Xi^*} = 1532$, and $M_{\Omega^-} = 1672$, the agreement is stunning. In fact this mass formula allowed Gell-Mann to successfully predict both the existence and mass of the Ω baryon.

Many more details can be found in Gell-Mann and Ne'eman's *The Eightfold Way* (Benjamin, 1964), as well as in S. Okubo's Rochester *Lectures on Unitary Symmetry*.

Before leaving this section, we note that the electric charges of the particles in the decuplet are the same as in the octet. Hence the Gell-Mann–Nishijima charge formula is now valid *for all* $SU(3)$ representations, and the charge operator is a linear combination of $SU(3)$ generators. With such an assignment, the three "particles" in the triplet are the fractionally charged quarks, u, d, and s.

7

Classification of compact simple Lie algebras

Now that we have looked at several examples and established useful techniques, it is time to present the complete classification (see Humphreys [11]). We define Lie algebras as a finite collection of elements X^A, $A = 1, 2, \ldots, N$, that satisfies the following axioms.

(i) It is closed under the antisymmetric operation of commutation

$$[\, X^A ,\; X^B \,] = i\, f^{AB}_{\;\;\;C}\, X^C . \tag{7.1}$$

(ii) It obeys the *Jacobi identity*

$$[\,[\, X^A ,\; X^B \,],\; X^C \,] + \text{cyclic permutations} = 0. \tag{7.2}$$

The coefficients $f^{AB}_{\;\;\;C}$ are called the *structure constants* of the Lie algebra. They are constrained to be antisymmetric under the interchange of A and B,

$$f^{AB}_{\;\;\;C} = -f^{BA}_{\;\;\;C},$$

by the first axiom and to satisfy

$$f^{AB}_{\;\;\;C}\, f^{DC}_{\;\;\;F} \;+\; f^{DA}_{\;\;\;C}\, f^{BC}_{\;\;\;F} \;+\; f^{BD}_{\;\;\;C}\, f^{AC}_{\;\;\;F} = 0, \tag{7.3}$$

obtained from the Jacobi identity. We restrict ourselves to the cases where the elements of the Lie algebra are hermitian. Then by taking the hermitian conjugate of the first axiom, it follows that the structure functions are real. Let us think of the structure functions as the $(N \times N)$ matrices

$$f^{AB}_{\;\;\;C} \;\equiv\; i\, (T^{A}_{\text{adj}})^{B}_{\;\;C}, \tag{7.4}$$

acting on a N-dimensional real vector space, so that the Jacobi identity translates into

$$(T^{A}_{\text{adj}}\, T^{D}_{\text{adj}})^{B}_{\;\;F} - (T^{D}_{\text{adj}}\, T^{A}_{\text{adj}})^{B}_{\;\;F} = i\, f^{AD}_{\;\;\;C}\, (T^{C}_{\text{adj}})^{B}_{\;\;F},$$

after using the antisymmetry of the upper two indices. We recognize this equation as the (BF) matrix element of

$$[T^A_{\text{adj}},\, T^D_{\text{adj}}] = i\, f^{AD}{}_C\, T^C_{\text{adj}}.$$

This explicit representation of the Lie algebra in terms of $(N \times N)$ matrices, on a real space with the same dimension as the algebra, is called the *adjoint representation*. The second-rank symmetric matrix

$$g^{AB} \;\equiv\; -f^{AC}{}_D\, f^{BD}{}_C = \text{Tr}\,(T^A_{\text{adj}}\, T^B_{\text{adj}}), \tag{7.5}$$

is called the Cartan–Killing form. We restrict ourselves to the study of *semi-simple* Lie algebras for which its determinant does not vanish. In that case we can define its inverse with lower indices, such that

$$g^{AB}\, g_{BC} = \delta^A{}_C. \tag{7.6}$$

It acts like a metric, which can be used to lower and raise indices. In particular, we can define the structure function with all three upper indices

$$f^{ABC} \;\equiv\; f^{AB}{}_D\, g^{CD} \;= -f^{AB}{}_D\, f^{CE}{}_F\, f^{DF}{}_E,$$

which, after use of the Jacobi identity, can be rewritten as

$$f^{ABC} = -i\text{Tr}\,(T^A_{\text{adj}}\, T^B_{\text{adj}}\, T^C_{\text{adj}} - T^B_{\text{adj}}\, T^A_{\text{adj}}\, T^C_{\text{adj}}),$$

showing it to be manifestly antisymmetric in all three indices.

7.1 Classification

These axioms severely constrain the structure functions. In this section, we derive the allowed Lie algebras. Consider a Lie algebra with r commuting generators H^i,

$$[H^i,\, H^j] = 0, \tag{7.7}$$

with $i, j = 1, 2, \ldots, r$, that is $f^{ij}{}_C = 0$. Since they mutually commute, the H^i can be represented by diagonal matrices. In the adjoint representation, we set

$$-i\, f^{ia}{}_b = (T^i_{\text{adj}})^a{}_b \;\equiv\; \beta^i(a)\, \delta^a{}_b, \tag{7.8}$$

where $a, b, \ldots$ label the remaining $(N - r)$ elements of the Lie algebra. Since the generators are hermitian, $\beta^i(a)$ are real, but otherwise undetermined. It follows from the basic commutation relations that

$$[H^i,\, X^a] = \beta^i(a)\, \delta^a{}_b\, X^b = \beta^i(a)\, X^a. \tag{7.9}$$

Every X^a is characterized by the r coefficients $(\beta^1(a), \beta^2(a), \dots, \beta^r(a))$. It is a vector in an r-dimensional vector space, $\boldsymbol{\beta}(a)$, called the *root* associated with the element X^a. It is convenient to relabel these in terms of the roots:

$$X^a \equiv E_{\boldsymbol{\beta}(a)}, \tag{7.10}$$

assuming no two elements with the same root. This yields for any root vector $\boldsymbol{\beta}$,

$$[\, H^i,\, E_{\boldsymbol{\beta}}\,] = \beta^i\, E_{\boldsymbol{\beta}}. \tag{7.11}$$

The hermitian conjugate equation is

$$[\, H^i,\, E^\dagger_{\boldsymbol{\beta}}\,] = -\beta^i\, E^\dagger_{\boldsymbol{\beta}}, \tag{7.12}$$

so that the dagger belongs to the negative root and we set

$$E^\dagger_{\boldsymbol{\beta}} = E_{-\boldsymbol{\beta}}. \tag{7.13}$$

Since the negative of each root is also a root, $(N - r)$ is always even. A Lie algebra of rank r and dimension N contains $(N - r)$ elements, each of which is described by a non-zero vector in r-dimensional space, and its r commuting elements can be viewed as associated with r roots of zero length.

Consider the commutator $[\, E_{\boldsymbol{\beta}},\, E_{\boldsymbol{\gamma}}\,]$, where $\boldsymbol{\beta}$ and $\boldsymbol{\gamma}$ are roots. Commute it with H^i and use the Jacobi identity to find (if not equal to zero)

$$[\, H^i,\, [\, E_{\boldsymbol{\beta}},\, E_{\boldsymbol{\gamma}}\,]] = (\beta^i + \gamma^i)\, [\, E_{\boldsymbol{\beta}},\, E_{\boldsymbol{\gamma}}\,],$$

so that the commutator belongs to the root $\boldsymbol{\beta} + \boldsymbol{\gamma}$. Accordingly we write in all generality,

$$[\, E_{\boldsymbol{\beta}},\, E_{\boldsymbol{\gamma}}\,] = \mathcal{N}_{\boldsymbol{\beta},\,\boldsymbol{\gamma}}\, E_{\boldsymbol{\beta}+\boldsymbol{\gamma}}, \tag{7.14}$$

where the yet undetermined constants satisfy

$$\mathcal{N}_{\boldsymbol{\beta},\,\boldsymbol{\gamma}} = -\mathcal{N}_{\boldsymbol{\gamma},\,\boldsymbol{\beta}}. \tag{7.15}$$

But wait! There are only $(N - r)$ roots, so that not all linear combinations of roots can be roots. Since all elements of the Lie algebra are spoken for, it must be that whenever $\boldsymbol{\beta} + \boldsymbol{\gamma}$ is *not* a root, the commutator vanishes, that is

$$\mathcal{N}_{\boldsymbol{\beta},\,\boldsymbol{\gamma}} = 0, \qquad \text{if} \quad \boldsymbol{\beta} + \boldsymbol{\gamma} \quad \text{not a root}. \tag{7.16}$$

By taking the hermitian conjugate we see that

$$\mathcal{N}_{\boldsymbol{\alpha},\,\boldsymbol{\beta}} = -\mathcal{N}^*_{-\boldsymbol{\beta},-\boldsymbol{\alpha}} = -\mathcal{N}_{-\boldsymbol{\beta},-\boldsymbol{\alpha}}, \tag{7.17}$$

since the structure functions are real.

If $\boldsymbol{\beta} + \boldsymbol{\gamma} = 0$, the commutator with H^i vanishes, and must therefore reduce to a linear combination of the H^i,

$$[\, E_{\boldsymbol{\beta}},\, E_{-\boldsymbol{\beta}}\,] \;\equiv\; \beta_i\, H^i, \tag{7.18}$$

where β_i are undetermined, not to be confused with β^i, the components of the root vector. To relate them, multiply this equation by H^j, and take the trace. A few manipulations yield

$$\beta^j\, \mathrm{Tr}\left(E_{-\boldsymbol{\beta}}\, E_{\boldsymbol{\beta}} \right) = \beta_i\, \mathrm{Tr}\left(H^i H^j\right). \tag{7.19}$$

The matrix

$$g^{ij} \;\equiv\; \mathrm{Tr}\left(H^i H^j\right) \tag{7.20}$$

is the Cartan–Killing form of eq. (7.5). It is invertible for a semi-simple algebra, acts as a metric for the r-dimensional root space, relates covariant and contravariant components of the root vectors, and is used to define a scalar product. The trace on the left-hand side is arbitrarily set to one,

$$\mathrm{Tr}\left(E_{-\boldsymbol{\beta}}\, E_{\boldsymbol{\beta}} \right) = 1, \tag{7.21}$$

for each $\boldsymbol{\beta}$. Thus

$$\beta^j = \beta_i\, g^{ij}; \qquad \beta_i = g_{ij}\, \beta^j. \tag{7.22}$$

This leads us to define the scalar product in the real r-dimensional root space as

$$(\,\boldsymbol{\alpha},\ \boldsymbol{\beta}\,) = (\,\boldsymbol{\beta},\ \boldsymbol{\alpha}\,) \;\equiv\; \alpha^i\, \beta^j\, g_{ij} = \alpha_i\, \beta_j\, g^{ij}. \tag{7.23}$$

One can show that the length of all these vectors is positive definite.

We are now in a position to determine the coefficients $\mathcal{N}_{\boldsymbol{\beta},\,\boldsymbol{\gamma}}$, by clever repeated use of the Jacobi identity. For three roots related by

$$\boldsymbol{\alpha} + \boldsymbol{\beta} + \boldsymbol{\gamma} = 0,$$

we easily see that

$$[\, E_{\boldsymbol{\alpha}},\, [\, E_{\boldsymbol{\beta}},\, E_{\boldsymbol{\gamma}}\,]] = \alpha_i\, \mathcal{N}_{\boldsymbol{\beta},\boldsymbol{\gamma}}\, H^i,$$

Taking cyclic permutations, the Jacobi identity yields

$$\left(\alpha_i\, \mathcal{N}_{\boldsymbol{\beta},\,\boldsymbol{\gamma}} + \beta_i\, \mathcal{N}_{\boldsymbol{\gamma},\,\boldsymbol{\alpha}} + \gamma_i\, \mathcal{N}_{\boldsymbol{\alpha},\,\boldsymbol{\beta}}\right) H^i = 0. \tag{7.24}$$

Multiply by H^j and take the trace, to get the constraint

$$\alpha^i\, \mathcal{N}_{\boldsymbol{\beta},\,\boldsymbol{\gamma}} + \beta^i\, \mathcal{N}_{\boldsymbol{\gamma},\,\boldsymbol{\alpha}} + \gamma^i\, \mathcal{N}_{\boldsymbol{\alpha},\,\boldsymbol{\beta}} = 0, \tag{7.25}$$

which leads to the identities

$$\mathcal{N}_{\boldsymbol{\alpha},\,\boldsymbol{\beta}} = \mathcal{N}_{\boldsymbol{\beta},-\boldsymbol{\alpha}-\boldsymbol{\beta}} = \mathcal{N}_{-\boldsymbol{\alpha}-\boldsymbol{\beta},\,\boldsymbol{\alpha}}. \tag{7.26}$$

A pair of integers p and q can be naturally associated to any two roots in the following way: take E_α and commute it k times with E_β, obtaining the $\boldsymbol{\alpha}$-chain of roots

$$[\,E_{\boldsymbol{\beta}},\,[\,E_{\boldsymbol{\beta}},\,[\cdots,\,[\,E_{\boldsymbol{\beta}},\,E_{\boldsymbol{\alpha}}\,]\cdots]]],$$

which either yields an element belonging to the root $\boldsymbol{\alpha}+k\,\boldsymbol{\beta}$, or else zero. Since the Lie algebra contains a finite number of elements, there has to be a maximum value $k_{\text{max}}=q$, after which further commutation yields zero, so that $\boldsymbol{\alpha}+(q+1)\,\boldsymbol{\beta}$ is *not* a root. Similarly, we can repeatedly commute $E_{-\boldsymbol{\beta}}$ with $E_{\boldsymbol{\alpha}}$, and there must be an integer p such that $\boldsymbol{\alpha}-(p+1)\,\boldsymbol{\beta}$ is not a root, with the vanishing of the multiple commutator after $(p+1)$ steps.

Now we form the Jacobi identity for the three elements belonging to the roots $\boldsymbol{\beta}$, $-\boldsymbol{\beta}$, and $\boldsymbol{\alpha}+k\,\boldsymbol{\beta}$, (with $\boldsymbol{\alpha}\neq\boldsymbol{\beta}$), and evaluate the commutators with H^i. The result is

$$\beta_i\,(\alpha^i+k\,\beta^i)\,E_{\boldsymbol{\alpha}+k\boldsymbol{\beta}}=[\,E_{\boldsymbol{\beta}},\,[\,E_{-\boldsymbol{\beta}},\,E_{\boldsymbol{\alpha}+k\boldsymbol{\beta}}\,]]+[\,E_{-\boldsymbol{\beta}},\,[\,E_{\boldsymbol{\alpha}+k\boldsymbol{\beta}},\,E_{\boldsymbol{\beta}}\,]].\quad(7.27)$$

Evaluation of the double commutators yields

$$\beta_i\,(\alpha^i+k\,\beta^i)=\mathcal{N}_{\boldsymbol{\beta},\,\boldsymbol{\alpha}+(k-1)\,\boldsymbol{\beta}}\,\mathcal{N}_{-\boldsymbol{\beta},\,\boldsymbol{\alpha}+k\,\boldsymbol{\beta}}+\mathcal{N}_{-\boldsymbol{\beta},\,\boldsymbol{\alpha}+(k+1)\,\boldsymbol{\beta}}\,\mathcal{N}_{\boldsymbol{\alpha}+k\,\boldsymbol{\beta},\,\boldsymbol{\beta}}.\quad(7.28)$$

Using (7.26), and the antisymmetry of the $\mathcal{N}$ symbols, it is rewritten as

$$\beta_i\,(\alpha^i+k\,\beta^i)=\mathcal{F}(k)-\mathcal{F}(k-1),\quad(7.29)$$

where

$$\mathcal{F}(k)=\mathcal{N}_{\boldsymbol{\beta},\,\boldsymbol{\alpha}+k\,\boldsymbol{\beta}}\,\mathcal{N}_{-\boldsymbol{\beta},\,-\boldsymbol{\alpha}-k\,\boldsymbol{\beta}},\quad(7.30)$$

and $-p\le k\le q$. Since $\boldsymbol{\alpha}+(q+1)\boldsymbol{\beta}$ is not a root, it follows from (7.27) that

$$\mathcal{F}(q)=0,\qquad\mathcal{F}(q-1)\;=-\beta_i\,(\alpha^i+q\,\beta^i).\quad(7.31)$$

The recursion relation (7.29) is then easily solved

$$\mathcal{F}(k)=(k-q)\,\beta_i\,[\,\alpha^i+\frac{1}{2}(k+q+1)\beta^i\,].\quad(7.32)$$

At the lower end of the chain, the use of (7.26) yields

$$\mathcal{F}(-p-1)=0,\quad(7.33)$$

so that

$$(p+q+1)\,[\,\beta_i\alpha^i+\frac{1}{2}(q-p)\beta_i\beta^i\,]=0.\quad(7.34)$$

Since both p and q are positive integers, we deduce that

$$\frac{2\,\beta_i\,\alpha^i}{\beta_j\,\beta^j}=2\frac{(\boldsymbol{\beta},\,\boldsymbol{\alpha})}{(\boldsymbol{\beta},\,\boldsymbol{\beta})}=p-q\;\equiv\;n,\quad(7.35)$$

is always an integer. This equation, central to Lie algebra theory, has many consequences.

First, by setting $k = 0$ in (7.32), and using (7.17), we obtain a neat expression for the structure constants,

$$(\mathcal{N}_{\alpha,\,\beta})^2 = \frac{1}{2} q\,(p+1)\,(\boldsymbol{\beta},\, \boldsymbol{\beta}\,). \tag{7.36}$$

It also means that

$$\boldsymbol{\alpha}' = \boldsymbol{\alpha} - (p-q)\boldsymbol{\beta} = \boldsymbol{\alpha} - \frac{2\,\beta_i\,\alpha^i}{\beta_j\,\beta^j}\,\boldsymbol{\beta}, \tag{7.37}$$

is always a root. The component of $\boldsymbol{\alpha}'$ along $\boldsymbol{\beta}$, is the negative of the projection of $\boldsymbol{\alpha}$ along the same direction, since

$$(\,\boldsymbol{\alpha}',\, \boldsymbol{\beta}\,) = -(\,\boldsymbol{\beta},\, \boldsymbol{\alpha}\,). \tag{7.38}$$

This suggests a nice geometrical interpretation in root space: $\boldsymbol{\alpha}'$ is the Weyl reflection of $\boldsymbol{\alpha}$ about the hyperplane perpendicular to $\boldsymbol{\beta}$. Weyl reflections generate roots of the same length, since

$$(\,\boldsymbol{\alpha}',\, \boldsymbol{\alpha}'\,) = (\,\boldsymbol{\alpha},\, \boldsymbol{\alpha}\,). \tag{7.39}$$

A further consequence is a set of restrictive values for the angle between two roots. To see this, note that we could have interchanged $\boldsymbol{\alpha}$ and $\boldsymbol{\beta}$, and arrived at a result with the same structure, namely

$$\frac{2(\,\boldsymbol{\alpha},\, \boldsymbol{\beta}\,)}{(\,\boldsymbol{\alpha},\, \boldsymbol{\alpha}\,)} = p' - q' \;\equiv\; n',$$

where p' and q' are the relevant integers for the chain $\boldsymbol{\beta} + k\boldsymbol{\alpha}$. Clearly n and n' must have the same sign, since

$$\frac{n}{n'} = \frac{(\,\boldsymbol{\alpha},\, \boldsymbol{\alpha}\,)}{(\,\boldsymbol{\beta},\, \boldsymbol{\beta}\,)}. \tag{7.40}$$

In addition

$$\frac{2(\,\boldsymbol{\beta},\, \boldsymbol{\alpha}\,)}{(\,\boldsymbol{\alpha},\, \boldsymbol{\alpha}\,)}\,\frac{2(\,\boldsymbol{\beta},\, \boldsymbol{\alpha}\,)}{(\,\boldsymbol{\beta},\, \boldsymbol{\beta}\,)} = n\,n', \tag{7.41}$$

from which

$$(\,\boldsymbol{\beta},\, \boldsymbol{\alpha}\,)^2 = (\,\boldsymbol{\alpha},\, \boldsymbol{\alpha}\,)\,(\,\boldsymbol{\beta},\, \boldsymbol{\beta}\,)\,\frac{n\,n'}{4}. \tag{7.42}$$

As a result, the angle $\Theta_{\alpha\,\beta}$ between the two roots is restricted by

$$\cos^2\Theta_{\alpha\,\beta} = \frac{n\,n'}{4} \;\leq 1. \tag{7.43}$$

If $\cos^2 \Theta_{\alpha\beta} = 1$, the two roots are proportional to one another, but this runs into the following theorem: if $\boldsymbol{\alpha}$ and $a\boldsymbol{\alpha}$ are roots, then $a = \pm 1$, but we assumed $\boldsymbol{\alpha} \neq \boldsymbol{\beta}$ from the onset. Hence, any two roots of a semi-simple Lie algebra must satisfy the constraints in the following table:

n	n'	Θ	Length ratio
0	0	$\pi/2$	not fixed
1	1	$\pi/3$	1
−1	−1	$2\pi/3$	1
1	2	$\pi/4$	$\sqrt{2}$
−1	−2	$3\pi/4$	$\sqrt{2}$
1	3	$\pi/6$	$\sqrt{3}$
−1	−3	$5\pi/6$	$\sqrt{3}$

so that always

$$\frac{4(\boldsymbol{\beta}, \boldsymbol{\alpha})^2}{(\boldsymbol{\alpha}, \boldsymbol{\alpha})(\boldsymbol{\beta}, \boldsymbol{\beta})} = 0, 1, 2, 3, \tag{7.44}$$

and since $|n|$, $|n'|$ are at most three, there are no chains with five roots or more.

7.2 Simple roots

The expressions derived above apply to any two arbitrary roots $\boldsymbol{\alpha}$ and $\boldsymbol{\beta}$. However, the non-zero $(N - r)$ roots live in an r-dimensional space, implying many linear dependencies. For $SU(3)$, we had split them into an equal number of positive and negative roots. Although our notion of positivity was basis-dependent, the final results were unaffected.

Repeating this procedure, let us order the components of any root along H^1, H^2, ..., H^r. A root is then said to be positive if its first non-zero component along the ordered axes is positive. This is fine since the negative of every non-zero root is also a root. Not all $(N - r)/2$ positive roots are independent. We introduce r linearly independent *simple* roots, $\boldsymbol{\alpha}_i$, $i = 1, 2, \ldots, r$, which are positive roots with the property that none can be expressed as linear combination of other simple roots with *positive* coefficients.

We now apply the results of the preceding section to chains of simple roots. Let $\boldsymbol{\alpha}_i$ and $\boldsymbol{\alpha}_j$ be two simple roots, and form their difference $\boldsymbol{\alpha}_i - \boldsymbol{\alpha}_j$, which is either positive, negative or zero. If positive, or negative, we see that the simple roots satisfy $\boldsymbol{\alpha}_i = (\boldsymbol{\alpha}_i - \boldsymbol{\alpha}_j) + \boldsymbol{\alpha}_j$, or $\boldsymbol{\alpha}_j = (\boldsymbol{\alpha}_j - \boldsymbol{\alpha}_i) + \boldsymbol{\alpha}_i$, contrary to their definition. Hence the difference of two simple roots must be zero.

It follows that any chain constructed out of two simple roots has $p = 0$, so that $(\boldsymbol{\alpha}_i, \boldsymbol{\alpha}_j) \leq 0$, and the angle between simple roots has only three possible values, $2\pi/3$, $3\pi/4$, and $5\pi/6$. This leads to only three chains built out of simple roots

$$\begin{array}{cc} \alpha_i & \alpha_i+\alpha_j \\ \circ & \circ \end{array}$$

$$\begin{array}{ccc} \alpha_i & \alpha_i+\alpha_j & \alpha_i+2\alpha_j \\ \circ & \circ & \circ \end{array}$$

$$\begin{array}{cccc} \alpha_i & \alpha_i+\alpha_j & \alpha_i+2\alpha_j & \alpha_i+3\alpha_j \\ \circ & \circ & \circ & \circ \end{array}$$

for $q = 1, 2, 3$, respectively. Only the first root is simple. For all three chains, we find

$$\frac{(\boldsymbol{\alpha}_i + \boldsymbol{\alpha}_j, \boldsymbol{\alpha}_i + \boldsymbol{\alpha}_j)}{(\boldsymbol{\alpha}_i, \boldsymbol{\alpha}_i)} = \frac{1}{q}, \tag{7.45}$$

for a chain starting at the simple root $\boldsymbol{\alpha}_i$ and ending at $\boldsymbol{\alpha}_i + q\,\boldsymbol{\alpha}_j$.

It is convenient to normalize each element of thc Lie algebra as

$$e_\alpha \equiv \sqrt{\frac{2}{(\boldsymbol{\alpha}_i, \boldsymbol{\alpha}_i)}}\, E_\alpha, \tag{7.46}$$

and write those corresponding to the simple roots as e_i, $i = 1, 2 \ldots, r$. Then

$$[\, e_i, \, e_{-i}\,] = \frac{2}{(\boldsymbol{\alpha}_i, \boldsymbol{\alpha}_i)} (\alpha_i)_j H^j, \tag{7.47}$$

where $(\alpha_i)_j$ is the jth component of $\boldsymbol{\alpha}_i$. In terms of

$$h_i = \frac{2}{(\boldsymbol{\alpha}_i, \boldsymbol{\alpha}_i)} (\alpha_i)_j H^j, \tag{7.48}$$

we have

$$[\, e_i, \, e_{-i}\,] = h_i, \tag{7.49}$$

and

$$[\, h_i, \, e_i\,] = 2\, e_i. \tag{7.50}$$

Thus each of the r sets e_i, e_{-i} and h_i, generates an $SU(2)$ algebra. However, their elements do not commute with one another:

$$[\,h_i,\,e_j\,] = \frac{2(\boldsymbol{\alpha}_i,\,\boldsymbol{\alpha}_j\,)}{(\boldsymbol{\alpha}_j,\,\boldsymbol{\alpha}_j\,)}\,e_j, \tag{7.51}$$

$$= A_{ji}\,e_j, \tag{7.52}$$

written in terms of the $(r \times r)$ Cartan matrix

$$A_{ji} \;\equiv\; \frac{2(\boldsymbol{\alpha}_i,\,\boldsymbol{\alpha}_j\,)}{(\boldsymbol{\alpha}_j,\,\boldsymbol{\alpha}_j\,)}, \tag{7.53}$$

with negative integer off-diagonal elements, -1, -2, and -3. We also have

$$[\,e_\alpha,\,e_\beta\,] = \mathcal{N}'_{\alpha,\beta}\,e_{\alpha+\beta}, \tag{7.54}$$

with

$$\mathcal{N}'_{\alpha,\beta} \;\equiv\; \sqrt{\frac{2(\boldsymbol{\alpha}+\boldsymbol{\beta},\,\boldsymbol{\alpha}+\boldsymbol{\beta}\,)}{(\boldsymbol{\alpha},\,\boldsymbol{\alpha}\,)(\boldsymbol{\alpha},\,\boldsymbol{\beta}\,)}}\,\mathcal{N}_{\alpha,\,\beta}. \tag{7.55}$$

Now, specialize this expression to the commutator of two simple roots, for which $p = 0$; use of eqs. (7.36) and (7.45) then yields the simple result

$$[\,e_i,\,e_j\,] \;= \pm\,e_{\alpha_i+\alpha_j}. \tag{7.56}$$

What of the commutator of elements which are not simple roots? It turns out that for any two roots $\boldsymbol{\alpha}$ and $\boldsymbol{\beta}$, eq. (7.45) generalizes to

$$\frac{(\boldsymbol{\alpha}+\boldsymbol{\beta},\,\boldsymbol{\alpha}+\boldsymbol{\beta}\,)}{(\boldsymbol{\alpha},\,\boldsymbol{\alpha}\,)} = \frac{p+1}{q}, \tag{7.57}$$

which leads to

$$[\,e_\alpha,\,e_\beta\,] = \pm\,(p+1)\,e_{\alpha+\beta}. \tag{7.58}$$

This is the Chevalley basis, where *all* structure constants are integers. The sign ambiguities can be determined with more work (see R. W. Carter, *Simple Groups of Lie Type* (John Wiley & Sons Ltd, London)). This is the Chevalley basis which we alluded to in the context of $SU(3)$; it is central to the construction of a plethora of finite simple groups.

7.3 Rank-two algebras

With the restrictions just derived, it is straightforward to build the root space of rank-two Lie algebras. The two zero roots are situated at the origin of the two-dimensional root space. The procedure is to align the first simple root along the horizontal positive axis. Its mirror image about the vertical negative represents its negative root. We then draw the second simple root at the prescribed length and angle. The rest of the roots are generated by Weyl reflections. Depending on the orientation of the second simple root, we have the following possibilities.

(i) Draw the second root at $\pi/2$ with arbitrary length. The Weyl reflections in this case produce the negative roots, and we obtain the following root diagram.

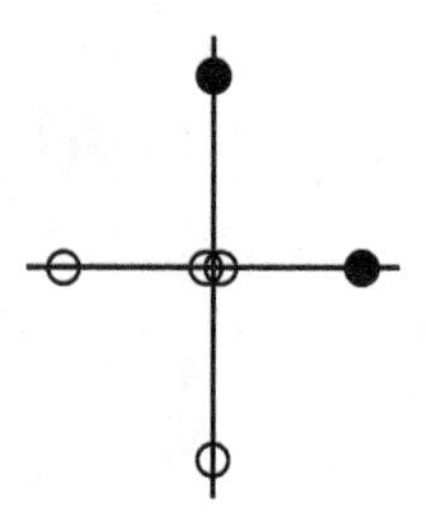

The two simple roots are indicated by the dark circles. Since Weyl reflections do not relate the roots on the vertical with those on the horizontal, we recognize two $SU(2)$ algebras, with their roots along the horizontal and vertical axes, respectively. This is the root diagram of the semi-simple algebra $SU(2) \times SU(2)$.

(ii) The second root is drawn at $2\pi/3$ from the first with the same length as the first. Draw their negatives, and start the Weyl reflections about the various perpendiculars. We generate three non-zero roots and their negatives, yielding the following $SU(3)$ root diagram.

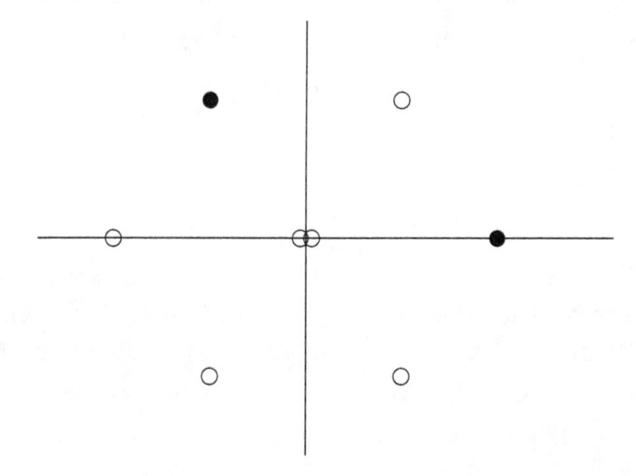

Note that it contains on its horizontal axis the three roots of the $SU(2)$ root spaces as well as a singlet root and two pairs of roots, one above, one below. This simply says that the $SU(3)$ adjoint representation contains an $SU(2)$ triplet (its adjoint), a singlet and two doublets. This diagram can be rotated by $\pi/3$ without affecting its structure.

(iii) The second simple root is now at an angle of $3\pi/4$ to the first, with length $1/\sqrt{2}$ that of the first. Add their negatives, and proceed by Weyl reflections. Reflect the small root about the vertical axis, then about the horizontal axis, then about the vertical axis. This yields four small roots. Further Weyl reflections generate no new roots. Now reflect the large root about the hyperplane perpendicular to the small roots. This yields a large root on the vertical axis. This process yields two more large roots on the axes. Hence we have obtained by Weyl reflection eight non-zero roots. This corresponds, adding the two zero-length roots, to a ten-dimensional algebra, called B_2 or C_2. If we had started with the small root on the horizontal axis, we would have obtained the same root structure, rotated by $\pi/4$.

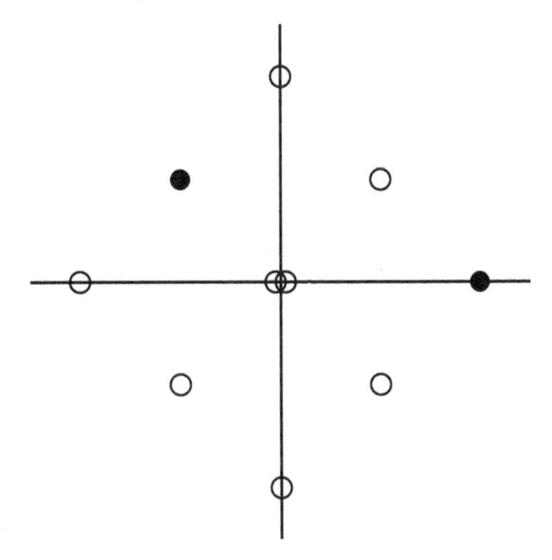

We recognize that this root diagram contains the diagram of two independent $SU(2)$. It also contains roots that look like doublets under *both* $SU(2)$s. Hence we infer that this algebra contains the six elements of $SU(2) \times SU(2)$ and four elements which transform as a doublet-doublet, that is a doublet under each!

(iv) Finally, draw the second root at an angle of $5\pi/6$ to the first. Take it to be the shorter one, with length $(3)^{-1/2}$ times that of the first. Draw their negatives, and proceed to build the root diagram by a judicious succession of Weyl reflections. With two zero roots, the result is the 14-dimensional Lie algebra G_2.

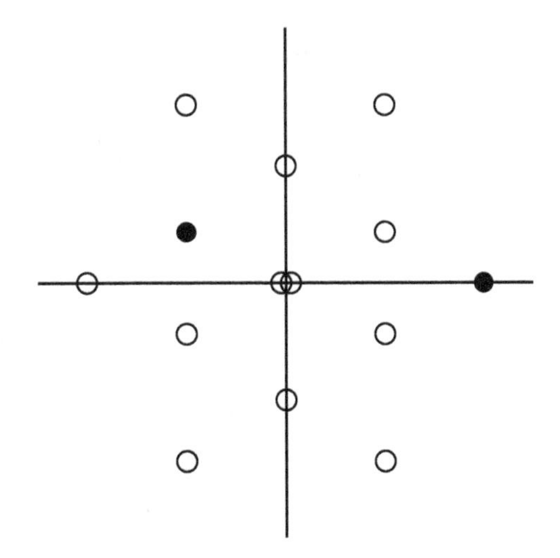

We recognize that the long roots form the $SU(3)$ root diagram, so that G_2 contains $SU(3)$, but we also discern two $SU(2)$s, so that it also contains $SU(2) \times SU(2)$.

This exhausts all possible rank-two semi-simple Lie algebras. All have roots of at most two different non-zero lengths. This is a general feature: the three possible length ratios, $1 : 1$, $1 : \sqrt{2}$, $1 : \sqrt{3}$ imply that an algebra with *all three* lengths would contain roots in the ratio $\sqrt{2} : \sqrt{3}$, contrary to our general results.

7.4 Dynkin diagrams

For higher-rank algebras, it is clearly not practical to describe the roots by drawing them in r-dimensional space. Fortunately, all the relevant information can be displayed in two-dimensional patterns for simple Lie algebras of any rank.

A *Dynkin diagram* is an assembly of r dots, one for each simple (positive) root. When the algebra contains roots of different lengths, short roots are indicated by an open dot, long roots by a filled dot.

It is convenient to work with the largest roots normalized to unit length

$$(\hat{\boldsymbol{\alpha}}_i, \hat{\boldsymbol{\alpha}}_i) = 1, \qquad i = 1, 2, \ldots, r.$$

As we have seen, the angle between two simple roots can take on four possible values. The rules for their construction are as follows.

(i) *Dots representing roots which satisfy*

$$4(\hat{\boldsymbol{\alpha}}_i, \hat{\boldsymbol{\alpha}}_j)^2 = 0,\ 1,\ 2,\ 3,\ \ i \neq j,$$

are connected by zero, one, two or three lines, respectively.

In the latter two cases, they must have different lengths, as indicated below.

These are the Dynkin diagrams of the rank-two algebras we just constructed.

(ii) *Removing dots (roots) from a Dynkin diagram generates Dynkin diagram(s) associated with lesser-rank Lie algebra(s) (simple or semi-simple).*

These simple rules are sufficient to derive the Dynkin diagrams of all simple Lie algebras.

Consider the vector

$$\boldsymbol{\alpha} = \sum_i^r \hat{\boldsymbol{\alpha}}_i .$$

Its norm

$$(\boldsymbol{\alpha}, \boldsymbol{\alpha}) = \sum_i^r (\hat{\boldsymbol{\alpha}}_i, \hat{\boldsymbol{\alpha}}_i) + 2\sum_{i<j}^r (\hat{\boldsymbol{\alpha}}_i, \hat{\boldsymbol{\alpha}}_j),$$

is strictly positive, that is

$$2\sum_{i<j}^r (\hat{\boldsymbol{\alpha}}_i, \hat{\boldsymbol{\alpha}}_j) < \sum_i^r (\hat{\boldsymbol{\alpha}}_i, \hat{\boldsymbol{\alpha}}_i) = r,$$

and the largest possible number of connected pairs by *one* line is $(r-1)$.

- *The number of connected pairs of roots is at most* $(r-1)$.
  Assume a Dynkin diagram with a closed cycle. Remove all dots outside the cycle, generating by rule (ii) a Dynkin diagram with s roots on the cycle, and s connected pairs, describing a rank s algebra! But we just proved that the number of connected pairs must be less than the rank. Hence we reach the following conclusion.
- *There are no Dynkin diagrams with closed cycles.*
  Let $\hat{\boldsymbol{\alpha}}$ be a root connected to n roots $\hat{\boldsymbol{\beta}}_i$, $i = 1, 2, \ldots, n$, so that

$$(\hat{\boldsymbol{\alpha}}, \hat{\boldsymbol{\alpha}}) = (\hat{\boldsymbol{\beta}}_i, \hat{\boldsymbol{\beta}}_i) = 1, \qquad (\hat{\boldsymbol{\alpha}}, \hat{\boldsymbol{\beta}}_i) < 0.$$

The absence of closed cycles implies

$$(\hat{\boldsymbol{\beta}}_j, \hat{\boldsymbol{\beta}}_i) = 0, \quad i \neq j.$$

The vector orthogonal to all $\hat{\boldsymbol{\beta}}_i$s

$$\hat{\boldsymbol{\alpha}} - \sum_i^n (\hat{\boldsymbol{\alpha}}, \hat{\boldsymbol{\beta}}_i)\hat{\boldsymbol{\beta}}_i,$$

has a positive norm, yielding the inequality

$$\sum_i^n (\hat{\boldsymbol{\alpha}}, \hat{\boldsymbol{\beta}}_i)^2 < 1,$$

but the left-hand side is simply one-quarter the number of lines emanating from the root $\boldsymbol{\alpha}$. Hence

$$\text{number of lines} = 4\sum_i^n (\hat{\boldsymbol{\alpha}}, \hat{\boldsymbol{\beta}}_i)^2 < 4.$$

It implies the following.

- *No more than three lines can originate from one root.*
  This limits the allowed multiple connectors to four types.

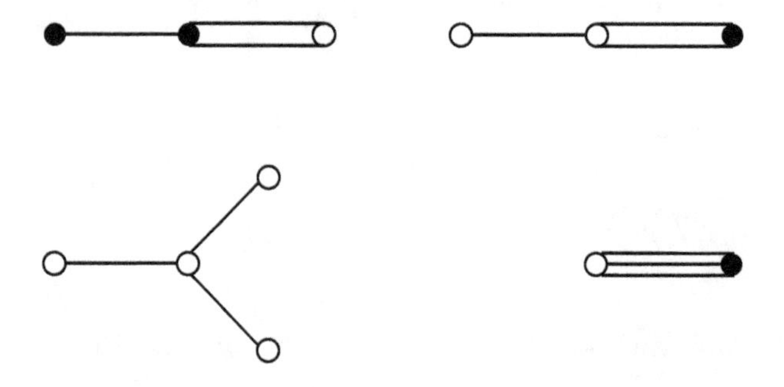

When two roots are connected by three lines, they cannot be further connected to other roots. Hence this last diagram stands by itself as the Dynkin diagram of the rank two algebra G_2

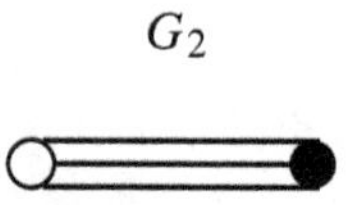

with Cartan matrix

$$A_{G_2} = \begin{pmatrix} 2 & -3 \\ -1 & 2 \end{pmatrix}.$$

Consider a Dynkin diagram with a linear chain of roots, $\{\hat{\boldsymbol{\beta}}_1, \hat{\boldsymbol{\beta}}_2, \ldots, \hat{\boldsymbol{\beta}}_n, \}$, connected to one another by one line, that is

$$(\hat{\boldsymbol{\beta}}_i, \hat{\boldsymbol{\beta}}_j) = -\frac{1}{2}, \quad i \neq j = 1, 2, \ldots, n.$$

Remove these n roots from the diagram, and replace them with one root described by the unit vector

$$\boldsymbol{\beta} = \sum_i^n \hat{\boldsymbol{\beta}}_i.$$

Let $\hat{\boldsymbol{\alpha}}$ be a root that was connected to the chain in the original Dynkin diagram; it was linked to at most one of $\hat{\boldsymbol{\beta}}_k$ of the chain, otherwise there would have been a cycle, that is

$$(\hat{\boldsymbol{\alpha}}, \hat{\boldsymbol{\beta}}) = (\hat{\boldsymbol{\alpha}}, \hat{\boldsymbol{\beta}}_k),$$

and we see that $\hat{\boldsymbol{\beta}}$ has the same projection along its connected root as the original root $\hat{\boldsymbol{\beta}}_k$, and thus satisfies all the axioms to be a root of the same length as those in the original chain.

- *Replacing a linear chain of roots by one root generates a Dynkin diagram.*
  As a result, there is no Dynkin diagram with a linear chain connecting to two different roots, each connected by two lines to the rest of the diagram; otherwise we could shrink the chain and put the two pairs together, resulting in a contradiction of a root with four lines coming out of it.

  Consider a Dynkin diagram with two roots $\hat{\boldsymbol{\alpha}}_n$ and $\hat{\boldsymbol{\beta}}_m$ connected by two lines, so that

$$(\,\hat{\boldsymbol{\alpha}}_n,\,\hat{\boldsymbol{\beta}}_m\,)^2 = 1/2.$$

Each is connected with its own linear chain, one of length n with roots $\hat{\boldsymbol{\alpha}}_i$, the other of length m, with roots $\hat{\boldsymbol{\beta}}_p$. For $i < n$, $p < m$,

$$(\,\hat{\boldsymbol{\alpha}}_i,\,\hat{\boldsymbol{\alpha}}_{i+1}\,) = (\,\hat{\boldsymbol{\beta}}_p,\,\hat{\boldsymbol{\beta}}_{p+1}\,) = -\frac{1}{2}, \qquad (\,\hat{\boldsymbol{\alpha}}_i,\,\hat{\boldsymbol{\beta}}_p\,) = 0.$$

The vectors

$$\boldsymbol{\alpha} = \sum_{j=1}^{n} j\,\hat{\boldsymbol{\alpha}}_j, \qquad \boldsymbol{\beta} = \sum_{p=1}^{m} p\,\hat{\boldsymbol{\beta}}_p,$$

have norms

$$(\,\boldsymbol{\alpha},\,\boldsymbol{\alpha}\,) = \sum_{j}^{n} j^2 - \sum_{i}^{n-1} j(j+1) = \frac{n(n+1)}{2},$$
$$(\,\boldsymbol{\beta},\,\boldsymbol{\beta}\,) = \sum_{p}^{m} p^2 - \sum_{p}^{m-1} p(p+1) = \frac{m(m+1)}{2}.$$

Since the roots in the two chains are orthogonal except for those at the end, we have

$$(\,\boldsymbol{\alpha},\,\boldsymbol{\beta}\,)^2 = m^2n^2\,(\,\hat{\boldsymbol{\alpha}}_n,\,\hat{\boldsymbol{\beta}}_m\,)^2 = \frac{m^2n^2}{2},$$

which, by Schwartz's inequality, must be less than the product of their norm,

$$\frac{m^2n^2}{2} < \frac{n\,m\,(n+1)\,(m+1)}{4},$$

that is

$$(n-1)(m-1) < 2.$$

This equation has three solutions.

– When $n = m = 2$, we find F_4, the rank-four algebra with the following Dynkin diagram.

F_4

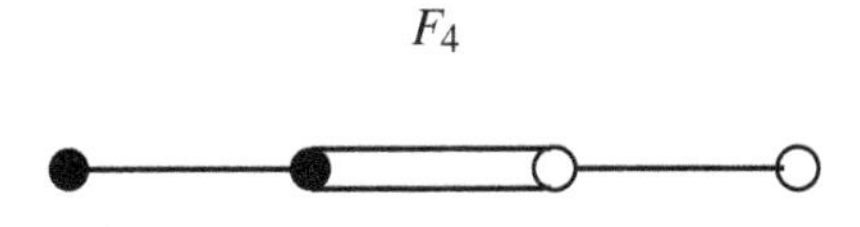

- When $n = 1$, and m arbitrary, there is only one linear chain connected to the long root. This generates the Lie algebra B_r or SO_{2r+1} with the following Dynkin diagram.

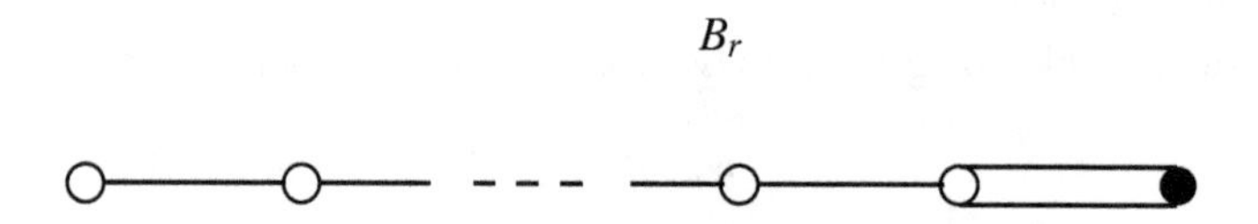

- With $m = 1$, and n arbitrary, the chain is connected to the short root. The resulting Lie algebra is called C_r, or Sp_{2r} with the following Dynkin diagram.

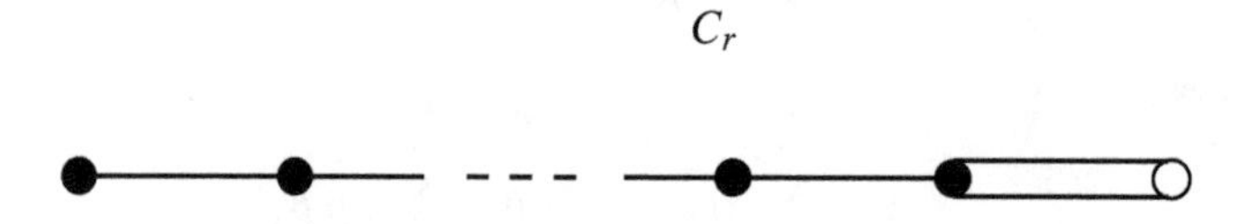

The two algebras obviously have the same Dynkin diagram when $r = 2$, as we have previously noted.

Now consider a Dynkin diagram with a root connected to three linear chains of lengths m, n, p with roots $\hat{\boldsymbol{\alpha}}_i$, $\hat{\boldsymbol{\beta}}_j$, and $\hat{\boldsymbol{\gamma}}_k$. The three chain vectors

$$\boldsymbol{\alpha} = \sum_i^{n-1} i\,\hat{\boldsymbol{\alpha}}_i, \qquad \boldsymbol{\beta} = \sum_j^{m-1} j\,\hat{\boldsymbol{\beta}}_j, \quad \boldsymbol{\gamma} = \sum_k^{p-1} k\,\hat{\boldsymbol{\gamma}}_k,$$

have norms

$$(\boldsymbol{\alpha}, \boldsymbol{\alpha}) = \frac{n(n-1)}{2}, \quad (\boldsymbol{\beta}, \boldsymbol{\beta}) = \frac{m(m-1)}{2}, \quad (\boldsymbol{\gamma}, \boldsymbol{\gamma}) = \frac{p(p-1)}{2}.$$

Let $\hat{\boldsymbol{\delta}}$ be the root that is connected by one line to $\hat{\boldsymbol{\alpha}}_{n-1}$, $\hat{\boldsymbol{\beta}}_{m-1}$ and $\hat{\boldsymbol{\gamma}}_{p-1}$. Now,

$$(\hat{\boldsymbol{\delta}}, \boldsymbol{\alpha})^2 = (n-1)^2(\hat{\boldsymbol{\delta}}, \hat{\boldsymbol{\alpha}}_{n-1})^2 = \frac{(n-1)^2}{4},$$

so that the angle between $\hat{\boldsymbol{\delta}}$ and $\boldsymbol{\alpha}$ is restricted to be

$$\cos^2 \Theta_{\hat{\delta}\alpha} = \frac{1}{2} - \frac{1}{2n}.$$

Similarly for the other chain vectors,

$$\cos^2 \Theta_{\hat{\delta}\beta} = \frac{1}{2} - \frac{1}{2m}, \qquad \cos^2 \Theta_{\hat{\delta}\gamma} = \frac{1}{2} - \frac{1}{2p}.$$

The vector perpendicular to all three *normalized* chain vectors $\hat{\boldsymbol{\alpha}}$, $\hat{\boldsymbol{\beta}}$ and $\hat{\boldsymbol{\gamma}}$ is

$$\hat{\boldsymbol{\delta}} - (\hat{\boldsymbol{\delta}}, \hat{\boldsymbol{\alpha}})\hat{\boldsymbol{\alpha}} - (\hat{\boldsymbol{\delta}}, \hat{\boldsymbol{\beta}})\hat{\boldsymbol{\beta}} - (\hat{\boldsymbol{\delta}}, \hat{\boldsymbol{\gamma}})\hat{\boldsymbol{\gamma}},$$

and the positivity of its norm requires

$$(\hat{\delta},\, \hat{\alpha}\,)^2 + (\,\hat{\delta},\, \hat{\beta}\,)^2 + (\,\hat{\delta},\, \hat{\gamma}\,)^2 \;<\; 1.$$

Putting it all together, the three integers m, n, p are restricted by the inequality

$$\frac{3}{2} - \frac{1}{2n} - \frac{1}{2m} - \frac{1}{2p} \;<\; 1,$$

or

$$\frac{1}{n} + \frac{1}{m} + \frac{1}{p} \;>\; 1. \tag{7.59}$$

This inequality has four types of solutions, one where two of the integers have the same value and three others where all assume different values.

- One solution leaps to the eye: if $m = p = 2$, this equation has solutions for arbitrary n, leading to the following Dynkin diagram of rank $n+2$ algebras, D_{n+2}, SO_{2n+4} to physicists.

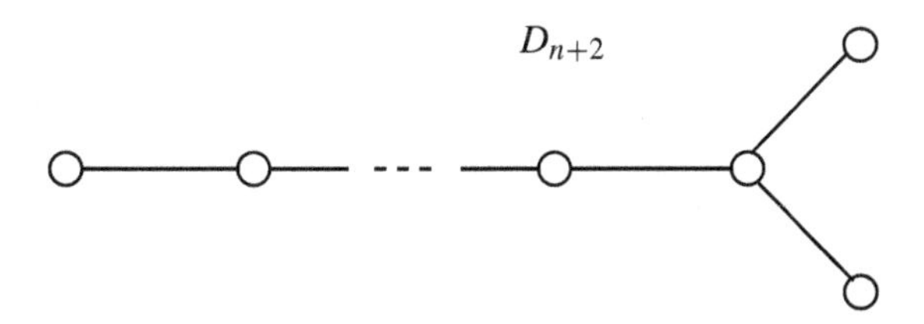

  It is symmetric under the interchange of the two antennas at the end.
- When $p = 2$, the inequality reduces to the very famous Diophantine equation

$$\frac{1}{n} + \frac{1}{m} \;>\; \frac{1}{2}. \tag{7.60}$$

  It has a remarkable geometrical meaning: let n be the number of m-polygons that meet at any vertex. Each m-gon comes with an angle $\pi - 2\pi/m$, since the sum of all the angles they make at the vertex must be less than 2π, we get the inequality

$$n\left(\pi - \frac{2\pi}{m}\right) \;<\; 2\pi,$$

  which is the same as eq. (7.60), and its solutions are in one-to-one correspondence with the platonic solids.
- The tetrahedron case, with $m = n = 2$, yields the following Dynkin diagram of the rank-six exceptional algebra E_6.

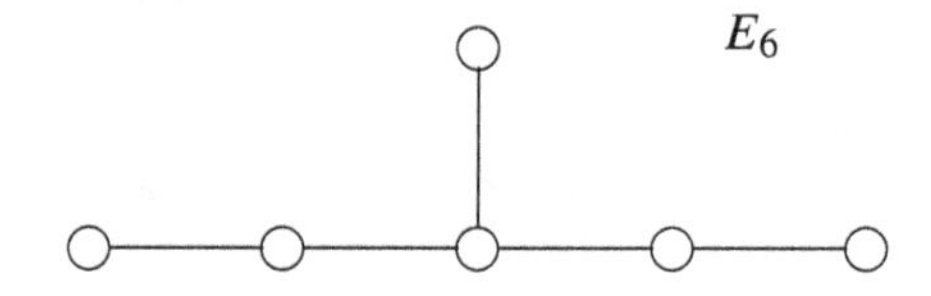

– The octahedron and cube correspond to $n = 3$, $m = 4$, resulting in the following Dynkin diagram of the rank-seven exceptional algebra E_7.

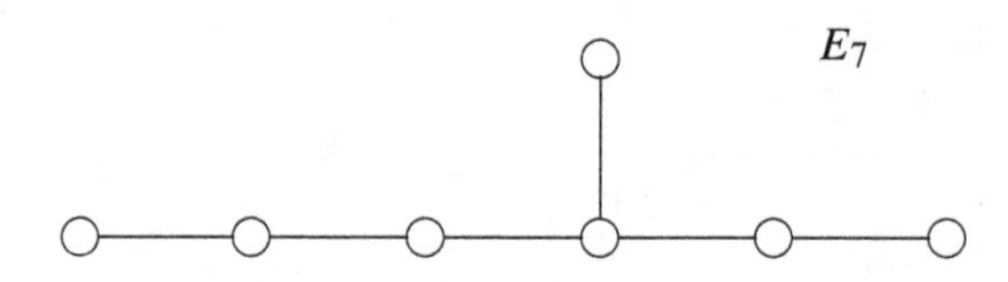

– Finally the icosahedron, $n = 5$, $m = 3$ yields the rank-eight exceptional Lie algebra E_8, with the following Dynkin diagram.

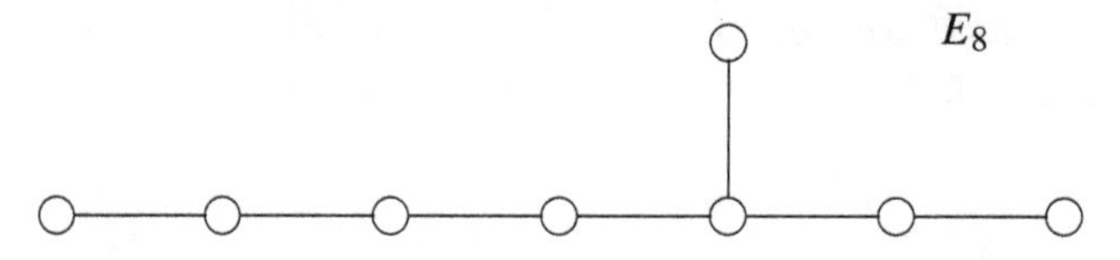

Lastly, roots of equal length can always be connected by one line, producing the Dynkin diagram of A_r or SU_{r+1}.

This exhausts all possible Dynkin diagrams: Lie algebras come as members of the four infinite families A_n, B_n, C_n, and D_n, $n = 1, 2 \ldots$, or one of the five exceptional algebras, G_2, F_4, E_6, E_7, and E_8.

7.5 Orthonormal bases

The information included in Dynkin diagrams and/or Cartan matrices can also be summarized by expressing the roots in terms of an orthonormal basis in the root space. Introduce an orthonormal basis, traditionally denoted by l vectors n_i, satisfying

$$(n_i, n_j) = \delta_{ij}. \tag{7.61}$$

The roots of all the simple Lie algebras are then expressed as linear combinations.

- $A_n = SU(n+1)$. In this case the roots all have the same length with the simple roots satisfying

$$(\boldsymbol{\alpha}_i, \boldsymbol{\alpha}_i) = 2, \qquad (\boldsymbol{\alpha}_i, \boldsymbol{\alpha}_{i+1}) = -1, \quad i = 1, 2, \ldots, n,$$

which suggest that we define $(n+1)$ orthonormal vectors e_i for $i, j = 1, 2, \ldots, n+1$, and set

$$\text{simple roots:} \quad n_i - n_{i+1}.$$

All the roots lie in the plane orthogonal to the vector $\sum n_i$. There are $(n^2 - n - 1)$ non-zero roots, half of them positive

$$\begin{aligned} \text{roots:} &\quad (n_i - n_j),\ i \neq j; \\ \text{positive roots:} &\quad (n_i - n_j),\ i < j. \end{aligned}$$

We define (for later use) the Weyl vector as half the sum of all positive roots

$$\rho \equiv \frac{1}{2} \sum_{\text{pos roots}} \alpha = (n_1 - n_{n+1}).$$

- $B_n = SO(2n+1)$. The roots all have the same length but one which is shorter. They can be written in terms of n orthonormal unit vectors n_i, $i = 1, 2, \ldots, n$ with

$$\text{roots:} \quad \begin{cases} \pm n_i, \\ \pm(n_i \pm n_j), \end{cases} \tag{7.62}$$

$$\text{simple roots:} \quad n_n,\ n_{i-1} - n_i,$$

with the short root of unit length, the long roots of length $\sqrt{2}$. An easy way to get this form is to start from the linear chain of $n-1$ long roots, which are given by the A-series, to which we add one root of unit length that is perpendicular to all but the last one in the chain, and that is n_n.
- $C_n = Sp(2n)$. This case is the same as the previous one except that the short and longer roots are interchanged, resulting in

$$\text{roots:} \quad \begin{cases} \pm 2\, n_i, \\ \pm(n_i \pm n_j), \end{cases} \tag{7.63}$$

$$\text{simple roots:} \quad 2\, n_n,\ n_{i-1} - n_i,\ \ i = 2, \ldots, n,$$

where the long roots are of length $\sqrt{2}$, while the short root has unit length.
- $D_n = SO(2n)$. We have n orthonormal vectors n_i, $i = 1, 2, \ldots, n$, with

$$\text{roots:} \quad \pm(n_i \pm n_j);$$

$$\text{simple roots:} \quad n_{n-1} + n_n,\ n_{i-1} - n_i.$$

This set of simple roots is obtained from the linear chain of the A-series, to which one must add a root of the same length that is perpendicular to all but the next to last root in the chain, and that is $n_{n-1} + n_n$.

- G_2. Its roots are best expressed in terms of three unit vectors n_i, $i = 1, 2, 3$, with

$$\text{roots:} \quad \begin{cases} \pm(n_i - n_j), \\ \pm(2n_i - n_j - n_k), i \neq j \neq k; \end{cases}$$
$$\text{simple roots:} \quad (n_1 - n_2), -(2n_1 - n_2 - n_3).$$

- F_4. In this case we need only four unit vectors to express the roots

$$\text{roots:} \quad \begin{cases} \pm n_i, \\ \pm\frac{1}{2}(n_1 \pm n_2 \pm n_3 \pm n_4), \\ \pm(n_i \pm n_j), \quad i \neq j = 1, 2, 3, 4; \end{cases}$$
$$\text{simple roots:} \quad (n_2 - n_3), (n_3 - n_4), n_4, \tfrac{1}{2}(n_1 - n_2 - n_3 - n_4).$$

In the above, we recognize the simple roots of B_3, to which we have added the short root $(n_1 - n_2 - n_3 - n_4)/2$ that is perpendicular to all but n_4.

- E_8. It is easy to express its simple roots, if we notice that by deleting one root at the right end of its Dynkin diagram, we obtain that of D_7, for which we already know how the simple roots are expressed. All we have to do is to add one of the same length that is perpendicular to all but the last one, $n_6 - n_7$ in our notation. The result is

$$\text{simple roots: } n_6 + n_7, \; n_{i-1} - n_i, \; i = 2, \ldots, 7; \tfrac{1}{2}\left(n_8 + n_7 - \sum_{i=1}^{6} n_i\right).$$

- E_7. This one is obtained simply from the previous case by chopping off the root at the left of the Dynkin, resulting in

$$\text{simple roots: } n_6 + n_7, \; n_{i-1} - n_i, \; i = 3, \ldots, 7; \tfrac{1}{2}\left(n_8 + n_7 - \sum_{i=1}^{6} n_i\right).$$

These are still in an eight-dimensional root space of E_8. If one wants to express the roots in seven dimensions, the last root is given by $n_8/\sqrt{2} + (n_7 - \sum_{i=2}^{6} n_i)/2$, which does not have rational coefficients. Hence it is best to leave its orthonormal basis in eight dimensions.

- E_6. Take out one more root to get

$$\text{simple roots: } n_6 + n_7, \; n_{i-1} - n_i, \; i = 4, \ldots, 7; \tfrac{1}{2}\left(n_8 + n_7 - \sum_{i=1}^{6} n_i\right).$$

As in E_7, we need to be in eight dimensions to express its last root with rational coefficients.

8

Lie algebras: representation theory

Now that we have determined all simple and semi-simple Lie algebras, the next step is to represent them in Hilbert spaces, where connections with physics may be established. The technique for building irreducible representations is rather simple to state: identify a highest weight state and then generate its other states by the repeated application of lowering operators until the process terminates, and you run out of steam. We have seen that this process is very efficient for $SU(2)$ which has only one lowering operator. However, a rank-r Lie algebra has r independent non-commuting annihilation operators, and the process becomes rather unwieldy, and a systematic approach must be adopted.

8.1 Representation basics

We work in the Dynkin basis where all states are labeled by r integers (positive, negative or zero), denoted by the ket

$$|\,\boldsymbol{\lambda}\,\rangle,$$

where $\boldsymbol{\lambda}$ is a vector whose components are the Dynkin labels of the state

$$\boldsymbol{\lambda} = (\,a_1\, a_2\, \dots\, a_n\,). \tag{8.1}$$

To each simple root, associate a lowering operator $T_-^{(i)}$ which, together with its hermitian conjugate, generates an $SU(2)$ algebra

$$[\,T_+^{(i)},\; T_-^{(i)}\,] = T_3^{(i)}, \qquad i = 1, 2 \dots, r. \tag{8.2}$$

We have

$$T_3^{(i)}\,|\,\boldsymbol{\lambda}\,\rangle = a^i\,|\,\boldsymbol{\lambda}\,\rangle. \tag{8.3}$$

The action of the lowering operators on this state yields another state with label $\boldsymbol{\lambda} - \boldsymbol{\alpha}_i$

$$T_-^{(i)} \mid \boldsymbol{\lambda} \rangle \sim \mid \boldsymbol{\lambda} - \boldsymbol{\alpha}_i \rangle. \tag{8.4}$$

When all a_i are positive, the state is said to be inside the *Weyl chamber* of the algebra's lattice, or on its wall if any of the entries are zero. A state inside or on the walls of the Weyl chamber is said to be *dominant*; denote it a capitalized vector

$$\boldsymbol{\Lambda} = (a_1\, a_2\, \ldots\, a_n), \qquad a_i \geq 0.$$

It is annihilated by the r raising operators

$$T_+^{(i)} \mid \boldsymbol{\Lambda} \rangle = 0, \qquad i = 1, 2 \ldots, r, \tag{8.5}$$

since the commutator of two raising operators is also a raising operator. Any dominant weight generates a representation of the algebra by repeated application of the lowering operators. For this reason, it is called the *highest weight state* of the representation. Representations of the Lie algebra are in one-to-one correspondence with the states inside or on the walls of the Weyl chamber. This is in contrast with the physics way to label representations in terms of the Casimir operators of the algebra.

Consider a highest weight state with entry a_i. We can subtract the simple root α_i a_i times, by applying $T_-^{(i)} = E(-\boldsymbol{\alpha}_i)$ as many times. The simple roots are expressed in Dynkinese by the rows of the Cartan matrix.

The number of applications of the annihilation operators to generate all states of a representations is called the *level* or *height* of the representation. It is given by the dot product of the Dynkin indices of the highest weight state with the level vector $\mathbf{R} = (R_1, R_2, \ldots, R_r)$, whose components are given by twice the sum of the rows of the inverse Cartan matrix

$$R_i = 2 \sum_j A^{-1}_{ij}. \tag{8.6}$$

The *fundamental representations* are those which are labeled by the Dynkin labels $(0^p\, 1\, 0^{n-p-1})$, where $p = 1, 2, \ldots\, n$, and 0^p is shorthand for p zero entries.

8.2 A_3 fundamentals

In this section, we use this technique to build the fundamental representations of A_3. There are three fundamental representations, since its Dynkin diagram is a linear chain with three nodes. Its three simple roots, expressed in the Dynkin basis, are the rows of the Cartan matrix ($\bar{1}$ stands for -1)

$$\boldsymbol{\alpha}_1 = (2\,\bar{1}\,0), \qquad \boldsymbol{\alpha}_2 = (\bar{1}\,2\,\bar{1}), \qquad \boldsymbol{\alpha}_3 = (0\,\bar{1}\,2).$$

From its Cartan matrix, we find the level vector

$$\mathbf{R}_{SU(4)} = (3,\ 4,\ 3), \tag{8.7}$$

whose entries give the levels of the fundamental representations.

The highest weight state of the first fundamental representation is $(1\,0\,0)$. With only one positive entry in the first spot, the only way to lower it is by application of the operator associated with $\boldsymbol{\alpha}_1$. This produces the state labeled by $(\bar{1}\,1\,0) = (1\,0\,0) - (2\,\bar{1}\,0)$. This new state has a positive entry in the second spot: we apply $T_-^{(2)}$ which subtracts the second root $\boldsymbol{\alpha}_2$. The result is a new state $(0\,\bar{1}\,1) = (\bar{1}\,1\,0) - (\bar{1}\,2\,\bar{1})$, with positive unit entry in the third spot. Then we apply $T_-^{(3)}$ which subtracts the third simple root to get $(0\,0\,\bar{1}) = (0\,\bar{1}\,1) - (0\,\bar{1}\,2)$. We are done because this state has no positive entries: application of any lowering operator gives zero. We have generated the four-dimensional representation of A_3. We observe that it has three levels, as expected.

Let us apply the same procedure to $(0\,1\,0)$. We first subtract $\boldsymbol{\alpha}_2$ to get $(1\,\bar{1}\,1)$. We now have positive entries in the first and third spots, so we need to subtract with $\boldsymbol{\alpha}_1$ and $\boldsymbol{\alpha}_3$, to produce two states. Then, as shown in the figure, the next level of subtractions produces the same state, so that the diagram is spindle-shaped. This is a general feature of Lie algebra representations. At the fourth level we finally arrive at a state without positive entries and the representation terminates. The height of this representation is four and it contains six states.

The highest weight state of the third fundamental is $(0\,0\,1)$. The first level is obtained by subtracting $\boldsymbol{\alpha}_3$, yielding $(0\,1\,\bar{1})$. Subtraction by $\boldsymbol{\alpha}_2$ produces $(1\,\bar{1}\,0)$; finally subtraction by $\boldsymbol{\alpha}_1$ yields the lowest weight state $(\bar{1}\,0\,0)$; as expected it has three levels.

These constructions are summarized by the following diagrams.

$SU(4)$ fundamental representations

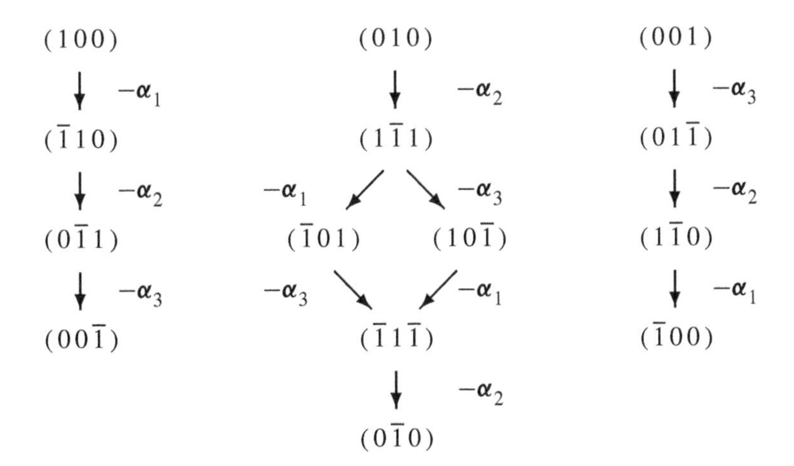

We see that the states in (0 0 1) have opposite Dynkin labels from those in (1 0 0). These representations are complex conjugates of one another, a consequence of the Dynkin diagram's reflection symmetry. Also the weights in the (0 1 0) are all accompanied by their negatives: this representation is self-conjugate, as expected from the Dynkin diagram.

Another method, more familiar to some physicists, is to represent the quadruplet (1 0 0) by a complex vector with upper index, the sextet (0 1 0) by a complex antisymmetric tensor with two upper indices, and the antiquadruplet (0 0 1) with a complex antisymmetric tensor with three upper indices

$$(1\,0\,0) \sim T^{a}, \qquad (0\,1\,0) \sim T^{ab}, \qquad (0\,0\,1) \sim T^{abc}.$$

We could have gone the other way, associating to the antiquadruplet a complex vector with one *lower* index, with the sextet represented by a tensor with two lower indices, and the quadruplet with a tensor with three lower indices

$$(0\,0\,1) \sim T_{a}, \qquad (0\,1\,0) \sim T_{ab}, \qquad (1\,0\,0) \sim T_{abc}.$$

This dual tensor description of the same representation is resolved by using the Levi–Civita tensor; for $SU(4)$, its four upper(lower) indices enable to trade one lower(upper) index for three upper(lower) indices and vice-versa. It allows us to write

$$T_{a} = \frac{1}{3!}\epsilon_{abcd}T^{bcd}. \tag{8.8}$$

This is called *Poincaré duality*. The sextet which sits in the middle of the Dynkin diagram is represented in two ways by a complex two-form (twelve real parameters), with upper or lower indices related by complex conjugation

$$\overline{T}_{ab} = T_{ab}. \tag{8.9}$$

The same duality relation

$$T^{ab} = \frac{1}{4!}\epsilon^{abcd}T_{cd}, \tag{8.10}$$

cuts the number of real parameters in half to six, the required number needed to describe the real sextet.

This pattern readily generalizes to all $SU(n+1)$ algebras, for which the basic representations correspond to the antisymmetric tensor products of its fundamental $(n+1)$-plet. One moves along the Dynkin diagram by adding an index and antisymmetrizing. Half way through the Dynkin diagram, one can use the totally antisymmetric Levi–Civita tensor with $(n+1)$ indices

$$\epsilon^{a_1 a_2 \ldots a_{n+1}},$$

to lower the indices.

The dimensions of the fundamental representations of A_n are summarized in the following table.

Dynkin	Symbol	Dimension
$(1\,0\,\cdots\cdot\,0)$	T^{a}	$\binom{n+1}{1}$
$(0\,1\,\cdots\cdot\,0)$	$T^{[a_1\,a_2]}$	$\binom{n+1}{2}$
$(0\,\cdot\cdot\,1\,\cdot\cdot\,0)$	$T^{[a_1\,a_2 \ldots a_k]}$	$\binom{n+1}{k}$
$(0\,0\,\cdots\,1\,0)$	$T_{[a_1\,a_2]}$	$\binom{n+1}{2}$
$(0\,0\,\cdots\,0\,1)$	T_{a}	$\binom{n+1}{1}$

In this table we have used duality in the last two entries. Each of these representations corresponds to one node in the Dynkin diagram.

The adjoint representation $(1\,0\,1)$ is real but is not a fundamental representation. Since the highest weight has two positive entries, there are two states at the next level. These produce two weights each of which have two positive entries, resulting again in two paths to the next level. There, redundancy occurs and there are only three states at this level, each of which has only one positive entry. The next level has three states, each of which has zero Dynkin labels; they correspond to the rank of the algebra. Then the process continues in reverse, ending with the lowest state $(\,\overline{1}\,0\,\overline{1}\,)$. This produces fifteen states, arranged in a self-conjugate spindle-shaped arrangements, with the conjugate weights appearing in the bottom half of the spindle.

$SU(4)$ adjoint representation

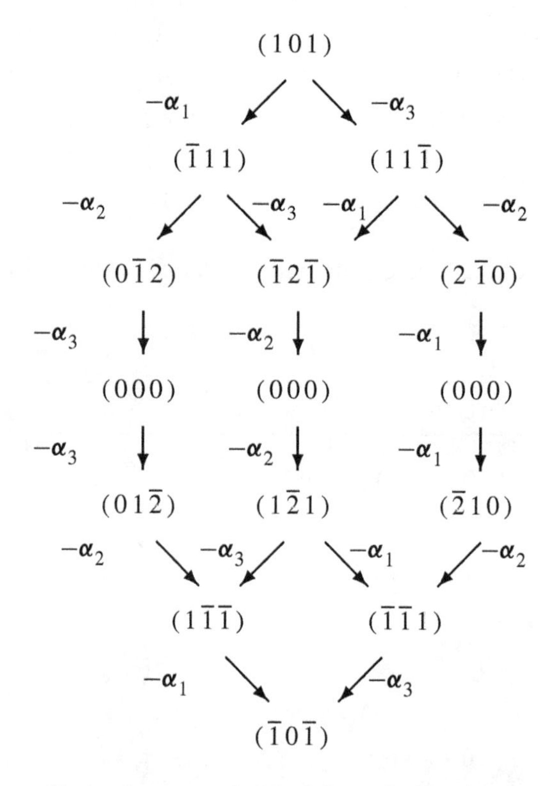

The positive roots lie in the upper half of the spindle; they are easily identified as $(\boldsymbol{\alpha}_1 + \boldsymbol{\alpha}_2 + \boldsymbol{\alpha}_3)$, $(\boldsymbol{\alpha}_2 + \boldsymbol{\alpha}_3)$, $(\boldsymbol{\alpha}_1 + \boldsymbol{\alpha}_2)$, and the three simple roots. Half the sum of all positive roots, the *Weyl vector*, is then

$$\boldsymbol{\rho} = \frac{1}{2}\,(3\boldsymbol{\alpha}_1 + 4\boldsymbol{\alpha}_2 + 3\boldsymbol{\alpha}_3).$$

Does this ring a bell? Yes you noticed it, the numerical factors are the same as in the level vector. This turns out to be a feature of the Lie algebras A_n, D_n, and $E_{6,7,8}$ which have a symmetric Cartan matrix, with roots of the same length. For the others, it is not true. The Weyl vector looks particularly simple in the Dynkin basis

$$\boldsymbol{\rho} = \boldsymbol{\omega}_1 + \boldsymbol{\omega}_2 + \boldsymbol{\omega}_3 = (1\,1\,1). \tag{8.11}$$

This is a general feature: the components of the Weyl vector of *any* simple Lie algebra are all equal to one in the Dynkin basis.

The height of the adjoint representation, l_{adj} (six for $SU(4)$), is related to the *Coxeter number* of the algebra

$$h = \frac{1}{2}\,l_{\text{adj}} + 1, \tag{8.12}$$

which relates D, the dimension of the algebra, to its rank r,

$$D = r\,(h+1). \tag{8.13}$$

8.3 The Weyl group

We have already seen that if $\boldsymbol{\alpha}$ and $\boldsymbol{\beta}$ are roots, then

$$\boldsymbol{\alpha}' = \boldsymbol{\alpha} - \frac{2\,\beta_i\,\alpha^i}{\beta_j\,\beta^j}\,\boldsymbol{\beta}, \tag{8.14}$$

is always a root. The root $\boldsymbol{\alpha}'$ is the reflection of the root $\boldsymbol{\alpha}$ about the hyperplane perpendicular to the direction of $\boldsymbol{\beta}$, with the same length, and its component along $\boldsymbol{\beta}$ is the negative of the projection of $\boldsymbol{\alpha}$ along the same direction:

$$(\boldsymbol{\alpha}',\,\boldsymbol{\alpha}') = (\boldsymbol{\alpha},\,\boldsymbol{\alpha}), \qquad (\boldsymbol{\alpha}',\,\boldsymbol{\beta}) = -(\boldsymbol{\alpha},\,\boldsymbol{\beta}). \tag{8.15}$$

Repeated applications of these reflections generate a finite group. A simple way to see this is by associating a basic reflection r_i with each simple root, so there are as many such reflections as the rank. For example, in $SU(3)$, the action of the basic reflections about the simple roots are

$$\begin{aligned} r_1\,\boldsymbol{\alpha}_1 &= -\boldsymbol{\alpha}_1, & r_1\,\boldsymbol{\alpha}_2 &= \boldsymbol{\alpha}_1 + \boldsymbol{\alpha}_2, \\ r_2\,\boldsymbol{\alpha}_2 &= -\boldsymbol{\alpha}_2, & r_2\,\boldsymbol{\alpha}_1 &= \boldsymbol{\alpha}_1 + \boldsymbol{\alpha}_2. \end{aligned} \tag{8.16}$$

These can be elegantly written in terms of the Cartan matrix

$$r_i\,\boldsymbol{\alpha}_j = \boldsymbol{\alpha}_j - A_{ji}\boldsymbol{\alpha}_i. \tag{8.17}$$

Reflections are elements of order two which generate all other positive and negative non-zero roots of $SU(3)$ through the three operations

$$r_2\,r_1, \qquad r_1\,r_2, \qquad r_1\,r_2\,r_1.$$

All other operations are redundant, for example one checks that $r_2\,r_1\,r_2$ is the same as $r_1\,r_2\,r_1$. Together with the identity, these six operations generate a six-dimensional *finite* group, called the *Weyl group*. The Weyl group of $SU(4)$ is $\mathcal{S}_3$, the permutation group on three objects. We can also express the action of these reflections in the ω-basis

$$\begin{aligned} r_1\,\boldsymbol{\omega}_2 &= \boldsymbol{\omega}_2, & r_1\,\boldsymbol{\omega}_1 &= \boldsymbol{\omega}_2 - \boldsymbol{\omega}_1, \\ r_2\,\boldsymbol{\omega}_1 &= \boldsymbol{\omega}_1, & r_2\,\boldsymbol{\omega}_2 &= \boldsymbol{\omega}_1 - \boldsymbol{\omega}_2. \end{aligned} \tag{8.18}$$

Each rank n Lie algebra has its own Weyl group, generated by n basic reflections r_i, $i = 1, 2, \ldots$. Their action on the simple roots are elegantly expressed in terms of the Cartan matrix

$$r_i \boldsymbol{\alpha}_j = \begin{cases} -\boldsymbol{\alpha}_i & j = i \\ \boldsymbol{\alpha}_j - A_{ji}\boldsymbol{\alpha}_i & j \neq i \end{cases}, \tag{8.19}$$

with no sum over i. The action of the Weyl group on the Weyl chamber generates the full group lattice. We note that the action of the basic reflections on the Weyl vector is simply to subtract their root

$$r_i \boldsymbol{\rho} = \boldsymbol{\rho} - \boldsymbol{\alpha}_i, \quad i = 1, 2, \ldots, n. \tag{8.20}$$

The Weyl groups of the simple Lie algebras are all finite, and they have an irreducible representation of dimension equal to the rank of the group. Below we list the Weyl groups for the simple Lie algebras.

Algebra	Weyl group	Order
$SU(n+1)$	$\mathcal{S}_{n+1}$	$(n+1)!$
$SO(2n+1)$	$\mathcal{S}_n \rtimes \underbrace{(\mathcal{Z}_2 \times \cdots \times \mathcal{Z}_2)}_{n}$	$2^n n!$
$Sp(2n)$	$\mathcal{S}_n \rtimes \underbrace{(\mathcal{Z}_2 \times \cdots \times \mathcal{Z}_2)}_{n}$	$2^n n!$
$SO(2n)$	$\mathcal{S}_n \rtimes \underbrace{(\mathcal{Z}_2 \times \cdots \times \mathcal{Z}_2)}_{n-1}$	$2^{n-1} n!$
G_2	$\mathcal{D}_6 = \mathcal{S}_3 \times \mathcal{Z}_2$	12
F_4	$\mathcal{S}_3 \times \mathcal{S}_4 \times \mathcal{Z}_2 \times \mathcal{Z}_2 \times \mathcal{Z}_2$	$2^7 \cdot 3^2$
E_6	$U_4(2) \cdot 2$	$2^7 \cdot 3^4 \cdot 5$
E_7	$Sp_6(2) \cdot 2$	$2^{10} \cdot 3^4 \cdot 5 \cdot 7$
E_8	$2 \cdot O_8^+(2) \cdot 2$	$2^{14} \cdot 3^5 \cdot 5^2 \cdot 7$

The Weyl groups of the classical algebras are easy to understand. For instance, the $A_n = SU(n+1)$ Weyl group permutes the n vectors e_i of its orthonormal basis. For $SO(2n+1)$ and $Sp(2n)$ their Weyl groups also permute their n orthonormal basis elements e_i, but also include the n reflections $e_i \to \pm e_i$. For $SO(2n)$, the reflections are restricted further with one less parity.

For G_2, $\mathcal{S}_3$ permutes its three orthonormal vectors e_i, and $\mathcal{Z}_2$ changes their sign.

The Weyl group of E_6 is almost simple, since it has only one normal subgroup, $\mathcal{Z}_2$. The quotient group is the simple $U_4(2)$, the simple group of unitary (4×4) matrices over the Galois field $GF(2)$.

The same applies for E_7's Weyl group, with a $\mathcal{Z}_2$ normal subgroup and the simple $Sp_6(2)$ of (6×6) symplectic matrices over $GF(2)$.

For E_8, the Weyl group has a more complicated structure. It is the double cover of $O_8^+(2)\cdot 2$, which has a $\mathcal{Z}_2$ normal subgroup with the simple $O_8^+(2)\cdot 2$ as quotient group (see Chapter 9 for notation).

Weyl groups are examples of the more general *Coxeter groups*, defined by the presentation

$$< r_1, r_2, \dots r_n \,|\, (r_i r_j)^{m_{ij}} = 1 >,$$

where the symmetric matrix elements m_{ij} are either positive integers or ∞

$$m_{ij} = \begin{cases} 1 & i = j \\ \geq 2 & i \neq j \end{cases},$$

and $m_{ij} = \infty$ if there is no such relation between r_i and r_j.

Finally, we note that irreducible representations can be described in terms of Weyl orbits, the collection of states obtained by repeated application of Weyl transformations. Their size is limited by the number of elements of the Weyl group: $SU(2)$ Weyl orbits are at most two-dimensional since its Weyl group is the two-element $\mathcal{Z}_2$. Its Weyl orbits contain the two states $|\, a\,\rangle$ and $|\,-a\,\rangle$, where a is the Dynkin weight, while the lone one-dimensional orbit is $|\,0\,\rangle$.

$SU(4)$ Weyl orbits come in different sizes: 1, 4, 6, 12, and 24, the most allowed by its Weyl group $\mathcal{S}_4$, and there can be many different orbits of the same size. Irreducible representations can then be viewed as a collection of Weyl orbits; for example, **4** and $\overline{\mathbf{4}}$ each contains one different size-four orbit, and the **6** is made up of a single orbit. On the other hand, **15** splits up into three orbits of unit length (for the three zero weights), and one twelve-dimensional orbit. For more details, the reader should consult *Tables of Dominant Weight Multiplicities For Representations of Simple Lie Algebras*, M. R. Bremner, R. V. Moody, and J. Patera (Marcel Dekker, Inc., New York and Basel).

8.4 Orthogonal Lie algebras

We have noted from the Dynkin diagrams several peculiar relations between unitary and orthogonal algebras. In particular, $SU(2)$ has the same algebra as $SO(3)$, as do $SU(4)$ and $SO(6)$.

In order to understand these types of relations, let us take a closer look at algebras which generate rotations in space. Start from an N-dimensional Hilbert space in which the largest algebra is generated by the $(N^2 - 1)$ Gell-Mann matrices. Some matrices have purely imaginary entries; in the original construction they are $\boldsymbol{\lambda}_2$, $\boldsymbol{\lambda}_5$, $\boldsymbol{\lambda}_7$, $\boldsymbol{\lambda}_{10}$, $\boldsymbol{\lambda}_{12}$, etc., one for each off-diagonal entry in the defining $(N \times N)$ matrix. There are $N(N-1)/2$ such entries. When exponentiated into group elements of the form

$$\exp\left(i\omega_a \frac{\lambda_a}{2}\right),$$

they form *real* matrices. As we have seen, the $SU(N)$ generators act on *complex* numbers z_i, $i = 1, 2, \ldots, N$. On the other hand the group elements generated by the matrices with imaginary entries act *independently* on the real and imaginary parts of these complex numbers. While the $SU(N)$ algebra preserves the norm of a complex N-dimensional vector, these matrices form a Lie algebra $SO(N)$ which preserves the norm of a real N-dimensional vector: they generate rotations.

From a geometrical point of view, we can think of rotations as transformations which rotate the ith axis into the jth axis. Denote their generators by X_{ij}. Clearly

$$X_{ij} = -X_{ji}. \tag{8.21}$$

Their commutator with another rotation is easy to visualize; if the i–j plane is not connected to the k–l plane, they must commute

$$[\, X_{ij},\, X_{kl} \,] = 0, \tag{8.22}$$

as long as all indices are different. If the two planes have a common axis, they no longer commute but their commutator generates another rotation, viz.

$$[\, X_{ij},\, X_{jk} \,] = i X_{ik}. \tag{8.23}$$

When acting on a real space with real coordinates x_i, $i = 1, 2, \ldots, N$, they are given by

$$X_{ij} = i \left(x_i \frac{\partial}{\partial x_j} - x_j \frac{\partial}{\partial x_i} \right). \tag{8.24}$$

Suppose $N = 2n$ is even. It is easy to see that the n generators

$$X_{12}, X_{34}, \ldots, X_{N-3\,N-2}, X_{N-1\,N},$$

commute with one another since they generate rotations in planes with no axis in common. Thus $SO(2n)$ has rank n.

The rotation generators in $(2n + 1)$ dimensions are the same as the $SO(2n)$ generators plus those of the form $X_{2n+1\,i}$ for $i = 1, 2, \ldots, 2n$. This does not add any generator to the previous commuting set, so that the rank of $SO(2n+1)$ is also n. We have seen that the Dynkin diagrams of the orthogonal groups in even and odd dimensions are different, although their tensor structures are much the same, except for the range of their indices. Their differences become evident in the study of their spinor representations. The best known example is the two-dimensional representation of $SO(3)$.

8.5 Spinor representations

The link between the two are the (2×2) Pauli matrices (found by F. Klein fifty years earlier). They satisfy

$$\sigma_a \sigma_b = \delta_{ab} + i\epsilon_{abc}\sigma_c,$$

from which one can obtain the $SO(3)$ Lie algebra and the anticommutator *Clifford* algebra

$$\{\sigma_a, \sigma_b\} = 2\,\delta_{ab}. \tag{8.25}$$

Also, the product of any two, say σ_1 and σ_2 generates the third one

$$\sigma_3 = -i\,\sigma_1\sigma_2. \tag{8.26}$$

In the geometrical notation, its generators are X_{12}, X_{23}, X_{31}, which suggests the rewriting

$$X_{12} = -\frac{i}{2}\sigma_1\sigma_2, \qquad X_{23} = -\frac{i}{2}\sigma_2\sigma_3, \qquad X_{31} = -\frac{i}{2}\sigma_3\sigma_1. \tag{8.27}$$

To derive their commutation relations, *it suffices to know that the Pauli matrices satisfy a Clifford algebra*. This gives a recipe for constructing generators of orthogonal groups from Clifford algebras by expressing the $SO(n)$ generators as bilinear products of Clifford elements. Building these representations of orthogonal algebras is reduced to the construction of all possible Clifford algebras and the spaces on which they act.

It is convenient to use the previously introduced fermionic harmonic oscillator operators. Recall they generate a finite-dimensional Hilbert space spanned by states of the form

$$b^\dagger_{i_1} b^\dagger_{i_2} \cdots b^\dagger_{i_k} |\Omega\rangle \;\equiv\; |\,[i_1 i_2 \cdots i_k]\,\rangle.$$

They can be thought of as complex totally antisymmetric tensors, and we know on general grounds that such tensors form representations of $SU(n)$. Indeed it is not too hard to see that the quadratic operators

$$T_j^{\;i} = b^\dagger_j b^i - \frac{1}{n}\delta_j^{\;i}\, b^\dagger_k b^k, \qquad T = \frac{1}{2n}\left(b^k b^\dagger_k - b^\dagger_k b^k\right),$$

form the $SU(n)$ and $U(1)$ compact Lie algebras, respectively, as

$$[\,T_j^{\;i},\, T_l^{\;k}\,] = \delta_l^{\;i} T^{\;k}_j - \delta_j^{\;k} T^{\;i}_l, \qquad [\,T,\, T_l^{\;k}\,] = 0.$$

Except for the vacuum state and the totally antisymmetric tensor with n indices, the states of the Hilbert space are thus the fundamental representations of $SU(n)$, one for each node in the Dynkin diagram.

Consider the remaining quadratic combinations

$$A^{ij} = b^i\, b^j, \qquad A_{ij} = b^\dagger_i\, b^\dagger_j,$$

which transform as antisymmetric two-forms of $SU(n)$,

$$[\,T_j{}^i,\ A^{kl}\,] = \delta_j{}^l A^{ik} - \delta_j{}^k A^{il},$$

and their conjugates. More computation shows that they close on the $SU(n)\times U(1)$ generators

$$[\,A^{ij},\ A_{kl}\,] = \delta_k{}^i T_l{}^j - \delta_k{}^j T_l{}^i - \delta_l{}^i T_k{}^j + \delta_l{}^j T_k{}^i - 2(\delta_k{}^i\delta_l{}^j - \delta_k{}^j\delta_l{}^i)\, T.$$

Taken together, these quadratic operators generate the larger $SO(2n)$ Lie algebra, with the embedding

$$SO(2n) \supset\ SU(n) \times U(1),$$

defined by the decomposition of its real vector representation

$$\mathbf{2n} = \mathbf{n} + \overline{\mathbf{n}},$$

as can be verified by forming the adjoint decomposition

$$\frac{\mathbf{2n(2n-1)}}{\mathbf{2}} = \mathbf{1} + (\mathbf{n^2-1}) + \frac{\mathbf{n(n-1)}}{\mathbf{2}} + \overline{\frac{\mathbf{n(n-1)}}{\mathbf{2}}}.$$

8.5.1 *SO*(2*n*) spinors

All states of the Hilbert space have a $U(1)$ charge, starting with the vacuum state,

$$T\,|\,\Omega\,\rangle = |\,\Omega\,\rangle,$$

with unit charge, and ending with the totally antisymmetric state made up of n oscillators

$$T|\,[i_1 i_2 \cdots i_n]\,\rangle = (1+n)|\,[i_1 i_2 \cdots i_n]\,\rangle.$$

We can build representations of $SO(2n)$ by repeated application of the double-raising operator A_{ij} on the vacuum state. This action generates all states with an *even number* of oscillators. A similar action starting on the one-oscillator state generates a second representation made up of states with an *odd number* of oscillators. These are the two *spinor representations* of $SO(2n)$.

Let us consider some simple examples. To build the spinors of $SO(6)$, we start from the $SU(3)$ Dynkin diagram to which we append the vacuum state and the totally antisymmetric state (open dots). We construct the two spinors by hopscotching across the Dynkin, starting with the vacuum state

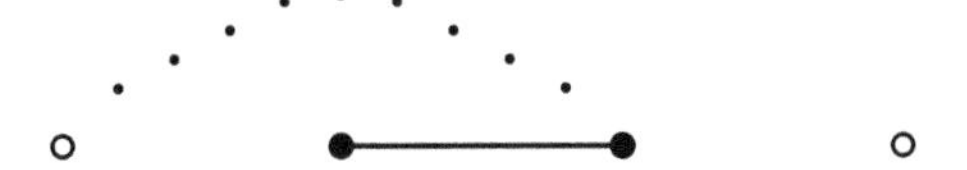

or the one-oscillator state

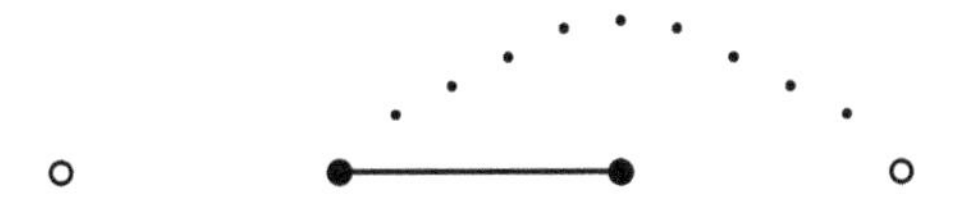

yielding the two spinor representations in terms of $SU(3)$ representations

$$\mathbf{1} + \overline{\mathbf{3}}, \qquad \mathbf{3} + \mathbf{1}.$$

They are evidently complex representations and also conjugate to one another, nothing but the $\overline{\mathbf{4}}$ and $\mathbf{4}$ of $SU(4) \sim SO(6)$, respectively.

Our second example is $SO(8)$. The first hopscotching

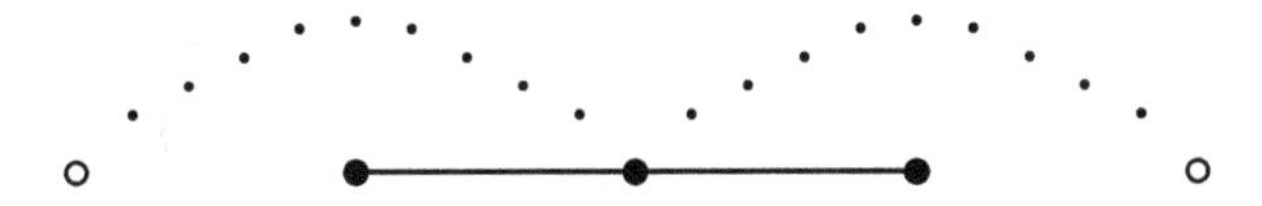

yields one real spinor representation with eight components, which contains the following $SU(4)$ representations

$$\mathbf{1} + \mathbf{6} + \mathbf{1}.$$

The second hopscotching

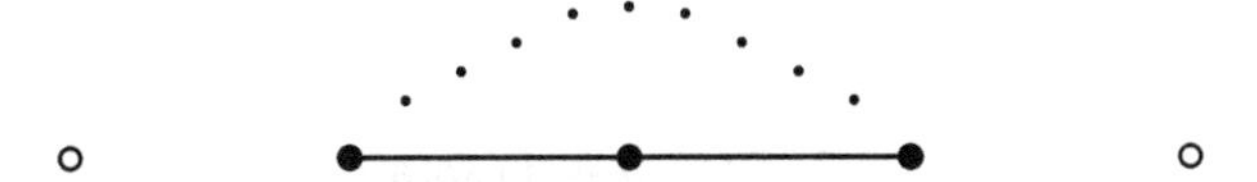

yields the second spinor representation made up of

$$\mathbf{4} + \overline{\mathbf{4}},$$

which is also eight-dimensional, real, and clearly different from the first one.

Judging from these examples, we infer that the structure of the spinor representations of $SO(2n)$ differ as to when n is even or odd. When n is even, the two spinor representations are complex and conjugates to one another; when n is odd, the two spinors are real and different from one another. In either case, both have the same dimensions. The two spinor representations of $SO(2n)$ correspond to the rabbit ears of its Dynkin diagram.

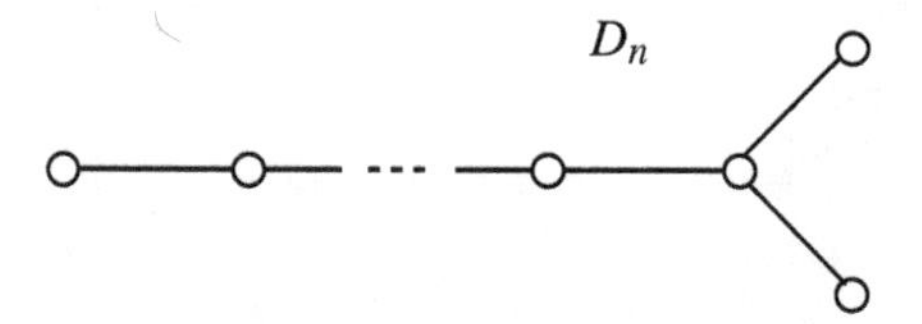

For orthogonal algebras in even dimensions, the first $(n-2)$ fundamental representations are antisymmetric tensors, and the last two are the spinor representations.

Dynkin	Tensor	Dimension
$(10\cdots0)$	T^a	$\binom{2n}{1}$
$(01\cdots0)$	$T^{[a_1 a_2]}$	$\binom{2n}{2}$
$(0\cdots1\cdots0)$	$T^{[a_1 a_2 \ldots a_k]}$	$\binom{2n}{k}$
$(00\cdots100)$	$T^{[a_1 a_2 \ldots a_{n-2}]}$	$\binom{2n}{n-3}$
$(00\cdots10)$	χ	$2^{(n-1)}$
$(00\cdots01)$	χ'	$2^{(n-1)}$

When n is even, the two spinor representations are complex and conjugates to one another; when n is odd, the two spinors are real and inequivalent.

8.5.2 $SO(2n+1)$ spinors

On the other hand, the spinor structure of $SO(2n-1)$ is very different as its Dynkin diagrams leads us to expect only one real spinor representation. This is exactly what happens for $SO(3)$ where we already found a (pseudoreal) spinor doublet. This can be seen by constructing the $SO(2n+1)$ generators in terms of the fermionic oscillators.

We begin with the expression of the $SO(2n)$ generators in terms of n fermionic oscillators. Then we add one more b and $b^\dagger$. According to the previous section, this would enable us to construct the generators of $SO(2n+2)$, but here we are interested in its $SO(2n+1)$ subgroup. In the embedding

$$SO(2n+1) \supset SO(2n), \qquad (\mathbf{2n+1}) = \mathbf{2n} + \mathbf{1},$$

the coset $SO(2n+1)/SO(2n)$ is just the **2n** representation of $SO(2n)$. In terms of $SU(n)$ it decomposes as **n** and $\overline{\mathbf{n}}$. Our extra fermionic oscillator allows us to write several quadratic expressions with the right $SU(n)$ transformation properties. The commutator of the combinations

$$\mathbf{n}: \; a^i = b^i(b + b^\dagger), \qquad \overline{\mathbf{n}}: \; a_i = b_i^\dagger(b^\dagger + b).$$

closes on the $SO(2n) \times U(1)$ algebra

$$[\, a^i ,\, a_j \,] = 2\, T_j{}^i + \frac{1}{n}\delta^i{}_j\, T.$$

They are the coset generators of $SO(2n+1)/SO(2n)$. These explicit expressions enable us to compute the spinor representation.

By adding one oscillator, we have doubled the size of the Hilbert space, which is now spanned by

$$|\, [i_1 i_2 \cdots i_k] \,\rangle, \qquad b^\dagger |\, [i_1 i_2 \cdots i_k] \,\rangle.$$

Each branch transforms by itself under the action of the $SO(2n)$ generators, but the a_i and a^i coset generators make transitions between them. To be explicit, consider their action on the vacuum state

$$a_i |\, \Omega \,\rangle = b_i^\dagger b^\dagger |\, \Omega \,\rangle.$$

Further action yields

$$a_j a_i |\, \Omega \,\rangle = -b_j^\dagger b_i^\dagger |\, \Omega \,\rangle,$$

and so on. This produces a crisscross pattern of states. It is manifestly self-conjugate, and yields half the states in the Hilbert space. We could have started with the action of the coset operators on the state $b^\dagger |\, \Omega \,\rangle$. The result would have been the same, that is a self-conjugate representation with half the state. Both representations are equivalent, and each is the real spinor representation of $SO(2n+1)$. Hence the fundamental representations of $SO(2n+1)$ differ from those of $SO(2n)$ by having only one real spinor representation (corresponding to the one dot of its Dynkin).

Dynkin	Symbol	Dimension
$(1 0 \cdots 0)$	T^a	$\binom{2n+1}{1}$
$(0 1 \cdots 0)$	$T^{[a_1 a_2]}$	$\binom{2n+1}{2}$
$(0 \cdots 1 \cdots 0)$	$T^{[a_1 a_2 \dots a_k]}$	$\binom{2n+1}{k}$
$(0 0 \cdots 1 0)$	$T^{[a_1 a_2]}$	$\binom{2n+1}{n-1}$
$(0 0 \cdots 0 1)$	χ	2^n

The lone spinor representation is of course real.

Let us now use this method to construct the spinor representation of $SO(9)$. We start from two copies of the $SU(4)$ Dynkin diagram, augmented by the two singlets. The leftmost dots represent the vacuum state. The diagonal dots represent the action of a_i, and link the states of the real spinor representation (not shown is the other equivalent way to construct the same representation).

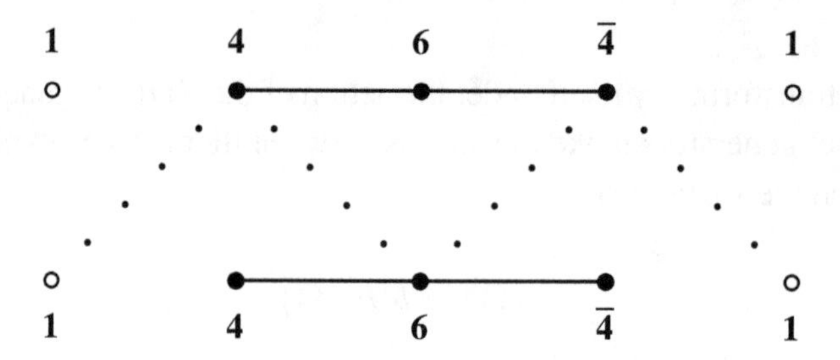

This expresses the real 16-dimensional $SO(9)$ spinor representation in terms of its $SU(4)$ subgroup

$$\mathbf{16} = \mathbf{1} + \mathbf{4} + \mathbf{6} + \bar{\mathbf{4}} + \mathbf{1}.$$

These graphical constructions show that there is only one real fundamental spinor representation in odd numbers of dimensions. As we saw, in even dimensions the two spinors can be either real and inequivalent, or else complex conjugates of one another.

We summarize these properties in a table of the spinor representations of the lowest orthogonal algebras.

Algebra	Spinor representation	Type
$SO(3)$	$\mathbf{2}$	Pseudoreal
$SO(3) \times SO(3)$	$(\mathbf{2}, \mathbf{1}), (\mathbf{1}, \mathbf{2})$	Pseudoreal
$SO(5)$	$\mathbf{4}$	Real
$SO(6)$	$\mathbf{4}, \overline{\mathbf{4}}$	Complex
$SO(7)$	$\mathbf{8}$	Real
$SO(8)$	$\mathbf{8}, \mathbf{8}'$	Real
$SO(9)$	$\mathbf{16}$	Real
$SO(10)$	$\mathbf{16}, \overline{\mathbf{16}}$	Complex
$SO(11)$	$\mathbf{32}$	Real
$SO(12)$	$\mathbf{32}, \mathbf{32}'$	Real

8.5.3 Clifford algebra construction

A more familiar way to build spinor representations is through the explicit construction of Clifford algebras. In the following we proceed in this way by using an iterative procedure which builds larger Clifford algebras starting from direct products of Pauli matrices. The direct product of matrices M and N, is the matrix $M \otimes N$, such that matrix multiplication satisfies

$$\left(M \otimes N\right)\left(P \otimes Q\right) = (M\,P) \otimes (N\,Q). \tag{8.28}$$

For example we can construct (4×4) matrices in this way, such as

$$\sigma_1 \otimes \sigma_3 = \begin{pmatrix} 0 & \sigma_3 \\ \sigma_3 & 0 \end{pmatrix}, \qquad \sigma_2 \otimes \sigma_3 = \begin{pmatrix} 0 & -i\,\sigma_3 \\ i\,\sigma_3 & 0 \end{pmatrix}, \tag{8.29}$$

and the like. Let us apply this construction to Clifford algebras.

- Start with the (2×2) matrices

$$\Gamma_1 = \sigma_1, \qquad \Gamma_2 = \sigma_2, \qquad \Gamma_3 = (-i)\sigma_1\sigma_2 = \sigma_3. \tag{8.30}$$

- By multiplying these by σ_1 and adding one along σ_2, we obtain the four (4×4) Clifford matrices

$$\begin{aligned} \Gamma_1 &= \sigma_1 \otimes \sigma_1, & \Gamma_2 &= \sigma_1 \otimes \sigma_2, \\ \Gamma_3 &= \sigma_1 \otimes \sigma_3, & \Gamma_4 &= \sigma_2 \otimes 1, \end{aligned} \tag{8.31}$$

  and their product

$$\Gamma_5 = \Gamma_1\Gamma_2\Gamma_3\Gamma_4 = \sigma_3 \otimes 1 = \begin{pmatrix} 1 & 0 & 0 & 0 \\ 0 & 1 & 0 & 0 \\ 0 & 0 & -1 & 0 \\ 0 & 0 & 0 & -1 \end{pmatrix}, \tag{8.32}$$

which is a diagonal matrix in this representation. Note that in this construction the Clifford matrices are all hermitian. We take their bilinear product to build the $SO(4)$ generators

$$
\begin{aligned}
X_{ij} &= -\frac{i}{2}\Gamma_i\,\Gamma_j = \frac{1}{2}\,1\otimes\sigma_k; \quad i,j,k = 1,2,3, \text{ cyclic},\\
X_{4k} &= -\frac{i}{2}\Gamma_4\,\Gamma_k = -\frac{1}{2}\sigma_3\otimes\sigma_k.
\end{aligned}
\tag{8.33}
$$

The combinations

$$
L_k^{\pm} = \frac{1}{2}\epsilon_{ijk}\,X_{ij} \pm X_k = \frac{1}{2}\,(1\mp\sigma_3)\otimes\sigma_k, \tag{8.34}
$$

commute with one another and form two $SU(2)$ Lie algebras. In block form

$$
L_k^{+} = \begin{pmatrix} 0 & 0\\ 0 & \sigma_k \end{pmatrix}, \qquad L_k^{-} = \begin{pmatrix} \sigma_k & 0\\ 0 & 0 \end{pmatrix}. \tag{8.35}
$$

The $SO(4)$ algebra is semi-simple, the sum of two $SU(2)$s. We can also define the operators

$$
P_{\pm} \;\equiv\; \frac{1}{2}(1\pm\Gamma_5), \tag{8.36}
$$

which satisfy the projection operator relations

$$
P_{\pm}\,P_{\pm} = P_{\pm}, \qquad P_{\pm}\,P_{\mp} = 0, \tag{8.37}
$$

so that they split the four-dimensional space into two parts. Since they clearly commute with the $SO(4)$ generators, the four-dimensional spinor space on which the generators act is reducible.

However, the full four-dimensional space is needed to realize the $SO(5)$ generators, since we need to add

$$
X_{45} = -\frac{i}{2}\Gamma_4\,\Gamma_5, \qquad X_{i5} = -\frac{i}{2}\Gamma_i\,\Gamma_5, \tag{8.38}
$$

showing that $SO(5)$ has a four-dimensional irreducible representation.

- We repeat the procedure to build (8×8) Clifford matrices, by multiplying the five previous matrices by σ_1, and adding one along σ_2

$$
\begin{aligned}
\Gamma_1 &= \sigma_1\otimes\sigma_1\otimes\sigma_1, & \Gamma_2 &= \sigma_1\otimes\sigma_1\otimes\sigma_2,\\
\Gamma_3 &= \sigma_1\otimes\sigma_1\otimes\sigma_3, & \Gamma_4 &= \sigma_1\otimes\sigma_2\otimes 1,\\
\Gamma_5 &= \sigma_1\otimes\sigma_3\otimes 1, & \Gamma_6 &= \sigma_2\otimes 1\otimes 1,
\end{aligned}
\tag{8.39}
$$

and one more by taking their product

$$
\Gamma_7 = (-i)\,\Gamma_1\,\Gamma_2\,\Gamma_3\,\Gamma_4\,\Gamma_5\,\Gamma_6 = \sigma_3\otimes 1\otimes 1. \tag{8.40}
$$

This means that we can represent both $SO(6)$ and $SO(7)$ algebras in this eight-dimensional space, using the projection operators

$$
P_{\pm} \;\equiv\; \frac{1}{2}(1\pm\Gamma_7), \tag{8.41}
$$

to separate the eight-dimensional spinor space into two four-dimensional spaces. Indeed, the fifteen $SO(6)$ generators

$$X_{ij} = -\frac{i}{2}\Gamma_i\Gamma_j, \qquad i \neq j = 1, 2, \ldots, 6, \tag{8.42}$$

are 8×8 matrices which split (in this representation) into two 4×4 block-diagonal matrices, since Γ_7 is diagonal. Thus, $SO(6)$ acts independently on two four-component spinors, and because $SO(6)$ has the same algebra as $SU(4)$, they must correspond to its four-dimensional *complex* representation and its conjugate. Write the infinitesimal changes in the form

$$\frac{i}{2}\omega_A \begin{pmatrix} \lambda_A & 0 \\ 0 & \hat{\lambda}_A \end{pmatrix} \begin{pmatrix} \psi \\ \chi \end{pmatrix}, \tag{8.43}$$

where $A = 1, 2, \ldots, 15$, ψ and χ are four-component spinors, and of course λ_A are the usual 4×4 Gell-Mann matrices. The hatted matrices are closely related, as it is only rotations containing Γ_6 that act on ψ and χ with different sign. For instance the diagonal generator corresponding to ω_3 is actually the same on both spinors

$$\frac{1}{2}\left(X_{12} + X_{34}\right) = \begin{pmatrix} \lambda_3 & 0 \\ 0 & \lambda_3 \end{pmatrix}. \tag{8.44}$$

However, the generator corresponding to λ_8 is actually different

$$\begin{cases} \lambda_8 = \frac{1}{2\sqrt{3}}\left(X_{34} - X_{12} + 2X_{56}\right), & \text{on } \psi \\ \hat{\lambda}_8 = \frac{1}{2\sqrt{3}}\left(X_{34} - X_{12} - 2X_{56}\right), & \text{on } \chi. \end{cases} \tag{8.45}$$

Some computation shows that

$$\hat{\lambda}_A = -\mathbf{C}\lambda_A^*\mathbf{C}^{-1} = \mathbf{C}\tilde{\lambda}_A\mathbf{C}^{-1}, \tag{8.46}$$

where the tilde matrix is the representation matrix on the $\overline{\mathbf{4}}$ representation, and the (4×4) *charge conjugation* matrix is

$$\mathbf{C} = \begin{pmatrix} 0 & -1 & 0 & 0 \\ 1 & 0 & 0 & 0 \\ 0 & 0 & 0 & 1 \\ 0 & 0 & -1 & 0 \end{pmatrix} = i\,\sigma_3 \otimes \sigma_2, \tag{8.47}$$

or as an eight-dimensional matrix,

$$\mathbf{C} = \frac{i}{2}(1 - \sigma_3) \otimes \sigma_3 \otimes \sigma_2. \tag{8.48}$$

Is there a direct way to see how conjugation arises? Our Clifford matrices are manifestly hermitian, but not symmetric nor real. Note that the Clifford algebra is equally satisfied by the transposed and complex conjugated matrices

$$\pm\left(\Gamma_i\right)^T, \qquad \pm\left(\Gamma_i\right)^*.$$

Similarity transformations on Clifford matrices

$$\Gamma_i \;\rightarrow\; S\,\Gamma_i\,S^{-1} \tag{8.49}$$

preserve the Clifford algebra, and they can be used to change representations; in our case we have opted for a representation where Γ_{2n+1} is diagonal. This also means the existence of a similarity transformation which transposes or takes the complex conjugate of the matrix.

Conjugation of spinors is tricky. We have seen that the $SO(3)$ spinor representation is pseudoreal in the sense that it has a charge conjugation operation

$$C = \sigma_2, \tag{8.50}$$

which is a member of the algebra, so that the **2** and $\overline{\mathbf{2}}$ are the same representation.

These examples suggest a recursive construction of Clifford algebras in higher dimensions. Given the $2n-1$ Clifford $(2^{n-1}\times 2^{n-1})$ matrices, $\tilde{\Gamma}_i$, we build $(2^n \times 2^n)$ Clifford matrices as follows

$$\begin{aligned}\Gamma_i &= \sigma_1 \otimes \tilde{\Gamma}_i, \quad i = 1, 2, \ldots, 2n-1.\\ \Gamma_{2n} &= \sigma_2 \otimes 1,\\ \Gamma_{2n+1} &= (-i)^n\,\Gamma_1\,\Gamma_2\,\ldots\,\Gamma_{2n} = \sigma_3 \otimes 1.\end{aligned} \tag{8.51}$$

These $(2^n \times 2^n)$ matrices, Γ_a, satisfy the $(2n+1)$-dimensional Clifford algebra

$$\{\,\Gamma_a,\,\Gamma_b\,\} = 2\,\delta_{ab}, \quad a, b = 1, 2, \ldots, 2n+1. \tag{8.52}$$

The $SO(2n+1)$ generators can be written as $(2^n \times 2^n)$ matrices

$$X_{ab} = -\frac{i}{2}\Gamma_a\,\Gamma_b, \quad a \neq b. \tag{8.53}$$

By forming the projection operators out of Γ_{2n+1}, we split the 2^n spinor into two 2^{n-1} spinors that transform according to $SO(2n)$. Hence we see in this way that there are two spinor representations in even dimensions and only one in odd dimensions.

The product of two $SO(3)$ spinors yields a triplet which describes a vector in three-space, and a singlet. Mathematically, the outer product of two two-component spinors ψ, χ is a (2×2) matrix, which can be written as a linear combination of the three Pauli matrices and the unit matrix

$$\psi\,\chi^T = 1\,v_0 + \sigma_i\,v_i, \tag{8.54}$$

where the coefficients are obtained by taking the appropriate traces; for example

$$\mathrm{Tr}(\psi\,\chi^T\,\sigma_j) = \mathrm{Tr}(\sigma_i\,\sigma_j)\,v_i = 2\,v_j.$$

This is called the Fierz expansion of spinor products. Let us apply it to the spinor representations of $SO(6)$, where

$$\mathbf{4} \times \mathbf{4} = \mathbf{10} + \mathbf{6}, \qquad \mathbf{4} \times \bar{\mathbf{4}} = \mathbf{15} + \mathbf{1}, \qquad \bar{\mathbf{4}} \times \bar{\mathbf{4}} = \overline{\mathbf{10}} + \mathbf{6}. \tag{8.55}$$

Representations appearing on the right-hand side are easily identified as $SO(6)$ tensors: **6** is the vector, **15** is the second-rank antisymmetric tensor. The third-rank antisymmetric tensor T_{ijk} has twenty real components, while only **10** and $\overline{\mathbf{10}}$ are in the product. The real combination is of course their sum which has twenty components. The way to understand this is to note that in $SO(6)$, there is the totally antisymmetric Levi–Civita tensor on six indices, ϵ_{ijklmn}, which can be used to generate self- and antiself-dual third rank tensors

$$T^{\pm}_{ijk} = \frac{1}{2}\left(T_{ijk} \pm \tilde{T}_{ijk}\right), \qquad \tilde{T}_{ijk} \equiv \frac{1}{3!}\,\epsilon_{ijklmn}\,T_{lmn}. \tag{8.56}$$

Similarly, we can use the Levi–Civita symbol to rewrite a four-component antisymmetric tensor in terms of a second rank tensor.

The numbers check since we can express the $8 \cdot 8 = 2^6$ components break up as

$$2^6 = (1+1)^6 = \sum_1^6 \binom{6}{i} = 1 + 6 + 15 + 20 + 15 + 6 + 1,$$

each being interpreted as rank p antisymmetric tensors (called *p-forms* by mathematicians: a scalar is a zero-form, a vector is a one-form, etc.). Their construction is straightforward in terms of products of different anticommuting Clifford matrices

$$\Gamma_{i_1 i_2 \dots i_n} = \Gamma_{i_1}\,\Gamma_{i_2}\,\cdots\,\Gamma_{i_n}, \qquad i_1 \neq i_2 \neq \cdots \neq i_n. \tag{8.57}$$

Their antisymmetry is obvious since all the matrices anticommute with one another. They are also traceless

$$\mathrm{Tr}\left(\Gamma_{i_1 i_2 \dots i_n}\right) = 0, \tag{8.58}$$

as one can verify from our explicit construction. The Fierz expansion expresses the outer product of two spinors as a linear combination of all these matrices

$$\Psi\,\Omega^T = \sum_{j=1}^{6} \Gamma_{i_1 i_2 \dots i_j}\, v_{i_1 i_2 \dots i_j}. \tag{8.59}$$

The coefficients are obtained by taking the appropriate traces. For example

$$\Omega^T\,\Gamma_{ij}\,\Psi = 8\,v_{ij}, \tag{8.60}$$

where we have used

$$\mathrm{Tr}\left(\Gamma_{ij}\,\Gamma_{kl}\right) = 8\,(\delta_{ik}\delta_{jl} - \delta_{jk}\,\delta_{il}). \tag{8.61}$$

$C_n \sim Sp(2n)$

Symplectic algebras are similar to the orthogonal ones except that symmetrization and antisymmetrization are interchanged. The fundamental representations move along the Dynkin diagram; all, including the nth node, are represented by symmetric tensors.

Dynkin	Tensor	Dimension
$(1 0 \cdots 0)$	T^{a}	$\binom{2n}{1}$
$(0 1 \cdots 0)$	$T^{(a_1 a_2)}$	$\binom{2n}{2} - \binom{2n}{0}$
$(0 \cdots 1 \cdots 0)$	$T^{(a_1 a_2 \ldots a_k)}$	$\binom{2n+k-1}{k}$
$(0 0 \cdots 1 0)$	$T^{(a_1 \ldots a_{n-1})}$	$\binom{3n-2}{n-1}$
$(0 0 \cdots 0 1)$	$T^{(a_1 \ldots a_n)}$	$\binom{2n}{n} - \binom{2n}{n-2}$

8.6 Casimir invariants and Dynkin indices

We have seen that irreducible representations of simple Lie algebras of rank n are characterized by n (positive or zero) integers, the Dynkin indices of its highest weight state. There are other ways to specify these representations, in terms of n operators constructed out of the generators, called *Casimir* operators, which commute with all generators of the algebra. Lie algebras always have one quadratic Casimir element, but the degree of the higher order Casimir invariants depend on the algebra. If we denote the generators of the algebra by the $(d_r \times d_r)$ matrices T^A in the $\mathbf{r}$ representation, the quadratic Casimir is defined to be

$$C_{\mathbf{r}}^{(2)} = T^A\, T^A, \tag{8.62}$$

where we have omitted the $(d_r \times d_r)$ unit matrix on the right-hand side.

In the case of $SU(n)$, these invariants can be constructed out of traces of products of $(n \times n)$ matrices. Since the determinant of such an $(n \times n)$ matrix can be expressed in terms of the trace of powers of the matrix up to the nth power, it follows that we have only $(n-1)$ independent invariants of degree $1, \ldots, n$. Since the trace of the matrix vanishes for an irrep, $SU(n)$ has $(n-1)$ Casimir invariants which are polynomials in its generators of order $2, \ldots, n$. To see how this works, consider the algebra $A_2 = SU(3)$, where we have only two quantities to consider

$$\mathrm{Tr}\left(T^A\, T^B\right), \qquad \mathrm{Tr}\left(T^A\, T^B\, T^C\right).$$

Since there is only one invariant tensor with two adjoint indices, we set

$$\mathrm{Tr}\left(T^A\, T^B\right) \;\equiv\; I^{(2)}_{\mathbf{r}} \delta^{AB},$$

where $I^{(2)}_{\mathbf{r}}$ is the quadratic *Dynkin index* of the $\mathbf{r}$ representation.

The trace of the cube can be split into an antisymmetric third-rank tensor, which is nothing but the structure function f^{ABC}, while the totally symmetric part is written as

$$\mathrm{Tr}\left(T^A\, T^B\, T^C\right)_{Sym} = d^{ABC}\, I^{(3)}_{\mathbf{r}},$$

yields $I^{(3)}_{\mathbf{r}}$, the cubic Dynkin index. Here, d^{ABC} is a symmetric third-rank tensor made out of the structure functions. Supposing that T^A are matrices in the adjoint representation, a straightforward computation yields

$$d^{ABC} = f^{AMN}\, f^{BNP}\, f^{CPM},$$

normalized so that $I^{(3)} = 1$ for the adjoint representation. The cubic invariant exists only for those algebras with the property that the *symmetric* product of its adjoint contains the adjoint. This is true only for the $SU(n)$ ($n \geq 3$) algebras.

To relate the quadratic Dynkin index to the quadratic Casimir operator, take the trace over A, B

$$I^{(2)}_{\mathbf{r}} \delta^{AA} = \mathrm{Tr}\left(T^A\, T^A\right) = C^{(2)}_{\mathbf{r}}\, \mathrm{Tr}(1) = C^{(2)}_{\mathbf{r}}\, d_{\mathbf{r}}, \tag{8.63}$$

so that

$$I^{(2)}_{\mathbf{r}} = \frac{d_{\mathbf{r}}}{D}\, C^{(2)}_{\mathbf{r}}, \tag{8.64}$$

where D is the dimension of the algebra. A similar computation expresses the third degree Dynkin index in terms of the cubic Casimir operator.

By taking the trace of arbitrary products of representation matrices,

$$C^{A_1 \cdots A_k} \;\equiv\; \mathrm{Tr}\left(T^{A_1}\, T^{A_2}\, \cdots\, T^{A_k}\right),$$

we generate quantities which are expressed as invariant kth-degree Dynkin indices multiplied by the appropriate invariant tensors.

In general a rank n Lie algebra has n independent invariants which are polynomials in its generators. For the simple Lie algebras, the orders of the invariant polynomials are as follows.

$$A_n: \ 2, 3, \ldots, n+1$$
$$B_n: \ 2, 4, 6, \ldots, 2n-1$$
$$C_n: \ 2, 4, 6, \ldots, 2n-1$$
$$D_n: \ 2, 4, 6, \ldots, 2n-4, n-2$$
$$G_2: \ 2, 6$$
$$F_4: \ 2, 5, 8, 12$$
$$E_6: \ 2, 5, 6, 8, 9, 12$$
$$E_7: \ 2, 6, 8, 10, 12, 14, 18$$
$$E_8: \ 2, 8, 12, 14, 18, 20, 24, 30$$

The irreducible representations of most Lie algebras can be further characterized by a *congruence number*, with the important property that it is additive under Kronecker products, modulo an integer specified for each Lie algebra.

Also, we can define the Chang number of a representation which is equal to its dimension mod$(h+1)$, where h is the Coxeter number (sum of the marks).

It is very useful to develop a graphical representation for the T^A matrices in arbitrary representations, inspired by Feynman diagram techniques. Following P. Cvitanovič (*Bird Tracks, Lie's and Exceptional Groups*, Princeton University Press, 2008), we assign to each state of an arbitrary d-dimensional representation a solid "propagator" line, with labels $a, b = 1, 2, \ldots, d$. Special status is given to the adjoint representation in the form of a dotted line.

In this language, we associate a "propagator line" to the Kronecker δ_{ab} and δ_{AB}.

δ^a_b 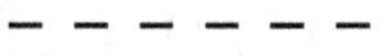 δ_{AB}

Closing the lines is the same as summing over the indices, so that a solid circle is d, the dimension of the representation.

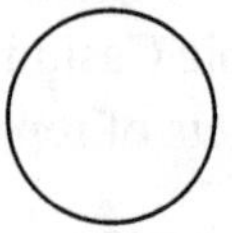

$= d$

The representation matrices in the d-dimensional representations are depicted as follows, respectively.

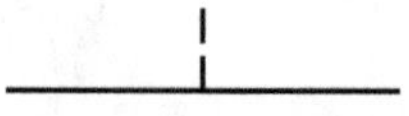

$= T^A_{ab}$

Tracelessness of these matrices is then shown as follows.

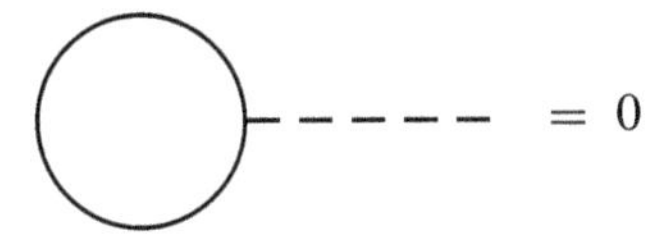

The second Dynkin index of the representation **r** is then represented by the following diagram.

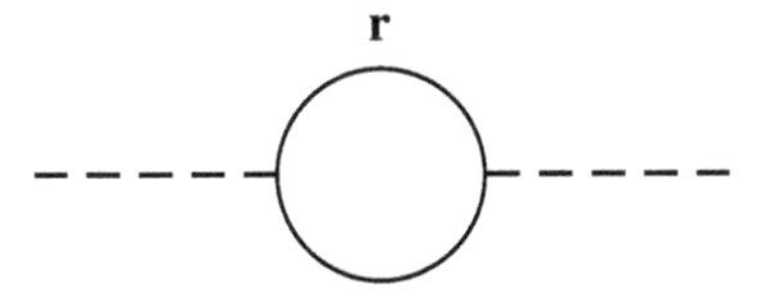

This makes it easy to develop a way to generate the quadratic Dynkin indices for any representation, using the Kronecker products. Label the reducible representation $\mathbf{r} \times \mathbf{s}$ by two solid lines,

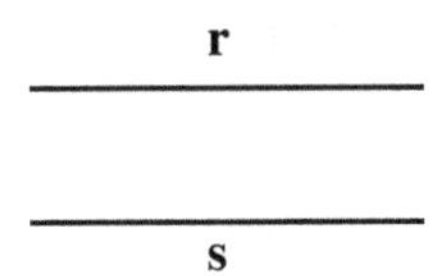

and compute its quadratic Dynkin index, the sum of the quadratic indices of the irreducible representations in $\mathbf{r} \times \mathbf{s}$. Since the representation matrices are traceless, they connect to the lines in pairs, leading to the sum of the following diagrams.

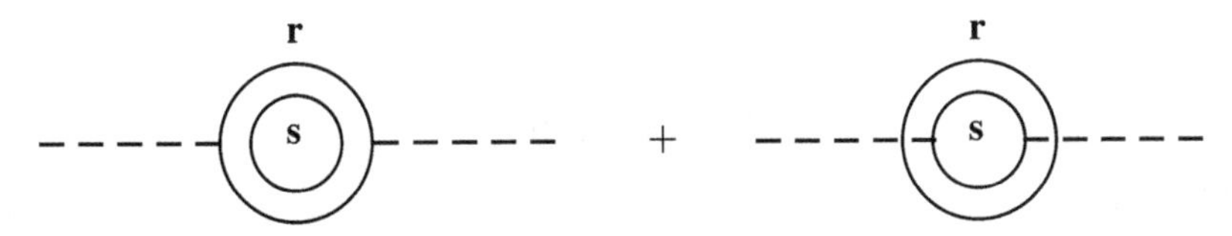

It follows that

$$I^{(2)}_{\mathbf{r}\times\mathbf{s}} = d_{\mathbf{s}}\, I^{(2)}_{\mathbf{r}} + d_{\mathbf{r}}\, I^{(2)}_{\mathbf{s}},$$

which gives us an easy way to compute its value for low-lying representations. When $\mathbf{r} = \mathbf{s}$, we can even specialize the rule diagrammatically by considering the symmetric and antisymmetric products in terms of the lines

and derive, just by counting diagrams, the following equation

$$I^{(2)}_{(\mathbf{r}\times\mathbf{r})_{s,a}} = (d_{\mathbf{r}} \pm 2)\, I^{(2)}_{\mathbf{r}}.$$

A similar reasoning can be applied to the cubic Dynkin index which arises in the following diagram.

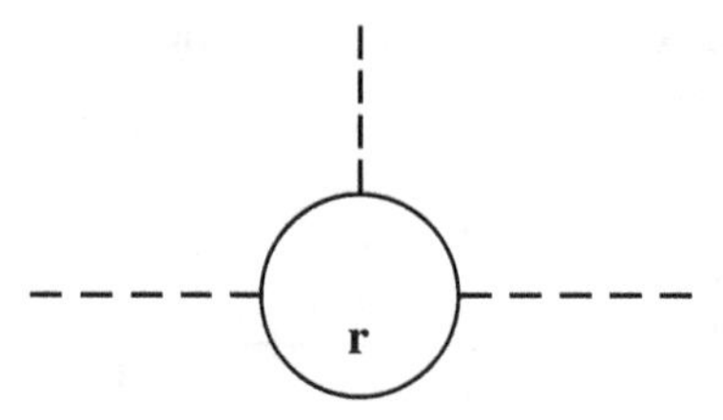

Using the same techniques we outlined for the quadratic Dynkin index, we derive the same composition law

$$I^{(3)}_{\mathbf{r}\times\mathbf{s}} = d_{\mathbf{s}}\, I^{(3)}_{\mathbf{r}} + d_{\mathbf{r}}\, I^{(3)}_{\mathbf{s}},$$

but the counting of possible diagrams yields for the symmetric and antisymmetric products a slightly different result

$$I^{(3)}_{(\mathbf{r}\times\mathbf{r})_{s,a}} = (d_{\mathbf{r}} \pm 4)\, I^{(3)}_{\mathbf{r}}.$$

The composition laws for the higher-order indices are complicated by the fact that two and more adjoint lines can hook to the same solid line, and we do not list them here.

The numerical values of the Dynkin indices for low-lying representations of choice Lie groups can be found in Appendix 2.

8.7 Embeddings

It is kind of obvious that small Lie algebras can live inside bigger ones: rotations in N dimensions contain rotations in $(N-1)$ dimensions, so we would write

$$SO(N) \supset SO(N-1).$$

This is an example of embedding. How can we systematize all embeddings inside an algebra of rank n? That is the subject of this section.

The analysis is based on two remarks. If a Lie algebra $\mathcal{H}$ lives inside another $\mathcal{G} \supset \mathcal{H}$, then

- the adjoint representation of $\mathcal{G}$ contains the adjoint representation of $\mathcal{H}$;
- the smallest representation of $\mathcal{G}$ is the sum of non-trivial representation(s) of $\mathcal{H}$.

This suggests a two-step method to catalog all embeddings: express the smallest irrep of the bigger algebra in terms of representation(s) of the smaller one; then take its Kronecker product to form the adjoint of the larger algebra and check if it contains the adjoint of the smaller one; if it does we have an embedding. We begin with several simple examples.

First example: inside $SU(3)$

This case is easy because there is only one simple algebra smaller than $SU(3)$: $SU(2) \sim SO(3)$ of dimension three and one semi-simple algebra, $SU(2) \times SU(2)$ with dimension six. Consider the defining fundamental **3**. First we note that there is not enough room for $SU(2) \times SU(2)$ in the triplet, since its smallest representation, $(\mathbf{2}, \mathbf{1}) + (\mathbf{1}, \mathbf{2})$, or $(\mathbf{2}, \mathbf{2})$ is four-dimensional. This leaves two ways to express the triplet in terms of representations of the $SU(2) \sim SO(3)$ algebras

$$\mathbf{3} = \mathbf{2} + \mathbf{1}; \qquad \mathbf{3} = \mathbf{3}.$$

Start with the first embedding $\mathbf{3} = \mathbf{2} + \mathbf{1}$, and form the adjoint

$$\begin{array}{ccc} \mathbf{3} \times \overline{\mathbf{3}} & = & (\mathbf{2} + \mathbf{1}) \times (\mathbf{2} + \mathbf{1}), \\ \| & & \| \\ \mathbf{8} + \mathbf{1} & = & \mathbf{3} + \mathbf{2} + \mathbf{2} + \mathbf{1} + \mathbf{1}. \end{array}$$

On the left-hand side are $SU(3)$ representations, and on the right-hand side are $SU(2)$ representations. Taking out a singlet on both sides, the $SU(3)$ adjoint is

$$\mathbf{8} = \mathbf{3} + \mathbf{2} + \mathbf{2} + \mathbf{1}.$$

We recognize the triplet as the $SU(2)$ adjoint, so this is a *bona fide* embedding, but there is more since the adjoint octet also contains a singlet, which generates a $U(1)$ Abelian algebra. Hence we really have

$$SU(3) \supset SU(2) \times U(1).$$

The $U(1)$ means that we can assign a charge to the members of the fundamental triplet, as long as they sum to zero. We indicate the value of the charge by a subscript, and choose an arbitrary normalization. The first embedding now reads

$$\mathbf{3} = \mathbf{2}_{-1} + \mathbf{1}_{2},$$

so that the charges in the adjoint are the sum of those in the fundamental

$$\mathbf{8} = \mathbf{3}_0 + \mathbf{2}_{+3} + \mathbf{2}_{-3} + \mathbf{1}_0.$$

The charge of the triplet is zero as it should be; otherwise the $U(1)$ would not commute with $SU(2)$.

Before leaving this embedding, we note that there are three ways to choose the $SU(2)$ singlet inside the triplet, leading to three equivalent embeddings, corresponding to the Weyl symmetry of the $SU(3)$ root diagram. These embeddings occur in elementary particle physics with $SU(2)$ as isospin and $SU(3)$ as the classification algebra of mesons and baryons, the eight-fold way of Gell-Mann and Ne'eman. The other $SU(2)$ are called U- and V-spins.

Now for the second embedding $\mathbf{3} = \mathbf{3}$, we form the adjoint

$$\begin{array}{ccc} \mathbf{3} \times \overline{\mathbf{3}} & = & \mathbf{3} \times \mathbf{3}, \\ \| & & \| \\ \mathbf{8} + \mathbf{1} & = & \mathbf{3} + \mathbf{5} + \mathbf{1}, \end{array}$$

so that

$$\mathbf{8} = \mathbf{3} + \mathbf{5}.$$

This time there are no singlets, and thus no $U(1)$. The embedding is just

$$SU(3) \supset SO(3),$$

written with $SO(3)$ rather than $SU(2)$ since only tensor representations appear. We saw that the Elliott model uses this embedding, where $SO(3)$ are space rotations, and $SU(3)$ classifies nuclear levels.

Second example: inside $SO(8)$

This is much less trivial because $SO(8)$ has rank 4 and also displays triality; its fundamental can be any one of the three real octets $\mathbf{8}_i$, where i stands for vector (v), and the two real spinors (s, s'). It does not matter which we start from, so we take the vector. The $\mathbf{28}$ adjoint is to be found in its antisymmetric product

$$(\mathbf{8}_v \times \mathbf{8}_v)_a = \mathbf{28}.$$

We consider all possible embeddings by listing the ways in which the vector representation can be expressed in terms of representations of algebras with rank equal or lower to 4. Simple algebras with representations smaller than eight are $SU(2) \sim SO(3) \sim Sp(2)$, $SU(3)$, $SU(4) \sim SO(6)$, $SO(5) \sim Sp(4)$, $SO(7)$, $Sp(6)$, and G_2. $SU(5)$ does not appear in the list although its fundamental quintet is small enough, but since the $\mathbf{8}_v$ is real, complex representations of sub-algebras must appear together with their conjugate. We can also have direct products of these algebras if their representations fit in the vector octet. We list the possibilities in terms of the partitions of eight into the largest integers, that is $8 = 8, 7+1, 6+2, 5+3, 4+4$, and identify these integers as possible representations of the sub-algebras. Then we form the adjoint and express it in terms of representations of the (assumed) sub-algebras.

(1) 8 = 8. There two possible sub-algebras with real octet representations, the $SU(3)$ adjoint octet,

$$SO(8) \supset SU(3)$$

$$\mathbf{8}_v = \mathbf{8}; \qquad \mathbf{28} = \mathbf{8} + \mathbf{10} + \overline{\mathbf{10}},$$

and the $SO(7)$ spinor.

$$SO(8) \supset SO(7)$$

$$\mathbf{8}_v = \mathbf{8}\,; \qquad \mathbf{28} = \mathbf{21} + \mathbf{7}.$$

These embeddings of $SU(3)$ and $SO(7)$ inside $SO(8)$ are peculiar as they seem to rely on numerical coincidences, the equality of the dimensions of the $SU(3)$ adjoint and the $SO(7)$ spinor to the $SO(8)$ vector representation.

(2) $8 = 7 + 1$. Since $SO(7)$ has a seven-dimensional representation, this case leads to

$$SO(8) \supset SO(7)$$

$$\mathbf{8}_v = \mathbf{7} + \mathbf{1}; \qquad \mathbf{28} = \mathbf{21} + \mathbf{7}.$$

Since the adjoint representation is the same as in the previous case, they must be the same embedding. This is explained by $SO(8)$ triality; after triality, one is the spinor, the other is the vector embedding.

(3) $8 = 7 + 1$. The seven can be interpreted as the **7** of the exceptional group G_2, leading to a perfectly good embedding with

$$SO(8) \supset G_2$$

$$\mathbf{8}_v = \mathbf{7} + \mathbf{1}; \qquad \mathbf{28} = \mathbf{14} + \mathbf{7} + \mathbf{7}.$$

By comparing with the previous case, we see that it also defines the embedding of G_2 inside $SO(7)$. This is not a maximal embedding as $SO(7)$ is in between $SO(8)$ and G_2 $(SO(8) \supset SO(7) \times G_2)$.

(4) $8 = 6 + 2$. We write the vector in the suggestive form

$$\mathbf{8}_v = \mathbf{6} + \mathbf{2} = (\mathbf{6},\ \mathbf{1}) + (\mathbf{1},\ \mathbf{2})$$

since it looks like $SO(6)$ and $SU(2)$. The adjoint is easily worked out

$$\mathbf{28} = (\mathbf{15},\ \mathbf{1}) + (\mathbf{1},\ \mathbf{1}) + (\mathbf{6},\ \mathbf{2})$$

but we see that, while it includes the $SU(4)$ adjoint, it does not contain the $SU(2)$ adjoint: there is no $SU(2)$ although there is a $U(1)$ factor corresponding to the singlet. This embedding is then

$$SO(8) \supset SU(4) \times U(1) \ \sim\ SO(6) \times SO(2).$$

Putting the $U(1)$ charges, we get

$$\mathbf{8}_v = \mathbf{6}_0 + \mathbf{1}_{+1} + \mathbf{1}_{-1}; \qquad \mathbf{28} = \mathbf{15}_0 + \mathbf{1}_0 + \mathbf{6}_{+1} + \mathbf{6}_{-1}.$$

(5) $8 = 5 + 3$. This looks like

$$\mathbf{8}_v = (\mathbf{5},\ \mathbf{1}) + (\mathbf{1},\ \mathbf{3}),$$

which yields the adjoint

$$\mathbf{28} = (\mathbf{10},\ \mathbf{1}) + (\mathbf{1},\ \mathbf{3}) + (\mathbf{5},\ \mathbf{3})$$

showing that we are dealing with

$$SO(8) \supset SO(5) \times SO(3)\ \sim\ Sp(4) \times Sp(2).$$

(6) $8 = 4 + 4$. The possible embedding

$$\mathbf{8}_v = (\mathbf{4},\ \mathbf{2})$$

can be interpreted either in terms of $SO(4) \times SU(2)$ or $Sp(4) \times Sp(2)$. To see which, we form the adjoint (the (a)sym subscript stands for the (anti)symmetrized product),

$$\begin{aligned} \mathbf{28} &= ([\mathbf{4} \times \mathbf{4}]_{sym},\ [\mathbf{2} \times \mathbf{2}]_a) + ([\mathbf{4} \times \mathbf{4}]_a,\ [\mathbf{2} \times \mathbf{2}]_{sym}), \\ &= ([\mathbf{4} \times \mathbf{4}]_{sym},\ \mathbf{1}) + ([\mathbf{4} \times \mathbf{4}]_a,\ \mathbf{3}). \end{aligned}$$

The adjoint of each sub-algebra must be a singlet of the other; but the $SO(4)$ adjoint is in the antisymmetric product which transforms as an $SU(2)$ triplet, and so cannot be $SO(4)$. On the other hand, the $Sp(4)$ adjoint lies in the symmetric product of its fundamental and the symmetric product contains a singlet. Hence we conclude that the embedding is

$$SO(8) \supset Sp(4) \times Sp(2)\ \sim\ SO(5) \times SO(3).$$

The adjoint representation is the same as that previously obtained; this is another example of triality as one embedding started from the vector, and the other (by triality) from the spinor. These two embeddings are equivalent.

(7) $8 = 4 + 4$. The decomposition

$$\mathbf{8}_v = (\mathbf{4},\ \mathbf{1}) + (\mathbf{1},\ \mathbf{4})$$

implies either $Sp(4) \times Sp(4)$ or $SO(4) \times SO(4)$. From the adjoint

$$\mathbf{28} = ([\mathbf{4} \times \mathbf{4}]_a,\ \mathbf{1}) + (\mathbf{1},\ [\mathbf{4} \times \mathbf{4}]_a) + (\mathbf{4},\ \mathbf{4}),$$

it is obvious that the embedding is

$$SO(8) \supset SO(4) \times SO(4),$$

since the $Sp(4)$ adjoint lies in the symmetric product of its fundamental, while the $SO(4)$ adjoint is to be found in the antisymmetric product.

(8) $8 = 4 + 4$. This can also be interpreted as

$$8_v = \mathbf{4} + \overline{\mathbf{4}},$$

leading to

$$\mathbf{28} = \mathbf{15} + \mathbf{1} + \mathbf{6} + \mathbf{6}.$$

It contains $SU(4)$ and a $U(1)$ factor, as indicated by the singlet. Hence the embedding

$$SO(8) \supset SU(4) \times U(1)$$

$$\mathbf{8}_v = \mathbf{4}_{+\frac{1}{2}} + \overline{\mathbf{4}}_{-\frac{1}{2}}; \qquad \mathbf{28} = \mathbf{15}_0 + \mathbf{1}_0 + \mathbf{6}_{+1} + \mathbf{6}_{-1}.$$

We have chosen the normalization of the $U(1)$ charges so as to get the same for the previously worked out (case (4)) embedding of $SU(4)$. This is yet another example of triality at work.

(9) $8 = 4 + 4$. This can be interpreted in terms of $SO(4)$ for the first four and two copies of one doublet for an additional $SU(2)$. We set

$$\mathbf{8}_v = (\mathbf{2},\ \mathbf{2},\ \mathbf{1})_0 + (\mathbf{1},\ \mathbf{1},\ \mathbf{2})_{+1} + (\mathbf{1},\ \mathbf{1},\ \mathbf{2})_{-1},$$

where we have anticipated a $U(1)$ charge since there are two copies of the doublet. The adjoint is then

$$\mathbf{28} = (\mathbf{3}, \mathbf{1}, \mathbf{1})_0 + (\mathbf{1}, \mathbf{3}, \mathbf{1})_0 + (\mathbf{1}, \mathbf{1}, \mathbf{3})_0 + (\mathbf{1}, \mathbf{1}, \mathbf{1})_0 + (\mathbf{2}, \mathbf{2}, \mathbf{2})_1 + (\mathbf{2}, \mathbf{2}, \mathbf{2})_{-1}.$$

It contains three $SU(2)$s and a $U(1)$ factor, as indicated by the singlet. This embedding is therefore

$$SO(8) \supset SU(2) \times SU(2) \times SU(2) \times U(1).$$

Finally we note that the possible decomposition

$$\mathbf{8}_v = (\mathbf{2},\ \mathbf{2},\ \mathbf{2}),$$

which looks like $SU(2) \times SU(2) \times SU(2)$, is not an embedding: when we form the adjoint,

$$\mathbf{28} = (\mathbf{1}, \mathbf{1}, \mathbf{1}) + (\mathbf{3}, \mathbf{3}, \mathbf{1}) + (\mathbf{1}, \mathbf{3}, \mathbf{3}) + (\mathbf{3}, \mathbf{1}, \mathbf{3}),$$

we do not find any adjoint $SU(2)$ (triplet). Rather, the $SU(2)$ adjoints are found in the *symmetric* product of two real octets, which means that this decomposition can be viewed as an embedding of the symplectic algebra $Sp(8)$

$$Sp(8) \supset SU(2) \times SU(2) \times SU(2)\,.$$

We can summarize the embeddings we have found in terms of rank, as shown below.

$SO(8)$ maximal embeddings	Remarks
$SU(2) \times SU(2) \times SU(2) \times SU(2)$	Semi-simple; rank 4
$SU(4) \times U(1)$	Not semi-simple; rank 3
$SU(2) \times SU(2) \times SU(2) \times U(1)$	Not semi-simple; rank 3
$SO(5) \times SO(3)$	Semi-simple; rank 3
$SO(7)$	Simple; rank 3
$SU(3)$	Simple; rank 2

We have derived the largest embeddings into $SU(3)$ and $SO(8)$ to provide examples and motivation for our method. Is there a systematic way to understand these embeddings, particularly in terms of Dynkin diagrams?

We note that by removing one dot from the $SO(8)$ Dynkin, we produce either $SU(4)$ or $SU(2) \times SU(2) \times SU(2)$; both of which appear on our list with an extra $U(1)$ factor. The same thing happens for the embeddings into $SU(3)$ as removing one dot yields $SU(2)$ which is embedded with a $U(1)$ factor as well. This leads us to the following rule.

- Start with the Dynkin diagram of an algebra $\mathcal{G}$ of rank n. Remove a dot from it, yielding the Dynkin of a smaller algebra $\mathcal{H}$ of rank $n-1$. This results in the *regular maximal non-semi-simple* embedding with a $U(1)$ factor

$$\mathcal{G} \supset \mathcal{H} \times U(1).$$

This accounts for two of the $SO(8)$ embeddings, but we also found one embedding of products of simple algebras without loss of rank. To account for this embedding, Dynkin invented the *extended Dynkin diagram* which has one *more* node, and then removed a node from it to obtain the Dynkin diagram of the sub-algebra! If the extra node is the negative of the highest root of the original algebra, the difference of that root with any other simple root is not a root, so that the removal of any node from the extended Dynkin diagram produces a Dynkin diagram of the same rank as the original algebra. Hence the second rule:

- Remove a dot from the extended Dynkin diagram of an algebra $\mathcal{G}$ of rank n to generate the Dynkin diagram of an algebra $\mathcal{H}$ of the same rank, resulting in the *regular maximal semi-simple* embedding

$$\mathcal{G} \supset \mathcal{H}.$$

Let us see how this works for $SO(8)$ by building its extended Dynkin diagram. We start from the $SO(8)$ Cartan matrix and its inverse

$$A_{D_4} = \begin{pmatrix} 2 & -1 & 0 & 0 \\ -1 & 2 & -1 & -1 \\ 0 & -1 & 2 & 0 \\ 0 & -1 & 0 & 2 \end{pmatrix}; \qquad A_{D_4}^{-1} = \frac{1}{2}\begin{pmatrix} 2 & 2 & 1 & 1 \\ 2 & 4 & 2 & 2 \\ 1 & 2 & 2 & 1 \\ 1 & 2 & 1 & 2 \end{pmatrix}.$$

In this numbering, the adjoint representation $(0\,1\,0\,0)$, corresponds to the second row of these matrices. It follows that the highest root, the highest weight of the adjoint representation, is read from the inverse Cartan matrix

$$\boldsymbol{\theta} = \boldsymbol{\alpha}_1 + 2\boldsymbol{\alpha}_2 + \boldsymbol{\alpha}_3 + \boldsymbol{\alpha}_4.$$

It is straightforward to compute its inner product with the simple roots to find

$$(\boldsymbol{\theta},\, \boldsymbol{\alpha}_1) = 0; \qquad (\boldsymbol{\theta},\, \boldsymbol{\alpha}_2) = 1,$$
$$(\boldsymbol{\theta},\, \boldsymbol{\alpha}_3) = 0; \qquad (\boldsymbol{\theta},\, \boldsymbol{\alpha}_4) = 0, \tag{8.65}$$

so that $\boldsymbol{\theta}$ has the same length as the other simple roots and is connected by one line to the second simple root. The negative of the highest root satisfies all the axioms to be part of a Dynkin diagram. This yields the $SO(8)$ *extended Dynkin diagram.*

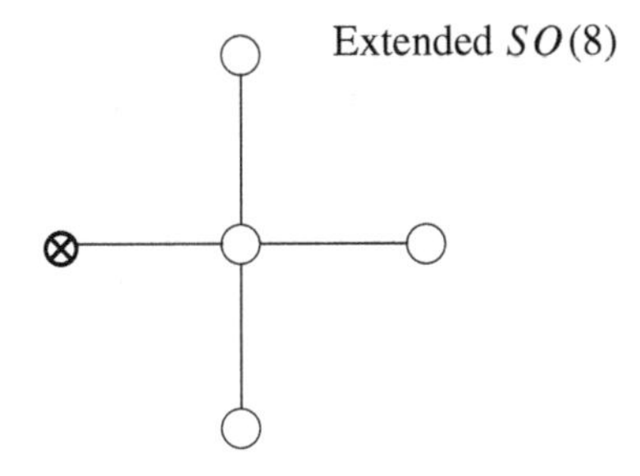

This diagram shows manifest quadrality, and a higher degree of symmetry. We can associate with it a (5×5) matrix

$$A_{D_4}^{(1)} = \begin{pmatrix} 2 & 0 & -1 & 0 & 0 \\ 0 & 2 & -1 & 0 & 0 \\ -1 & -1 & 2 & -1 & -1 \\ 0 & 0 & -1 & 2 & 0 \\ 0 & 0 & -1 & 0 & 2 \end{pmatrix}, \tag{8.66}$$

which satisfies all the properties to be a Cartan matrix, except that its determinant vanishes

$$\det A_{D_4}^{(1)} = 0,$$

so that it has one zero eigenvalue. We will see later how this matrix can be used to generate infinite-dimensional Kac–Moody algebras, but for the present we are more interested in using it to generate embeddings.

How do we know which dot to remove to obtain, say the maximal subgroup? There is a simple way to label each dot. The extended matrix has vanishing determinant, and one eigenvector with zero eigenvalue. Its components $\check{c}_i$ are the *co-marks*, which satisfy

$$\sum A^{(1)}_{ij} \check{c}_j = 0.$$

The *marks* c_i are the components of the zero eigenvector of the transposed matrix

$$\sum A^{(1)}_{ji} c_j = 0.$$

In this case, all roots are of equal length, and the marks and co-marks are the same, and equal to $(1, 1, 2, 1, 1)$.

One dot is singled out by this labeling. Removing the central dot from the extended diagram yields four unconnected nodes corresponding to $SU(2) \times SU(2) \times SU(2) \times SU(2)$, while removing any one of the outer nodes brings us back to the original diagram.

These two rules account for the so-called *regular maximal embeddings*. There are some notable exceptions where this rule with the extended Dynkin diagram does not yield maximal sub-algebras, all having to do with the exceptional algebras F_4, E_7 and E_8.

All other maximal embeddings are dubbed *special*. The three embeddings of $SO(5) \times SO(3)$, $SO(7)$, and $SU(3)$ are special. They seem to pop up because of apparent numerical coincidences relating dimensions of their representations. For instance, the embedding

$$SO(2n) \supset SO(2n-1)$$

is special because it relies on the numerical equality between the dimension of either the $SO(2n)$ spinor or the $SO(2n-1)$ spinor. The $SU(3)$ embedding is also special to $SO(8)$ because of the numerical equality with the $SU(3)$ adjoint. Finally,

$$SO(2n) \supset Sp(4) \times SO(2n-5)$$

is also a special embedding, because of the dual role of $Sp(4)$ as a symplectic and orthogonal algebra. It is clear that the special embeddings have to be worked out on a case by case basis, although there are also some general patterns as we have just seen.

As a final example, we construct the G_2 extended Dynkin. The G_2 Cartan matrix and its inverse are

$$A_{G_2} = \begin{pmatrix} 2 & -3 \\ -1 & 2 \end{pmatrix}; \qquad A_{G_2}^{-1} = \begin{pmatrix} 2 & 3 \\ 1 & 2 \end{pmatrix}.$$

In this numbering, the long root is in the first Dynkin label, so the adjoint $(1\,0)$ corresponds to the first row. Hence the highest root

$$\boldsymbol{\theta} = 2\boldsymbol{\alpha}_1 + 3\boldsymbol{\alpha}_2.$$

Since

$$(\boldsymbol{\theta},\, \boldsymbol{\alpha}_1) = 1; \qquad (\boldsymbol{\theta},\, \boldsymbol{\alpha}_2) = 0,$$

the negative highest root is a long root connected to the $\boldsymbol{\alpha}_1$ node by one line. This leads to the extended Dynkin diagram for G_2.

Extended G_2

From it we obtain the extended Cartan matrix

$$A_{G_2}^{(1)} = \begin{pmatrix} 2 & -1 & 0 \\ -1 & 2 & -3 \\ 0 & -1 & 2 \end{pmatrix},$$

whose determinant vanishes. The marks are $(1, 2, 3)$, and the co-marks are different and equal to $(1, 2, 1)$. The co-marks single out one node, and its removal produces the maximal embedding. Removal of the short node yields the $SU(3)$ Dynkin, with embedding $G_2 \supset SU(3)$:

$$\mathbf{7} = \mathbf{3} + \overline{\mathbf{3}} + \mathbf{1}; \qquad \mathbf{14} = \mathbf{8} + \mathbf{3} + \overline{\mathbf{3}}.$$

Removal of the middle node yields the Dynkin diagram of $SU(2) \times SU(2)$, that is the embedding $G_2 \supset SU(2) \times SU(2)$

$$\mathbf{7} = (\mathbf{2},\, \mathbf{2}) + (\mathbf{1},\, \mathbf{3})\,; \qquad \mathbf{14} = (\mathbf{1},\, \mathbf{3}) + (\mathbf{3},\, \mathbf{1}) + (\mathbf{2},\, \mathbf{4})\,.$$

Removal of the extended node yields of course the G_2 Dynkin, resulting in the trivial embedding. As we noted for $SO(8)$, the extended Dynkin yields only the maximal embedding without loss of rank. There is for example the anomalous embedding $G_2 \supset SU(2)$, with $\mathbf{7} = \mathbf{7}$.

The extended Dynkin diagrams can be constructed for any of the simple Lie algebras. In each case, the resulting extended Cartan matrix has one zero eigenvalue, since one row is linearly dependent.

8.8 Oscillator representations

The representation theory of Lie algebras is best understood in terms of the *defining* representation of the algebra, which generates all others by direct product. For example, the three fundamental representations of $SU(4)$, $(1\,0\,0) \equiv \mathbf{4}$, $(0\,1\,0) \equiv \mathbf{6}$, and $(0\,0\,1) \equiv \overline{\mathbf{4}}$ are generated by **4** since the other two can be obtained by the antisymmetric Kronecker products

$$(\mathbf{4}\times\mathbf{4})_a = \mathbf{6}, \qquad (\mathbf{4}\times\mathbf{4}\times\mathbf{4})_a = \overline{\mathbf{4}},$$

so that **4** is the defining representation.

The antisymmetrization means that we need *three* copies of the fundamental weights to generate all others. Note that, in $SO(6)$ language, the spinor representation is defining, not the vector: one can build a vector out of two spinors but not vice versa. This is a general feature of orthogonal groups, and for all Lie algebras, the defining representation is not necessarily the one with the smallest dimension, as the spinor is larger than the vector for $SO(n)$ with $n \geq 9$. The defining representation resides at the end of the Dynkin diagram.

We have already seen how Schwinger constructed all the $SU(2)$ representations, starting with creation and annihilation operators that transform as the **2** of $SU(2)$. His construction can be generalized, but with some important distinctions.

We start with four creation and annihilation operators, $A^\dagger_a$ and A^b, which satisfy the usual commutation relations

$$[\,A^b,\; A^\dagger_a\,] = \delta^b{}_a, \qquad a, b = 1, 2, 3, 4. \tag{8.67}$$

The $SU(4)$ generators are

$$X_a{}^b = i\,A^\dagger_a\,A^b, \tag{8.68}$$

so that the $A^\dagger_a$ transform like the defining quartet representation. The highest weight state of the quartet is then expressed as

$$A^\dagger_1\,|\,0\,\rangle,$$

and repeated application of this operator on the vacuum state

$$\left(A^\dagger_1\right)^a|\,0\,\rangle,$$

generate all irreducible representations of type $(a\,0\,0)$.

So far this is Schwinger's construction, but one type of harmonic oscillator is not sufficient to generate the whole $SU(4)$ lattice, which contains states of the form $(0\,b\,0)$, or $(0\,0\,c)$, obtained by antisymmetrizing the defining representation.

Two more copies of the quartet oscillators are required to generate *all* irreducible representations. For example, the **6** is made up of states of the form

$$\left(A^{(1)\dagger}_a\,A^{(2)\dagger}_b - A^{(1)\dagger}_b\,A^{(2)\dagger}_a\right)|\,0\,\rangle,$$

while the $\overline{\mathbf{4}}$ is made up of three different oscillators

$$\epsilon^{abcd}\,A^{(1)\dagger}_a\,A^{(2)\dagger}_b\,A^{(3)\dagger}_c|\,0\,\rangle,$$

using the Levi–Civita symbol. The Hilbert space generated by these three sets of oscillators contains *all* the $SU(4)$ irreducible representations, although it is no easy task to separate them out. Alternatively, one could have introduced one set of oscillators for each of the fundamental representations with the same result; either way, the resulting Hilbert space is large and redundant, and to be useful it must be organized into irreducible blocks of the algebra.

This oscillator representation of the group lattice can be generalized for any simple Lie algebra (T. Fulton, *J. Phys. A: Math. Gen.* **18**, 2863 (1985)).

A non-trivial example is the construction of all representations of the exceptional group F_4. We need three sets of oscillators transforming as **26**. Each copy of 26 oscillators is decomposed in terms of the $SO(9)$ subgroup as one singlet $A^{[\kappa]}_0$, nine vectors $A^{[\kappa]}_i$, $i = 1,\ldots,9$ and sixteen $SO(9)$ spinors $B^{[\kappa]}_a$, $a = 1,\ldots,16$, and their hermitian conjugates, and where $\kappa = 1,2,3$. They satisfy the commutation relations of ordinary harmonic oscillators

$$[\,A^{[\kappa]}_i,\,A^{[\kappa']\dagger}_j\,] = \delta_{ij}\,\delta^{\kappa\,\kappa'},\quad [\,A^{[\kappa]}_0,\,A^{[\kappa']\dagger}_0\,] = \delta^{\kappa\kappa'},\quad [\,B^{[\kappa]}_a,\,B^{[\kappa']\dagger}_b\,] = \delta_{ab}\,\delta^{\kappa\kappa'},$$

and commute with one another. The generators T_{ij} and T_a

$$T_{ij} = -i\sum_{\kappa=1}^{3}\left\{\left(A^{[\kappa]\dagger}_i A^{[\kappa]}_j - A^{[\kappa]\dagger}_j A^{[\kappa]}_i\right) + \frac{1}{2}\,B^{[\kappa]\dagger}\,\gamma_{ij}B^{[\kappa]}\right\},$$

$$T_a = -\frac{i}{2}\sum_{\kappa=1}^{3}\left\{(\gamma_i)^{ab}\left(A^{[\kappa]\dagger}_i B^{[\kappa]}_b - B^{[\kappa]\dagger}_b A^{[\kappa]}_i\right) - \sqrt{3}\left(B^{[\kappa]\dagger}_a A^{[\kappa]}_0 - A^{[\kappa]\dagger}_0 B^{[\kappa]}_a\right)\right\},$$

satisfy the F_4 algebra,

$$[\,T_{ij},T_{kl}\,] = -i\,(\delta_{jk}\,T_{il} + \delta_{il}\,T_{jk} - \delta_{ik}\,T_{jl} - \delta_{jl}\,T_{ik}),$$

$$[\,T_{ij},T_a\,] = \frac{i}{2}\,(\gamma_{ij})_{ab}\,T_b,$$

$$[\,T_a,T_b\,] = \frac{i}{2}\,(\gamma_{ij})_{ab}\,T_{ij}.$$

The structure constants are given by

$$f_{ij\,ab} = f_{ab\,ij} = \frac{1}{2}\,(\gamma_{ij})_{ab}.$$

The reader is encouraged to verify the commutation relations, with the use of the identity

$$3\,\delta^{ac}\delta^{db} + (\gamma^i)^{ac}\,(\gamma^i)^{db} - (a \leftrightarrow b) = \frac{1}{4}\,(\gamma^{ij})^{ab}\,(\gamma^{ij})^{cd},$$

satisfied by the (16×16) Dirac matrices. Similar constructions can be found for the exceptional groups, through the use of their orthogonal subgroups.

8.9 Verma modules

We have stated that irreducible representations of simple groups are characterized by their highest weight state, the state annihilated by all creation (raising) operators, and also have outlined their construction through the action of lowering operators. Physicists, who usually deal with low-lying representations, do not concern themselves with the construction of arbitrarily large representations. In this section, we introduce some mathematical machinery which is useful in building them.

While informal, the following presentation relies on the book by James E. Humphreys [11], and S. Sternberg's unpublished notes on Lie Algebras.

Algebras contain at least as many lowering operators as their rank, and the states obtained by their repeated application on the highest weight state form a highly redundant set, from which it is not a priori obvious how to single out the irreducible representations. Yet, this is the vehicle used by mathematicians.

A useful concept in this extraction is the *Verma module* $\mathrm{Verm}(\mathbf{\Lambda})$: the infinite number of states obtained by repeated application lowering operators to a state of highest weight (living in or on the walls of the Weyl chamber) $|\,\mathbf{\Lambda}\rangle$.

To see how it works, consider the $SU(2)$ Verma module associated with the highest weight state with Dynkin label (twice the magnetic quantum number) $|\,2\,\rangle$. Since $SU(2)$ has only one lowering operator T_-, the full Verma module is generated by its repeated application on $|\,2\,\rangle$. In this way we obtain the infinite set of states

$$\mathrm{Verm}(2): \quad \{\,|\,2\,\rangle,\ |\,0\,\rangle,\ |\,-2\,\rangle, |\,-4\,\rangle, \cdots\}, \tag{8.69}$$

where

$$|\,2-2k\,\rangle = (T_-)^k\,|\,2\,\rangle. \tag{8.70}$$

We are free to apply the raising operator to any one of these, and we find from the algebra that the state $|\,-4\,\rangle$ is annihilated by T_+

$$T_+|\,-4\,\rangle = T_+(T_-)^3\,|\,2\,\rangle = 0. \tag{8.71}$$

Thus $(T_-)^3|\,2\,\rangle$ is a highest weight state. The first three states form the three-dimensional irrep, and we have decomposed the Verma module into an irreducible representation and another Verma module.

$$\text{Verm}(2) = \{|\,2\,\rangle, |\,0\,\rangle, |\,-2\,\rangle\} \oplus \text{Verm}(-4). \tag{8.72}$$

We can then continue the decomposition of Verm(-4). It does not contain any more highest weight states, since

$$T_+|\,2-2k\,\rangle = [\,T_+,\ (T_-)^k\,]|\,2\,\rangle = k(3-k)|\,1\,\rangle, \tag{8.73}$$

vanishes only for $k = 0,\ 3$. Thus, if we take away the first three states, we obtain another Verma module with a state of highest weight with a negative Dynkin index, which contains no other state of highest weight. What do the states $|\,2\,\rangle$ and $|\,-4\,\rangle$ have in common?

This is a general feature: a Verma module with highest weight state in or on the walls of the Weyl chamber contains one finite-dimensional representation, and the rest assemble into infinite-dimensional Verma modules, each of which starts with a state outside the Weyl chamber.

In general things are not so simple: consider a rank r algebra, and form its Verma module associated with the highest weight state $|\,\mathbf{\Lambda}\,\rangle = |\,a_1\,a_2\,\cdots\,a_r\,\rangle$ in or on the walls of the Weyl chamber (all a_i positive or zero), that is

$$T_+^{(i)}\,|\,\mathbf{\Lambda}\,\rangle = 0, \qquad i = 1, 2, \ldots, r. \tag{8.74}$$

The Verma module Verm(Λ) is the collection of states

$$\left[T_-^{(i_1)}\right]^{n_1}\left[T_-^{(i_2)}\right]^{n_j}\cdots\left[T_-^{(i_n)}\right]^{n_r}|\,\Lambda\,\rangle, \tag{8.75}$$

for all values of $i_m = 1, 2, \ldots, r$, and arbitrary integers n_j. Among this infinite set of states lie the irreducible representation labelled by $\mathbf{\Lambda}$, and more. The systematic decomposition of the Verma module is what makes this concept useful.

Since the effect of $T_-^{(i)}$ is to subtract the simple root $\boldsymbol{\alpha}_i$, the Verma module consists of states of the form

$$|\,\boldsymbol{\mu}\,\rangle = |\,\mathbf{\Lambda} - \sum n_i\,\boldsymbol{\alpha}_i\,\rangle, \tag{8.76}$$

where the n_i are either zero or positive integers. In addition, since the lowering operators associated with the simple roots do not operate in any preassigned order, their commutator produces all the lowering operators associated with the positive roots. For instance, in $SU(4)$, the three lowering operators generate by commutations three more lowerers. As a result, there are many different ways to obtain a particular weight. The multiplicity of a weight $\boldsymbol{\mu}$ in Verm($\mathbf{\Lambda}$), $N^{\mathbf{\Lambda}}(\mu)$, is equal to the number of ways $\mathbf{\Lambda} - \boldsymbol{\mu}$ can be expressed as a sum of the positive roots with positive coefficients.

In order to determine this multiplicity, it is convenient to introduce an operator which does this counting. First, let us associate to any state $| \boldsymbol{\mu} \rangle$, an operator $e(\boldsymbol{\mu})$ which satisfies an exponential-like multiplication rule

$$e(\boldsymbol{\lambda}) \cdot e(\boldsymbol{\mu}) = e(\boldsymbol{\lambda} + \boldsymbol{\mu}), \tag{8.77}$$

for any two weights $\boldsymbol{\lambda}$ and $\boldsymbol{\mu}$. Introduce the Kostant counter

$$\mathcal{C} \equiv \prod_{\text{pos. roots}} \frac{1}{1 - e(\boldsymbol{\alpha})}, \tag{8.78}$$

where the product is over the positive roots. For $SU(3)$ it is the product of three terms

$$\frac{1}{1 - e(-\boldsymbol{\alpha}_1)} \frac{1}{1 - e(-\boldsymbol{\alpha}_2)} \frac{1}{1 - e(-\boldsymbol{\alpha}_1 - \boldsymbol{\alpha}_2)}. \tag{8.79}$$

For a Lie algebra of rank r and dimension D, it will be the product of $(N - r)/2$ terms. Expand each product as a power series

$$\frac{1}{1 - e(-\boldsymbol{\alpha}_1)} = 1 + e(-\boldsymbol{\alpha}_1) + e(-2\boldsymbol{\alpha}_1) + \cdots , \tag{8.80}$$

using the multiplication rule (8.77) and then multiply them together. A little algebra yields

$$\mathcal{C} = 1 + \sum_{\boldsymbol{\mu}} N(\boldsymbol{\mu})\, e(\boldsymbol{\mu}), \tag{8.81}$$

where the sum is over the weights of the form

$$\boldsymbol{\mu} = \sum n_i\, \boldsymbol{\alpha}_i, \tag{8.82}$$

with n_i positive integers or zero, and $N(\boldsymbol{\mu})$ simply counts the *sets* of sums over positive roots that yields $\boldsymbol{\mu}$. In $SU(3)$,

$$\mathcal{C} = 1 + e(-\boldsymbol{\alpha}_1) + e(-\boldsymbol{\alpha}_2) + 2\, e(-\boldsymbol{\alpha}_1 - \boldsymbol{\alpha}_2) + \cdots , \tag{8.83}$$

since there are two linear combinations of positive roots that yields that combination: one is by adding two roots $\boldsymbol{\alpha}_1$ and $\boldsymbol{\alpha}_2$ and the other is the positive root $\boldsymbol{\alpha}_1 + \boldsymbol{\alpha}_2$, and $\mathcal{C}$ is indeed a counting operator. It is closely related to the Weyl function

$$\mathcal{Q} \equiv \prod_{\text{pos. roots}} \Big(e(\boldsymbol{\alpha}/2) - e(-\boldsymbol{\alpha}/2) \Big), \tag{8.84}$$

since

$$\begin{aligned}\mathcal{Q} &= \prod_{\text{pos. roots}} e(\boldsymbol{\alpha}/2)\left(1 - e(-\boldsymbol{\alpha})\right) \\ &= e(\boldsymbol{\rho}) \prod_{\text{pos. roots}} \left(1 - e(-\boldsymbol{\alpha})\right) \\ &= e(\boldsymbol{\rho})\, \mathcal{C}^{-1},\end{aligned} \tag{8.85}$$

where $\boldsymbol{\rho}$ is the Weyl vector, half the sum of the positive roots, which we introduced earlier.

The *formal character* over the Verma module

$$\mathrm{Ch}_{\mathrm{Verm}(\boldsymbol{\Lambda})} \equiv \sum_{\boldsymbol{\mu}} N^{\boldsymbol{\Lambda}}(\boldsymbol{\mu})\, e(\boldsymbol{\mu}), \tag{8.86}$$

with the sum over all the weights in the module, is closely related to the counter

$$\mathrm{Ch}_{\mathrm{Verm}(\boldsymbol{\Lambda})} = \mathcal{C}\, e(\boldsymbol{\Lambda}). \tag{8.87}$$

We also have

$$\mathcal{Q}\, \mathrm{Ch}_{\mathrm{Verm}(\boldsymbol{\Lambda})} = \mathcal{Q}\mathcal{C}\, e(\boldsymbol{\Lambda}) = e(\boldsymbol{\Lambda} + \boldsymbol{\rho}). \tag{8.88}$$

Further understanding is gained by splitting up Verma modules into irreducible components. We have seen that our sample $SU(2)$ Verma module contains a state annihilated by creation operators. In order to identify such states in a general Verm(Λ), consider the quadratic Casimir operator

$$C_2 \equiv \sum_A T^A_{\mathbf{r}}\, T^A_{\mathbf{r}},$$

which commutes with all elements of the Lie algebra,

$$[\, C_2,\ T^A_{\mathbf{r}}\ ] = 0. \tag{8.89}$$

We rewrite it in the root notation as

$$C_2 \equiv g_{ij}\, H^i\, H^j + \sum_{\text{roots } \boldsymbol{\alpha}} E_{\boldsymbol{\alpha}}\, E_{-\boldsymbol{\alpha}}, \tag{8.90}$$

where the sum is over all non-zero roots. Its value on a highest weight state $|\, \boldsymbol{\Lambda}\, \rangle$ is particularly simple. Split the sum over positive and negative roots, and use for the positive roots

$$E_{\boldsymbol{\alpha}}\, |\, \boldsymbol{\Lambda}\, \rangle = 0,$$

to find

$$C_2 \mid \mathbf{\Lambda} \rangle = \Big(g_{ij}\, H^i\, H^j + \sum_{\text{pos. roots}} E_\alpha\, E_{-\alpha} \Big) \mid \mathbf{\Lambda} \rangle,$$

$$= \Big(g_{ij}\, H^i\, H^j + \sum_{\text{pos. roots}} [\, E_\alpha,\, E_{-\alpha}\,] \Big) \mid \mathbf{\Lambda} \rangle. \tag{8.91}$$

Using the commutation relations, we obtain

$$C_2 \mid \mathbf{\Lambda} \rangle = \Big(g_{ij}\, H^i\, H^j + \sum_{\text{pos. roots}} \alpha_i\, H^i \Big) \mid \mathbf{\Lambda} \rangle, \tag{8.92}$$

which can be rewritten as

$$C_2 \mid \mathbf{\Lambda} \rangle = \Big[(\, \mathbf{\Lambda} + \boldsymbol{\rho},\, \mathbf{\Lambda} + \boldsymbol{\rho}\,) - (\, \boldsymbol{\rho},\, \boldsymbol{\rho}\,) \Big] \mid \mathbf{\Lambda} \rangle, \tag{8.93}$$

by completing the squares, and using the inner product defined by the Killing form. Since the Casimir operator commutes with all generators, its value is the same on any state of the Verma module.

$$C_2 \mid \boldsymbol{\mu} \rangle = \Big[(\, \mathbf{\Lambda} + \boldsymbol{\rho},\, \mathbf{\Lambda} + \boldsymbol{\rho}\,) - (\, \boldsymbol{\rho},\, \boldsymbol{\rho}\,) \Big] \mid \boldsymbol{\mu} \rangle, \tag{8.94}$$

for $\mid \boldsymbol{\mu} \rangle$ in Verm($\mathbf{\Lambda}$). Furthermore, if $\boldsymbol{\mu}_k$ is a highest weight state in Verm($\mathbf{\Lambda}$), the above reasoning shows that also

$$C_2 \mid \boldsymbol{\mu}_k \rangle = \Big[(\, \boldsymbol{\mu}_k + \boldsymbol{\rho},\, \boldsymbol{\mu}_k + \boldsymbol{\rho}\,) - (\, \boldsymbol{\rho},\, \boldsymbol{\rho}\,) \Big] \mid \boldsymbol{\mu}_k \rangle, \tag{8.95}$$

so that the highest weight states in Verm($\mathbf{\Lambda}$) are those which satisfy

$$(\, \mathbf{\Lambda} + \boldsymbol{\rho},\, \mathbf{\Lambda} + \boldsymbol{\rho}\,) = (\, \boldsymbol{\mu}_k + \boldsymbol{\rho},\, \boldsymbol{\mu}_k + \boldsymbol{\rho}\,). \tag{8.96}$$

Since they have the same length, this suggests that $\mathbf{\Lambda} + \boldsymbol{\rho}$ and $\boldsymbol{\mu}_k + \boldsymbol{\rho}$ are related by Weyl transformations; in fact they are. Using our sample $SU(2)$ Verma module, with $\boldsymbol{\rho} = 1$, the two highest weight states yield $\mid 2 + 1 \rangle = \mid 3 \rangle$ and $\mid -4 + 1 \rangle = \mid -3 \rangle$, which are indeed on the same Weyl orbit.

This singles out a finite set of weight states $\boldsymbol{\mu}_k$, $k = 1, 2 \ldots n$, each standing at the top of an *irreducible module*, Irr($\boldsymbol{\mu}_k$), which contains no other state annihilated by the creation operators. Again in our $SU(2)$ example, there are two irreducible modules; one is finite-dimensional, with its highest weight in the Weyl chamber, the other is infinite-dimensional with its highest weight outside the Weyl chamber.

Verma modules split up as finite sums of irreducible modules; all are infinite-dimensional except for Irr($\mathbf{\Lambda}$), which is the finite-dimensional irrep of the Lie algebra.

The next step is to define a Verma module for each $\boldsymbol{\mu}_k$, and express it in terms of irreducible modules contained within. To that effect, we order the weights $\boldsymbol{\mu}_k$ inside Verm($\mathbf{\Lambda}$) so that $\boldsymbol{\mu}_1$ is the "largest" weight obtained by subtracting positive roots, $\boldsymbol{\mu}_2$ the second largest, etc. This enables us to write a relation among the Verma and irreducible characters

$$\mathrm{Ch}_{\mathrm{Verm}(\boldsymbol{\mu}_k)} = \mathrm{Ch}_{\mathrm{Irr}(\boldsymbol{\mu}_k)} + \sum_{j<k} a_j^{[\boldsymbol{\mu}_k]}\, \mathrm{Ch}_{\mathrm{Irr}(\boldsymbol{\mu}_j)}, \tag{8.97}$$

where the character of the irreducible module, labeled by the highest weight $\boldsymbol{\mu}_k$, is given by

$$\mathrm{Ch}_{\mathrm{Irr}(\boldsymbol{\mu}_k)} = \sum_{\boldsymbol{\lambda}} n^{[k]}(\boldsymbol{\lambda})\, e(\boldsymbol{\lambda}), \tag{8.98}$$

with the sum over the $n^{[k]}(\boldsymbol{\lambda})$ states of weight $\boldsymbol{\lambda}$ in the irreducible module.

Alternatively, we can invert this equation and express the character of the irreducible representations in terms of the character of the Verma module of highest weights which satisfy eq. (8.96). In so doing, we should be able to write the character of any irreducible representation in terms of those of the Verma modules associated with the weights which satisfy eq. (8.96)

$$\mathrm{Ch}_{\mathrm{Irr}(\boldsymbol{\Lambda})} = \mathrm{Ch}_{\mathrm{Verm}(\boldsymbol{\Lambda})} + \sum_{k=1} a_k^{[\boldsymbol{\Lambda}]}\, \mathrm{Ch}_{\mathrm{Verm}(\boldsymbol{\mu}_k)}, \tag{8.99}$$

where of course $\boldsymbol{\mu}_k < \boldsymbol{\Lambda}$. Therefore, using eq. (8.88),

$$\mathcal{Q}\,\mathrm{Ch}_{\mathrm{Irr}(\boldsymbol{\Lambda})} = e(\boldsymbol{\Lambda} + \boldsymbol{\rho}) + \sum_{k=1} a_k^{[\boldsymbol{\Lambda}]}\, e(\boldsymbol{\mu}_k + \boldsymbol{\rho}). \tag{8.100}$$

Further simplification is achieved by considering the effect of a Weyl transformation on this equation, as the character is invariant under Weyl transformations since irreducible transformations are a collection of weights on Weyl orbits.

In addition, the Weyl function transforms simply under the Weyl group: a simple reflection r_i will permute positive roots into other positive roots, except for the simple root $\boldsymbol{\alpha}_i$ which goes into its negative, that is

$$r_i\colon\ \Big[e(\boldsymbol{\alpha}_i/2) - e(-\boldsymbol{\alpha}_i/2)\Big] \ \rightarrow\ -\Big[e(\boldsymbol{\alpha}_i/2) - e(-\boldsymbol{\alpha}_i/2)\Big],$$

and all others in the product get permuted into one another. It follows that

$$r_i\colon\ \mathcal{Q} \rightarrow\ -\mathcal{Q}, \tag{8.101}$$

so that for any element w of the Weyl group

$$w\colon\ \mathcal{Q} \rightarrow\ (-1)^{l(w)}\,\mathcal{Q}, \tag{8.102}$$

where $l(w)$ is the number of reflections in the Weyl group element. Hence

$$\mathcal{Q}\,\mathrm{Ch}_{\mathrm{Irr}(\boldsymbol{\Lambda})} = (-1)^{l(w)}\left(e(w[\boldsymbol{\Lambda} + \boldsymbol{\rho}]) + \sum_{k} a_k^{[\boldsymbol{\Lambda}]}\, e(w[\boldsymbol{\mu}_k + \boldsymbol{\rho}])\right), \tag{8.103}$$

where $w[\boldsymbol{\mu}_k + \boldsymbol{\rho}]$ is the Weyl image of $\boldsymbol{\mu}_k + \boldsymbol{\rho}$. Thus, for any $w \in W$,

$$\begin{aligned} \mathcal{Q}\,\mathrm{Ch}_{\mathrm{Irr}(\boldsymbol{\Lambda})} &= e(\boldsymbol{\Lambda} + \boldsymbol{\rho}) + \sum a_k^{[\boldsymbol{\Lambda}]}\, e(\boldsymbol{\mu}_k + \boldsymbol{\rho}), \\ &= (-1)^{l(w)} \Big(e(w[\boldsymbol{\Lambda} + \boldsymbol{\rho}]) + \sum a_k^{[\boldsymbol{\Lambda}]}\, e(w[\boldsymbol{\mu}_k + \boldsymbol{\rho}]) \Big), \end{aligned} \tag{8.104}$$

which holds for *all* elements of the Weyl group. This implies that $e(\boldsymbol{\mu}_k + \boldsymbol{\rho}) \sim e(w[\boldsymbol{\Lambda} + \boldsymbol{\rho})])$ for some Weyl group element. It follows that

$$\mathcal{Q}\,\mathrm{Ch}_{\mathrm{Irr}(\boldsymbol{\Lambda})} = \sum_{w \in W} (-1)^{l(w)}\, e(w[\boldsymbol{\Lambda} + \boldsymbol{\rho}]), \tag{8.105}$$

as long as that *all* weights of the form $|\, \boldsymbol{\mu}_k + \boldsymbol{\rho}\, \rangle$ are on the Weyl orbit of $\boldsymbol{\Lambda} + \boldsymbol{\rho}$.

To see that it is indeed the case, assume the existence of a state, say $|\, \boldsymbol{\mu}_m + \boldsymbol{\rho}\, \rangle$, which is *not* on the orbit of $\boldsymbol{\Lambda} + \boldsymbol{\rho}$. By repeated application of Weyl transformations, we arrive at a new state in the Weyl chamber or on its wall. This would imply that both this state and $|\, \boldsymbol{\Lambda} + \boldsymbol{\rho}\, \rangle$ live in the Weyl chamber.

It is easy to show that with the same length, and differing by a sum of positive roots with positive coefficients, they must be equal to one another. Let $\boldsymbol{\delta} = \boldsymbol{\Lambda} - \boldsymbol{\mu}_m$ be their difference, and write

$$\begin{aligned} 0 &= (\, \boldsymbol{\Lambda} + \boldsymbol{\rho},\ \boldsymbol{\Lambda} + \boldsymbol{\rho}\,) - (\, \boldsymbol{\mu}_m + \boldsymbol{\rho},\ \boldsymbol{\mu}_m + \boldsymbol{\rho}\,), \\ &= (\, \boldsymbol{\Lambda} + \boldsymbol{\rho},\ \boldsymbol{\delta}\,) + (\, \boldsymbol{\delta},\ \boldsymbol{\mu}_m + \boldsymbol{\rho}\,), \\ &\geq (\, \boldsymbol{\Lambda} + \boldsymbol{\rho},\ \boldsymbol{\delta}\,) \geq 0, \end{aligned}$$

since it is assumed to be in the Weyl chamber $(\, \boldsymbol{\delta},\ \boldsymbol{\mu}_m + \boldsymbol{\rho}\,) \geq 0$. Hence, it can only be zero and $\boldsymbol{\mu}_m = \mathbf{0}$. The two states are the same and eq. (8.105) is exact. For $\boldsymbol{\Lambda} = 0$, it reduces to

$$\mathcal{Q} = \sum_{w \in W} (-1)^{l(w)}\, e(w[\boldsymbol{\rho}]), \tag{8.106}$$

leading us to the *Weyl character formula*

$$\mathrm{Ch}_{\mathrm{Irr}(\boldsymbol{\Lambda})} = \frac{\sum_w (-1)^w\, e(w[\boldsymbol{\Lambda} + \boldsymbol{\rho}])}{\sum_w (-1)^w\, e(w[\boldsymbol{\rho}])}. \tag{8.107}$$

This formula appears to be quite unwieldy: for $SU(4)$ the sum is over the $4! = 24$ different Weyl elements, but for $SU(2)$, it is easy to describe. We have only one simple positive root $\boldsymbol{\alpha}$, and its Weyl group of one reflection, $\mathcal{Z}_2$, which changes its sign. All weight vectors are of the form $|\, n\boldsymbol{\alpha}/2\, \rangle$, $n = 0, \pm 1, \pm 2, \ldots$, and the Weyl vector is $\boldsymbol{\rho} = 1$. Hence

$$\sum_w (-1)^w\, e(w[\rho]) = e(\boldsymbol{\alpha}/2) - e(-\boldsymbol{\alpha}/2), \tag{8.108}$$

so that

$$\frac{1}{e(\boldsymbol{\alpha}/2) - e(-\boldsymbol{\alpha}/2)} = \frac{1}{e(\boldsymbol{\alpha}/2)}\left(1 + \sum_{p} e(-\boldsymbol{\alpha})\right). \tag{8.109}$$

The numerator is given by

$$\sum_{w} (-1)^{w}\, e(w[\boldsymbol{\Lambda} + \boldsymbol{\rho}]) = e((n+1)\boldsymbol{\alpha}/2) - e(-(n+1)\boldsymbol{\alpha}/2), \tag{8.110}$$

where

$$|\,\boldsymbol{\Lambda}\,\rangle = |\,\frac{n}{2}\boldsymbol{\alpha}\,\rangle. \tag{8.111}$$

Now if we divide one by the other, we obtain

$$\sum_{p=0}^{n} e(\frac{(n-2p)}{2}\boldsymbol{\alpha}), \tag{8.112}$$

which contains $(n+1)$ terms in the representation $\mathbf{n}$ $(n = 2j)$. This ratio of characters exactly reproduces the states in the $SU(2)$ irreducible representation labeled by $\boldsymbol{\Lambda} = \mathbf{n}$.

8.9.1 Weyl dimension formula

From the Weyl character formula to the dimensions of irreducible representations is but a small step, since the dimension is simply the character with the operators $e(\boldsymbol{\mu})$ replaced by one. Thus for any weight $\boldsymbol{\mu}$, we define

$$\mathcal{E}_{\mu}\Big(e(\boldsymbol{\lambda})\Big) \equiv \exp\Big\{(\,\boldsymbol{\mu},\,\boldsymbol{\lambda}\,)\,t\Big\}, \tag{8.113}$$

where t is some real parameter. Write the Weyl formula for convenience as

$$\mathrm{Ch}_{\mathrm{Irr}(\boldsymbol{\Lambda})} = \frac{A(\boldsymbol{\Lambda} + \boldsymbol{\rho})}{A(\boldsymbol{\rho})}, \tag{8.114}$$

with

$$A(\boldsymbol{\mu}) \equiv \sum_{w \in W} (-1)^{l(w)}\, e(w[\boldsymbol{\mu}]). \tag{8.115}$$

Then

$$\mathcal{E}_{\mu}(A(\boldsymbol{\nu})) = \sum_{w \in W} (-1)^{l(w)} \exp\Big\{(\,\boldsymbol{\mu},\, w[\boldsymbol{\nu}]\,)\,t\Big\}, \tag{8.116}$$

and observe that, since Weyl transformations do not change the length,

$$(\,\boldsymbol{\mu},\, w[\boldsymbol{\nu}]\,) = (\,w^{-1}[\boldsymbol{\mu}],\, \boldsymbol{\nu}\,),$$

after summing over $w' = w^{-1}$, we obtain

$$\mathcal{E}_\mu(A(\nu)) = \sum_{w\in W} (-1)^{l(w)} \exp\Big\{(\,\nu,\ w[\mu]\,)\,t\Big\},$$
$$= \mathcal{E}_\nu(A(\mu)). \tag{8.117}$$

Hence,

$$\mathcal{E}_\rho(A(\lambda)) = \mathcal{E}_\lambda(A(\rho)) = \mathcal{E}_\lambda\Big(\prod[e(\alpha/2) - e(-\alpha/2)]\Big),$$
$$= \prod\Big[\exp\Big\{(\lambda,\alpha/2)t\Big\} - \exp\Big\{-(\lambda,\alpha/2)t\Big\}\Big]. \tag{8.118}$$

Now expand about $t = 0$ to get

$$\mathcal{E}_\rho(A(\lambda)) \approx t^N \prod_{\text{pos. roots}} (\lambda,\alpha) + \cdots, \tag{8.119}$$

where N is the number of positive roots and the dots denote higher powers of t. It follows that

$$\mathcal{E}_\rho\left(\frac{A(\Lambda+\rho)}{A(\rho)}\right) = \prod_{\text{pos. roots}} \frac{(\Lambda+\rho,\alpha)}{(\rho,\alpha)} + \mathcal{O}(t). \tag{8.120}$$

We set $t = 0$ and note that the left-hand side is the dimension of the representation, the character of the irreducible representation at $e = 1$. Hence, Weyl's dimension formula:

$$\mathrm{Dim}_{\mathrm{Irr}(\Lambda)} = \prod_{\alpha>0}\left(1 + \frac{(\Lambda,\alpha)}{(\rho,\alpha)}\right). \tag{8.121}$$

This yields a convenient way to get the dimension as a polynomial in the Dynkin labels of the highest weight state. In $SU(4)$, with six positive roots, the dimension of the representation $(\,a_1\,a_2\,a_3\,)$ is a sixth-order polynomial

$$(1+a_1)\,(1+a_2)\,(1+a_3)\left(1+\frac{a_1+a_2}{2}\right)\left(1+\frac{a_2+a_3}{2}\right)\left(1+\frac{a_1+a_2+a_3}{3}\right),$$

corresponding to the six positive roots in the α-basis

$$(1\,0\,0)\quad(0\,1\,0)\quad(0\,0\,1)\quad(1\,1\,0)\quad(0\,1\,1)\quad(1\,1\,1).$$

8.9.2 Verma basis

It is possible to construct directly the linearly independent states of irreducible representations, using the Verma module construction.

We state the results without proof and refer the reader to S.-P. Ling, R. Moody, M. Nicolescu, and J. Patera, *J. Math. Phys.* **27**, 666 (1986).

For $SU(3)$, the (un-normalized) states in the Verma basis are of the form

$$\left[T_-^{(1)}\right]^{n_3}\left[T_-^{(2)}\right]^{n_2}\left[T_-^{(1)}\right]^{n_1}|\,a_1\,a_2\,\rangle, \tag{8.122}$$

with

$$\begin{aligned}
0 \le n_1 &\le a_1, \\
0 \le n_2 &\le a_2 + n_1, \\
0 \le n_3 &\le \min[a_2,\, n_2].
\end{aligned}$$

For G_2, the (un-normalized) states are given by products of the two lowering operators associated with the simple roots

$$\left[T_-^{(2)}\right]^{n_6}\left[T_-^{(1)}\right]^{n_5}\left[T_-^{(2)}\right]^{n_4}\left[T_-^{(1)}\right]^{n_3}\left[T_-^{(2)}\right]^{n_2}\left[T_-^{(1)}\right]^{n_1}|\,a_1\,a_2\,\rangle, \tag{8.123}$$

with

$$\begin{aligned}
0 \le n_1 &\le a_1, \\
0 \le n_2 &\le a_2 + 3n_1, \\
0 \le n_3 &\le \min[(a_2 + 2n_2)/3,\, n_2], \\
0 \le n_4 &\le \min[(a_2 + 3n_3)/2,\, 2n_3], \\
0 \le n_5 &\le \min[(a_2 + n_4)/3,\, n_4/2], \\
0 \le n_6 &\le \min[a_2,\, n_5].
\end{aligned}$$

While clumsy to write, these constructions are very easy to implement on computers, and as such are quite useful.

9

Finite groups: the road to simplicity

Earlier, we saw that finite groups can be taken apart through the composition series, in terms of *simple* finite groups, that is groups without normal subgroups. Remarkably the infinitude of simple groups is amenable to a complete classification. Indeed, most simple groups can be understood as finite elements of Lie groups, with parameters belonging to finite Galois fields. Their construction relies on the Chevalley basis of the Lie algebra, as well as on the topology of its Dynkin diagram. The remaining simple groups do not follow this pattern; they are the magnificent 26 *sporadic* groups. A singular achievement of modern mathematics was to show this classification to be complete.

So far, this beautiful subject has found but a few applications in physics. We feel nevertheless that physicists should acquaint themselves with its beauty. In this mostly descriptive chapter we introduce the necessary notions from number theory, and outline the construction of the Chevalley groups as well as that of some sporadic groups. We begin by presenting the two smallest non-Abelian simple finite groups.

$\mathcal{A}_5$ *is simple*

The 60 even permutations of the alternating group $\mathcal{A}_5$ are 3-ply or triply transitive. We can use this fact to prove something startling about alternating groups.

By definition, all even permutations are generated by the product of two transpositions, which can be reduced to three-cycles or the product of three-cycles. Indeed, the product of two transpositions is either

- the unit element if they are the same $(a\,b)(a\,b) = e$,
- a three-cycle if they have one element in common: $(a\,b)(b\,c) = (a\,b\,c)$,
- the product of two three-cycles if they have no elements in common: $(a\,b)(c\,d) = (a\,b)(c\,a)(c\,a)(c\,d) = (a\,b\,c)(c\,a\,d)$.

It follows that all even permutations are generated by three-cycles (any five-cycle can be written as the product of two three-cycles: $(a\,b\,c\,d\,e) = (a\,b\,c)(c\,d\,e)$). We can use this fact and the $(n-2)$-ply transitivity of $\mathcal{A}_n$ to prove that all alternating groups are simple as long as $n \geq 5$. Below we carry the proof for the case $n = 5$, but the generalization to larger n is easy.

Assume that $\mathcal{A}_5$ has a normal subgroup $\mathcal{B}$, with elements of the form (x x x x x), (x x x), or (x x)(x x).

Suppose first that $\mathcal{B}$ contains the five-cycle $b = (1\,2\,3\,4\,5)$. Consider the element $t = (3\,4\,5)$, and form its conjugate

$$\widetilde{b} = tbt^{-1} = (1\,2\,4\,5\,3). \tag{9.1}$$

Since $\mathcal{B}$ is assumed to be normal, $\widetilde{b}$ and its inverse are also in $\mathcal{B}$, as is their product

$$b(\widetilde{b})^{-1} = (1\,4\,3) \in \mathcal{B}. \tag{9.2}$$

If $\mathcal{B}$ contains the five-cycle, it must also contain a three-cycle.

Then suppose that $b = (1\,2)(3\,4)$. Let $t = (1\,2\,5)$. Form

$$\widetilde{b} = tbt^{-1} = (2\,5)(3\,4), \tag{9.3}$$

and

$$b\widetilde{b} = (1\,5\,2), \tag{9.4}$$

which must be the normal subgroup: any normal subgroup of $\mathcal{A}_5$ must contain at least one three-cycle.

Now for *l'envoi*: assume that $\mathcal{B}$ contains the three-cycle $(1\,2\,3)$. Since $\mathcal{A}_5$ is 3-ply transitive, any other three-cycle is related to it by conjugation, for example

$$(345) = t(1\,2\,3)t^{-1}, \qquad t = (1\,3\,5), \tag{9.5}$$

and $(3\,4\,5)$ must belong to the normal $\mathcal{B}$. Since $\mathcal{A}_5$ is generated by *all* three-cycles, it must be that $\mathcal{B} = \mathcal{A}_5$, and $\mathcal{A}_5$ is simple. The proof carries for arbitrary n along the same lines: first one shows that any normal subgroup of $\mathcal{A}_n$ must contain three-cycles (the case $n = 5$ is a bit easier to analyze), then use the triple transitivity of $\mathcal{A}_n$. Note that the proof does not carry for $n = 4$ which is at most doubly transitive, and in fact $\mathcal{A}_4$ is not simple. We have identified two infinite families of simple groups, $\mathcal{Z}_p$, p prime, and $\mathcal{A}_n$, $n \geq 5$. But there are many more!

Note that $\mathcal{A}_5$ has only one singlet representation; this is a consequence of the fact that it is a perfect group, equal to its derived subgroup.

$\mathcal{A}_5$	C_1	$12C_2$	$12C_3$	$15C_4$	$20C_5$
$\chi^{[1]}$	1	1	1	1	1
$\chi^{[3_1]}$	3	$\frac{1}{2}(1+\sqrt{5})$	$\frac{1}{2}(1-\sqrt{5})$	-1	0
$\chi^{[3_2]}$	3	$\frac{1}{2}(1-\sqrt{5})$	$\frac{1}{2}(1+\sqrt{5})$	-1	0
$\chi^{[4]}$	4	-1	-1	0	1
$\chi^{[5]}$	5	0	0	1	-1

It is the symmetry group of the icosahedron, the platonic solid with 12 vertices, 20 triangular faces, and 30 edges. Its classes describe rotations of the vertices ($12C_2$, $12C_3$), edges ($15C_4$) , and faces ($20C_5$) into themselves.

A closely related structure, the truncated icosahedron, is found in Nature, notably in the form of a football (soccer ball), Fullerene, and virus coating. A truncated icosahedron is simply an icosahedron where each vertex is chopped off and replaced by a pentagon.

Fullerene, C_{60} is a molecule with 60 carbon atoms. Its surface has $5 \cdot (12) = 60$ vertices. Its $20 + 12 = 32$ faces, split into 20 hexagons, and 12 pentagons which do not touch one another. The carbons are linked by two types of chemical bonds: 30 "double bonds" edge two hexagons, and 90 slightly shorter "single bonds" edge a pentagon and a hexagon, making up the 90 edges required by Euler's formula. Each carbon has one double bond and two single bonds.

$\mathcal{A}_5 = \mathcal{PSL}_2(5)$ (see later) acts on pairs of vertices in C_{60}, but does not take into account its intricate bond structure. Remarkably, by upgrading to $\mathcal{PSL}_2(11)$, the bond structure can be encoded as well, as shown by B. Kostant (*Notices of the American Mathematical Society*, **42**, number 9, 959 (1995)). $\mathcal{PSL}_2(11)$, the order 660 projective group of (2×2) matrices over GF_{11} (see later), contains the icosahedral group, and a conjugacy class of 60 elements. Crucial to this identification is the fact that its transformations can be written as $z\,h$, where $z \in \mathcal{Z}_{11}$, and $h \in \mathcal{A}_5$, as noted by Galois, the night before his death.

9.1 Matrices over Galois fields

In this section, we study groups as matrices whose elements are members of finite fields, called Galois fields. This approach has not yet proven useful for physical applications, but it is by far the most efficient way to present the landscape of finite groups. We begin by reminding the reader of some basic mathematical facts about finite fields.

Scholium. *Rings and fields*

A set of elements with two operations, addition $+$ and multiplication $\star$. A *ring* contains the zero element so that it is an Abelian group under addition. Under multiplication, a ring contains the unit element, and satisfies closure and associativity, **but** its elements need not have multiplicative inverses, so it is not a group under multiplication. The product law is not necessarily commutative. Octonions do not form a ring, since their multiplication is not associative.

A *field* is just a commutative ring, i.e. one where the multiplication law is commutative, and where each element has a unique inverse. Hence it forms an Abelian group under both addition and multiplication. It must contain a zero and a unit element, denoted by 0 and 1, respectively. The real and complex numbers, including zero and one form a field. Quaternions do not form a field because their multiplication law is not commutative: they form only a ring.

Scholium. *Galois fields*

A Galois field is a finite set of elements which satisfy the field axioms. The simplest example is the p elements $0, 1, 2, \ldots, p-1$, where p is a prime number. Take both addition and multiplication to be mod(p). They form the Galois field $GF(p)$. An example will suffice to illustrate its properties.

The Galois field $GF(7)$ has seven elements, 0, 1, 2, 3, 4, 5, 6. Their inverses are obtained by modular addition:

$$-1 \sim 6 : 1+6 = 0 \mod(7), \ldots, -6 \sim 1 : 6+1 = 0 \mod(7). \tag{9.6}$$

This makes perfect sense under ordinary multiplication mod(7), as $1 = (-1)^2 = (6)^2 = 36 = 1 + 35 = 1 \mod(7)$. Galois fields have two elements which are roots of one, except for the simplest Galois field $GF(2)$ with elements 0, 1, where 1 is its own inverse.

One might think that any set of integers with mod(n) addition and multiplication forms a field, but that is not so. It suffices to look at the mod(6) case: the element 2 clearly does not have an *integer* inverse: $? \cdot 2 = 1 \mod(6)$ has not an integer solution. To find out the allowed values of n, let us start from $GF(p)$ (p prime), with elements $u_0, u_1, \ldots, u_{p-1}$, and build bigger finite fields. Assume a finite field with q elements, where $q > p$, which contains $GF(p)$. Since $q > p$, it has at least one element v not in $GF(p)$. The p new elements $u_i v$, are in the field, but not in $GF(p)$ (since v is not). Together with u_i, we can form p^2 elements, the sums

$$u_i + u_j \star v. \tag{9.7}$$

If $q = p^2$, we have constructed a new field of order p^2. Otherwise, $q > p^2$, allowing us to repeat the procedure: pick a new element v', and generate p^3 elements, and so on until we run out of new elements, with $q = p^n$. We have shown how to build a Galois field of order p^n, which contains the elements of $GF(p)$. This procedure is general: finite fields are necessarily of order p^n. The prime p is called the *characteristic* of the field.

Having established its existence, we have yet to find its multiplication table. For $GF(p^n)$ to be a field, it must be an Abelian group of order $(p^n - 1)$ under its multiplication rule. Hence the order of any of its elements must divide $(p^n - 1)$, that is

$$\underbrace{(x \star x \star \cdots \star x)}_{(p^n-1)} = x_\star^{(p^n-1)} = 1, \tag{9.8}$$

under the yet to be defined multiplication rule for the field (this does not mean that x is a complex root of unity!). This equation simply says that the elements of $GF(p^n)$, $u_0, u_1, \ldots, u_a, \ldots, u_{p^n-1}$, are the roots of the polynomial

$$p(x) = x_\star^{p^n} - x = (x - u_0) \star (x - u_1) \star (x - u_2) \star \cdots \star (x - u_{p^n-1}). \tag{9.9}$$

To find the multiplication rule, Galois started with a polynomial of degree n over $GF(p)$ (he did not call it that, of course), that is

$$p(x) = v_0 x^n + v_1 x^{n-1} + \cdots + v_{p^n}, \tag{9.10}$$

where $v_a \in GF(p)$. There are p^2 such polynomials. Only a subset of these polynomials, $\pi_a(x)$, have no roots which belong to $GF(p)$. They are called *irreducible polynomials*.

The multiplication rule of the elements of $GF(p^n)$ are determined by setting one of these irreducible polynomials to zero, and it does not matter which we choose!

Some simple examples: $n = 2$, $p = 2$. To find the quadratic irreducible polynomials, we form all second order polynomials with roots over $GF(2)$: $(x - 1) \star (x - 1) = x^2 - 2x + 1 = x + 1$ and , $x \star (x - 1) = x^2 - 1 = x^2 + 1$ mod(2). Comparing with all possible second-order polynomials over $GF(2)$, we see that there is only one irreducible polynomial

$$\pi(x) = x^2 + x + 1, \tag{9.11}$$

leading to the multiplication rules: $x \star x = x^2 = -x - 1 = x + 1$ mod(π). We henceforth dispense with the $\star$, and build the elements of $GF(4)$ by successive additions $\{0, 1, x, x + 1\}$. Believe it or not, all are roots of unity under multiplication. For instance, $x^3 = xx^2 = x(x + 1) = x^2 + x = 2x + 1 = 1$, all mod(2). This leads to the multiplication table for $GF(4)$

$$\begin{matrix} 1 & x & x+1 \\ x & x+1 & 1 \\ x+1 & 1 & x \end{matrix} \qquad (9.12)$$

where of course the zero element does not appear.

A less trivial example with several possible irreducible polynomials is $GF(9)$. Start from the *integral* elements of $GF(3)$, 0, 1, 2, to which we add polynomials obtained by successive addition, leading to the nine (three integral plus four non-integral) elements

$$GF(9): \quad \{0, 1, 2, x, x+1, x+2, 2x, 2x+1, 2x+2\}. \qquad (9.13)$$

For convenience, we label them by their number, $a = 1, 2 \ldots, 8$. Their multiplication is determined from any irreducible polynomial. The reducible quadratic polynomials are $(x-1)^2$, $(x-2)^2$, $(x-1)(x-2)$, $x(x-1)$, $x(x-2)$, x^2 to be evaluated mod(3). This leaves three possible irreducible polynomials

$$\pi_1(x) = x^2 + 1, \qquad \pi_2(x) = x^2 + 2x + 2, \qquad \pi_3(x) = x^2 + x + 2. \quad (9.14)$$

It is simplest to use $\pi_1(x)$ to find the following multiplication table, that is $x^2 = -1 \sim 2 \bmod(\pi_1)$.

1	2	3	4	5	6	7	8
2	1	6	8	7	3	5	4
3	6	2	5	8	6	4	7
4	8	5	6	1	7	2	3
5	7	8	1	3	4	6	2
6	3	1	7	4	2	8	5
7	5	4	2	6	8	3	1
8	4	7	3	2	5	1	6

The reader can verify that the other choices for irreducible polynomials, π_2 and π_3, lead to the same multiplication table. Note the interesting fact that one of its elements 4, cannot be written as a square.

Finally, we note that a Galois field $GF(p^n)$, with p^n elements, p prime, sustains an important map into itself, called the Frobenius map. It is defined as

$$u \rightarrow u' = u^p, \qquad (9.15)$$

and it clearly has order n, since (symbolically)

$$(u')^{\prime n} = u^{p^n} = u. \qquad (9.16)$$

For special values of n, the Frobenius map plays a crucial role in the construction of some infinite families of simple finite groups.

We are now in a position to study the property of matrices whose elements are elements of finite Galois fields.

Matrices over finite fields

As long as they have inverses, matrices over the real and complex number fields generate infinite Lie groups, since they satisfy all the group axioms. Hence we can expect that non-singular matrices over finite fields also form (finite) groups.

We can immediately consider the *general linear group* of $(n \times n)$ matrices over the Galois field $GF(q)$, where q is a power of a prime. It is denoted by $\mathcal{GL}_n(q)$. These matrices form finite groups which we wish to catalog. An instructive example is $\mathcal{GL}_2(7)$, of (2×2) matrices over the Galois field $GF(7)$:

$$\mathcal{GL}_2(7)\colon \begin{pmatrix} a & b \\ c & d \end{pmatrix}, \quad a, b, c, d \in GF(7), \tag{9.17}$$

with

$$(ad - bc) \neq 0 \mod(7) \tag{9.18}$$

required because these matrices must have inverses to form a group. To determine how many such matrices there are, look at the first row: a and b can take on seven values each, but $a = b = 0$ is forbidden by the determinant condition. Hence it has $7^2 - 1$ possible entries. The second row c, d also has 7^2 possible values, but for each assignment of the first row, there are seven choices which give vanishing determinant, when both elements are zero and when the second row is a multiple of the first. Hence the second row can have $7^2 - 7 = 42$ assignments. Hence, $\mathcal{GL}_2(7)$ contains $48 \cdot 42 = 2016$ matrices.

We then specialize to the *special linear group*, the set of matrices in $\mathcal{GL}_2(7)$ with unit determinant:

$$\mathcal{SL}_2(7)\colon \begin{pmatrix} a & b \\ c & d \end{pmatrix}, \quad a, b, c, d \in GF(7); \qquad (ad - bc) = 1 \mod(7). \tag{9.19}$$

Since the determinant can have one of six values, we see that it has one sixth as many elements as $\mathcal{GL}_2(7)$: $\mathcal{SL}_2(7)$ is a group with $2016/6 = 336$ elements.

We can think of these matrices as acting on two-vectors

$$\begin{pmatrix} x \\ y \end{pmatrix} \to \begin{pmatrix} x' \\ y' \end{pmatrix} \begin{pmatrix} a & b \\ c & d \end{pmatrix} \begin{pmatrix} x \\ y \end{pmatrix}, \tag{9.20}$$

and define *projective transformations*

$$u \to u' = \frac{au + b}{cu + d}, \qquad u = \frac{x}{y}, \tag{9.21}$$

where u is called the homogeneous coordinate. These are generated by a smaller group, the *projective special linear group*, $\mathcal{PSL}_2(7)$. It is smaller because we can multiply the first row and the second row by the same element of $GF(7)$ without altering the transformation. In order to stay within the unit determinant requirement, the element must square to one. Hence, in $\mathcal{PSL}_2(7)$, the mappings

$$\begin{pmatrix} 1 & 0 \\ 0 & 1 \end{pmatrix}\begin{pmatrix} a & b \\ c & d \end{pmatrix}, \qquad \begin{pmatrix} 6 & 0 \\ 0 & 6 \end{pmatrix}\begin{pmatrix} a & b \\ c & d \end{pmatrix}, \tag{9.22}$$

lead to the same projective map, since

$$\begin{pmatrix} 6 & 0 \\ 0 & 6 \end{pmatrix}\begin{pmatrix} 6 & 0 \\ 0 & 6 \end{pmatrix} = \begin{pmatrix} 1 & 0 \\ 0 & 1 \end{pmatrix} \quad \text{mod}(7). \tag{9.23}$$

In $\mathcal{PSL}_2(7)$, these two elements are taken to be the same, resulting in $336/2 = 168$ elements. Technically speaking, these two diagonal matrices form the *center* of $\mathcal{SL}_2(7)$.

This construction can be generalized to any $(n \times n)$ matrices over any Galois field $GF(p^m)$. We have then $\mathcal{GL}_n(p^m)$, $\mathcal{SL}_n(p^m)$, and $\mathcal{PSL}_n(p^m)$. There are special cases where simplifications arise. Clearly $\mathcal{GL}_n(2) = \mathcal{SL}_n(2)$, since the determinant can only have one non-zero value, and it is also the same as $\mathcal{PSL}_n(2)$, since in $GF(2)$, $1 = -1$.

All the projective groups are simple, except for $\mathcal{PSL}_2(2)$. Sometimes you will see the notation $\mathcal{PSL}_n(p^m) = L_n(p^m)$. By counting their number of elements, we can see the isomorphisms

$$\mathcal{PSL}_2(2) \sim \mathcal{S}_3; \qquad \mathcal{PSL}_2(3) \sim \mathcal{A}_4. \tag{9.24}$$

The smallest simple group is also of this type, as

$$\mathcal{PSL}_2(4) \sim \mathcal{PSL}_2(5) \sim \mathcal{A}_5. \tag{9.25}$$

By specializing the matrices of $\mathcal{GL}_n(p^m)$, one can find several infinite families of simple groups. These are the groups of Lie type, whose construction will have to await our discussion of Dynkin diagrams.

9.1.1 $\mathcal{PSL}_2(7)$

$\mathcal{PSL}_2(7)$ is the second smallest simple group. It is not an alternating group, and it is a perfect group.

To start, we easily see that $\mathcal{PSL}_2(7)$ contains only three diagonal matrices

$$e = \begin{pmatrix} 1 & 0 \\ 0 & 1 \end{pmatrix}, \qquad b = \begin{pmatrix} 3 & 0 \\ 0 & 5 \end{pmatrix}, \qquad b^2 = \begin{pmatrix} -5 & 0 \\ 0 & -3 \end{pmatrix} = \begin{pmatrix} 5 & 0 \\ 0 & 3 \end{pmatrix}, \tag{9.26}$$

using mod(7) addition and multiplication, and remembering that projective transformations are the same as their negatives. Thus b generates a $\mathcal{Z}_3$ subgroup. We can identify another subgroup. The matrix

$$a = \begin{pmatrix} 1 & 1 \\ 0 & 1 \end{pmatrix}, \tag{9.27}$$

is easily seen to generate a $\mathcal{Z}_7$ subgroup; indeed, since

$$a^k = \begin{pmatrix} 1 & k \\ 0 & 1 \end{pmatrix}, \; k \in GF(7), \tag{9.28}$$

it follows that $a^7 = e$. Now let us form the product

$$b^{-1}ab = \begin{pmatrix} 5 & 0 \\ 0 & 3 \end{pmatrix} \begin{pmatrix} 1 & 1 \\ 0 & 1 \end{pmatrix} \begin{pmatrix} 3 & 0 \\ 0 & 5 \end{pmatrix} = \begin{pmatrix} 1 & 4 \\ 0 & 1 \end{pmatrix} = a^4. \tag{9.29}$$

It therefore has a semi-direct product structure, so that

$$\mathcal{Z}_7 \rtimes \mathcal{Z}_3 \subset \mathcal{PSL}_2(7). \tag{9.30}$$

This subgroup of order 21 is sometimes called the Frobenius group.

We can recognize yet another subgroup. Adjoin to the above $\mathcal{Z}_3$ subgroup generated by b, the matrix

$$c = \begin{pmatrix} 0 & 1 \\ 6 & 0 \end{pmatrix}, \qquad c^2 = e, \tag{9.31}$$

and we see that

$$cbc^{-1} = \begin{pmatrix} 2 & 0 \\ 0 & 4 \end{pmatrix} = \begin{pmatrix} 5 & 0 \\ 0 & 3 \end{pmatrix} = b^2, \tag{9.32}$$

which indicates a semi-direct product $\mathcal{Z}_3 \rtimes \mathcal{Z}_2 = \mathcal{S}_3$, the permutation group on three letters. A convenient presentation of $\mathcal{PSL}_2(7)$ in terms of the generators of $\mathcal{Z}_3$ and $\mathcal{Z}_7$ is given by $< a, b \mid a^7 = e, b^3 = e, (ba)^3 = (ba^4)^4 = e >$.

9.1.2 A doubly transitive group

Consider a set of seven points, imaginatively labeled as 1, 2, 3, 4, 5, 6, 7. One could use any label or names, for instance (Bashful, Doc, Dopey, Grumpy, Happy, Sleepy, Sneezy), or (E, G, A, L, O, I, S). We wish to construct the group formed by those permutations that permute the seven pillars (of wisdom).

$$\begin{bmatrix} 1 \\ 2 \\ 4 \end{bmatrix} \begin{bmatrix} 2 \\ 3 \\ 5 \end{bmatrix} \begin{bmatrix} 3 \\ 4 \\ 6 \end{bmatrix} \begin{bmatrix} 4 \\ 5 \\ 7 \end{bmatrix} \begin{bmatrix} 5 \\ 6 \\ 1 \end{bmatrix} \begin{bmatrix} 6 \\ 7 \\ 2 \end{bmatrix} \begin{bmatrix} 7 \\ 1 \\ 3 \end{bmatrix}$$

This group is not triply transitive since it does not contain transformations that map all sets of three points (for instance [1 2 4] does not map into [1 2 6]). We will see that it is doubly transitive.

We start by determining its subgroup $\mathcal{H}$ that leaves the first column invariant. It contains permutations that leave the three points in the first pillar invariant: $1 \to 1$, $2 \to 2, 4 \to 4$. Inspection of the other six columns yields only four elements, depending on the transformation of the 3 element:

$$(3)(5)(6)(7), \qquad (35)(67), \qquad (36)(57), \qquad (37)(56), \tag{9.33}$$

yielding a subgroup of order four. With three elements of order two, it is the Abelian dihedral group $\mathcal{D}_2 = \mathcal{Z}_2 \times \mathcal{Z}_2$. Next, we can also transpose 2 and 4, $a = (24)$ without changing the first column. This produces $8 = 2{\cdot}4$ transformations that leave the 1 element invariant

$$\mathcal{H}_2 : \quad \mathcal{D}_2 + \mathcal{D}_2\, a = \mathcal{D}_4. \tag{9.34}$$

The only other transformations that leave the first column invariant are in the three-cycle (1 2 4). One of these is the permutation of order three

$$b = (124)(3)(567), \qquad b^3 = e, \tag{9.35}$$

which we use to find $\mathcal{H}$ through the coset construction

$$\mathcal{H} = \mathcal{H}_2 + \mathcal{H}_2 b + \mathcal{H}_2 b^2. \tag{9.36}$$

This exhausts all transformations that leave the first column invariant. We see that $\mathcal{H}$ contains the $3 \cdot 8 = 24$ permutations of $\mathcal{S}_4$.

Finally, we consider the transformation that maps the columns into one another, such as the order seven permutation

$$c = (1234567), \qquad c^7 = e, \tag{9.37}$$

which maps the first column into the second one. Similarly, c^2 maps the first column into the third one, and so on. This produces the group we are seeking through the coset decomposition

$$\mathcal{G} = \mathcal{H} + \mathcal{H}c + \cdots + \mathcal{H}c^6. \tag{9.38}$$

This is a group of order $7 \cdot 24 = 168$. Geometrically, we can think of the seven pillars as lines, leading to the interpretation of a projective geometry with seven points and seven lines, such that two points are *only* on one line. It follows that this group is *doubly transitive*.

We can summarize this construction by means of the following diagram.

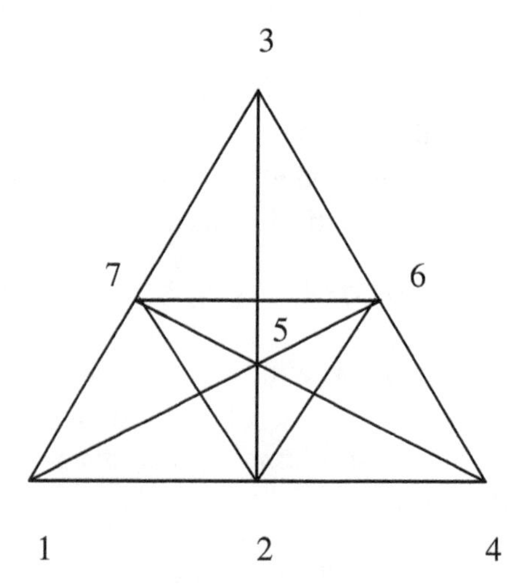

This picture is useful to identify the subgroup that leaves the line [1 2 4] invariant. We can permute the remaining points without changing the picture: the three transpositions (6 7), (5 7), and (3 7) do not alter the picture but relabel the points while keeping the base of the triangle invariant. For example, (6 7) simply interchanges the two outer edges of the triangle. All three generate the permutations on the four points [3, 5, 6, 7]: the subgroup that leaves one line invariant is nothing but $\mathcal{S}_4$.

This geometry is an example of a *Steiner system*, in this case $\mathcal{S}(2, 3, 7)$: **seven** points assembled in sets of **three**, such that any **two** points belong only to one three-set. In general, finite projective planes can be defined in terms of Steiner systems as follows: let p be a prime number, and consider a geometry with $p^2 + p + 1$ points, such that each line will contain $p + 1$ points, with two points lying on only one line. These form the Steiner systems $\mathcal{S}(2, p+1, p^2+p+1)$. Ours is the simplest non-trivial case with $p = 2$.

Sylow's theorems tell us that $\mathcal{PSL}_2(7)$ has elements of order 2, 3, 4 and 7. They are conveniently displayed in the presentation $< a, b \,|\, a^2 = e, b^3 = e, (ab)^7 = e, ([a, b])^4 = e >$, where $[a, b] = a^{-1}b^{-1}ab$.

The generators have simple expressions in terms of (2×2) matrices over $GF(7)$

$$a = \begin{pmatrix} 0 & -1 \\ 1 & 0 \end{pmatrix}, \qquad b = \begin{pmatrix} 0 & -1 \\ 1 & 1 \end{pmatrix}. \tag{9.39}$$

Note that a belongs to one class with 42 elements of order 2, while b gives rise to a second class with 56 elements of order 3. The element

$$[a, b] = \begin{pmatrix} 1 & 1 \\ 1 & 2 \end{pmatrix} \tag{9.40}$$

generates a class of 21 elements of order 4. Finally, the two elements of order 7

$$ab = \begin{pmatrix} 1 & 1 \\ 0 & 1 \end{pmatrix}, \qquad abb = \begin{pmatrix} 1 & 0 \\ 1 & 1 \end{pmatrix}, \tag{9.41}$$

generate two classes, each of 24 elements of order 7. It follows that there are six irreps. Only one of these is one-dimensional. To see this, note that in such irrep, a and b are simply commuting numbers, so that

$$a^2 = 1, \ b^3 = 1, \ (ab)^7 = ab = 1, \tag{9.42}$$

with unique solution $a = b = 1$. The rest of the character table is given as follows.

$\mathcal{PSL}_2(7)$	C_1	$21C_2^{[2]}(a)$	$56C_3^{[3]}(b)$	$42C_4^{[4]}([a,b])$	$24C_5^{[7]}(ab)$	$24C_6^{[7]}(ab^2)$
$\chi^{[1]}$	1	1	1	1	1	1
$\chi^{[3]}$	3	-1	0	1	$\frac{1}{2}\left(-1+i\sqrt{7}\right)$	$\frac{1}{2}\left(-1-i\sqrt{7}\right)$
$\chi^{[\bar{3}]}$	3	-1	0	1	$\frac{1}{2}\left(-1-i\sqrt{7}\right)$	$\frac{1}{2}\left(-1+i\sqrt{7}\right)$
$\chi^{[6]}$	6	2	0	0	-1	-1
$\chi^{[7]}$	7	-1	1	-1	0	0
$\chi^{[8]}$	8	0	-1	0	1	1

This group is one of an infinite family of finite simple groups, and there are many others. Thankfully, there is a systematic way to construct *all* finite simple groups. This will be the subject of the next section.

9.2 Chevalley groups

We have seen that there exists a basis for any Lie algebra where all the structure functions and matrix elements are integers, the Chevalley basis. Chevalley observed that a Lie group element in this basis with parameters belonging to a finite Galois field $GF(q)$ will generate a finite group. They are the Chevalley groups, most of which are simple. For details the reader is referred to Roger Carter's classic book [3].

- $A_n(q)$, generated by elements of the form

$$U_i(u) \equiv e^{u\,e_i}, \qquad U_{-i}(u) \equiv e^{u\,e_{-i}}, \tag{9.43}$$

where e_i are the generators associated with the simple roots in the Chevalley basis of A_n, $i = 1, 2, \ldots, n$, and where u is an element of the Galois field, $GF(q)$, with $u^q = 1$, q being any power of a prime.

Let us see how this works for $n = 2$, where we use the (3×3) representation of the two nilpotent matrices. The group elements are generated by the matrices

$$\begin{pmatrix} 1 & u & 0 \\ 0 & 1 & 0 \\ 0 & 0 & 1 \end{pmatrix}, \qquad \begin{pmatrix} 1 & 0 & 0 \\ 0 & 1 & u \\ 0 & 0 & 1 \end{pmatrix}, \tag{9.44}$$

as well as

$$\begin{pmatrix} 1 & 0 & 0 \\ u & 1 & 0 \\ 0 & 0 & 1 \end{pmatrix}, \qquad \begin{pmatrix} 1 & 0 & 0 \\ 0 & 1 & 0 \\ 0 & u & 1 \end{pmatrix}. \tag{9.45}$$

All have unit determinant, and generate $\mathcal{PSL}_{n+1}(q)$. Most are simple with few exceptions: $\mathcal{PSL}_2(5) \sim \mathcal{PSL}_2(4) \sim \mathcal{A}_5$, $\mathcal{PSL}_2(3) \sim \mathcal{A}_4$, and $\mathcal{PSL}_2(2) \sim \mathcal{S}_3$.

- $B_n(q)$, the classical orthogonal finite groups $\mathcal{O}_{2n+1}(q)$, also known as $\mathcal{P}\Omega_{2n+1}(q)$, which leave the quadratic form

$$(u, u) = u_0^2 + u_1^2 + \cdots + u_{2n}^2,$$

invariant, where the u_i belong to $GF(q)$.

- $C_n(q)$, the projective groups $\mathcal{P}Sp_{2n}(q)$, which leave the antisymmetric symplectic quadratic form

$$(u, v) = \sum_{i=1}^{n} (u_i\, v_{i+n} - v_i\, u_{i+n})$$

invariant. All but $\mathcal{P}Sp_2(2) \sim \mathcal{S}_3$, $\mathcal{P}Sp_2(3) \sim \mathcal{A}_3$, and $\mathcal{P}Sp_4(2) \sim \mathcal{S}_6$, are simple.

- $D_n(q)$ are classical finite groups of the orthogonal type. Here the situation is a little bit more complicated because we are dealing with finite fields. *Two* different quadratic invariants can be defined for finite orthogonal transformations. This is because some elements of finite fields may not have a square root, as we have seen.

Consider a symmetric quadratic form $(\mathbf{u}, \mathbf{v})$, where $\mathbf{u}$, $\mathbf{v}$ are m-dimensional vectors over $GF(q)$, q odd. Suppose that we have managed to find a basis in terms of m orthogonal vectors $\mathbf{e}_i$, that is

$$(\mathbf{e}_i, \mathbf{e}_j) = 0, \quad i \neq j. \tag{9.46}$$

So far so good. If the field elements $(\mathbf{e}_i, \mathbf{e}_i)$ are squares, a trivial redefinition normalizes them to the unit element. If this is true *for all* basis vectors, we arrive at the familiar quadratic invariant (with a + superscript)

$$(\mathbf{u}, \mathbf{u})^+ = \sum_{i=1}^{2n} u_i^2 = u_1^2 + \cdots + u_{2n}^2,$$

where

$$\mathbf{u} = \sum_{i=1}^{2n} u_i \, \mathbf{e_i}. \tag{9.47}$$

However, suppose there is one vector $\mathbf{e}_k$ for which

$$(\, \mathbf{e}_k, \, \mathbf{e}_k \,) = v,$$

where v is *not* a square element of the finite field. Then we have to define a new quadratic form

$$(\, u, \, u \,)^- = \sum_{i=1, \neq k}^{2n} u_i^2 + v \, u_k^2.$$

The case with two vectors whose norm is not a square, is reduced by judicious linear transformations to the plus case. Hence there are only two distinct possibilities.

The Chevalley groups $D_n(q)$ are the orthogonal groups defined in terms of the $+$ quadratic form. They are called $\mathcal{O}_{2n}^+(q)$, or $\mathcal{P}\Omega_{2n}^+(q)$. There are interesting isomorphisms, $O_4^+(q) = \mathcal{PSL}_2(q) \times \mathcal{PSL}_2(q)$, $O_6^+(q) = \mathcal{PSL}_4(q)$.

When q is even, a whole new discussion is needed.

The remaining Chevalley groups are constructed from the exceptional groups, yielding $G_2(q)$, $F_4(q)$, $E_6(q)$, $E_7(q)$, and $E_8(q)$.

Twisted Chevalley groups

The Chevalley construction can be further refined to produce more groups. We have seen that the Dynkin diagrams of the $A - D - E$ Lie algebras have a parity which maps the Dynkin diagram into itself. We also noted the Frobenius map of the elements of the Galois field into itself; in particular when $q = p^2$, it is also a parity, splitting the elements into even and odd parts.

The combined operation of both order two maps, one on the Dynkin diagram, the other on the Galois field, maps the Chevalley groups $A_n(p^2)$, $D_n(p^2)$, and $E_6(p^2)$ into themselves.

The beautiful thing is that each contains a subgroup, the twisted Chevalley groups, which are invariant under the combined mapping. If we split the Galois field elements into even and odd elements $u^\pm$, and the generators as $e^\pm$, even and odd under Dynkin parity, these groups are generated by

$$U_i^+(u^+) = e^{u^+ e_i^+}, \qquad U_i^-(u^-) = e^{u^- e_i^-}, \tag{9.48}$$

and their conjugates.

Finally, to match the order three $\mathcal{Z}_3$ automorphism of the $SO(8)$ Dynkin diagram, we consider a Galois field $GF(p^3)$ which has an order-three Frobenius automorphism. In this way we obtain the following.

- ${}^2A_n(p^2) \equiv U_{n+1}(p^2) = \mathcal{P}SU_{n+1}(p^2)$. Defined for $n \geq 2$, all but ${}^2A_2(2^2)$ are simple. They are the classical unitary groups, whose elements can be viewed as matrices acting on $(n+1)$-dimensional vectors with elements in the Galois field, and which leave invariant the quadratic form

$$\sum_{j=1}^{n+1} \overline{u}_j \, u_j, \tag{9.49}$$

where $\overline{u}_j = u_j^p$ is the Frobenius conjugate.

For ${}^2A_2(p^2)$, the Dynkin automorphism is a Weyl reflection about the line which links the origin to the $(1, 1)$ state. Thus we can readily identify the even (3×3) matrices in the Chevalley basis

$$\begin{pmatrix} 0 & 1 & 0 \\ 1 & 0 & 0 \\ 0 & 0 & 0 \end{pmatrix}, \quad \begin{pmatrix} 0 & 0 & 0 \\ 0 & 0 & 1 \\ 0 & 1 & 0 \end{pmatrix}. \tag{9.50}$$

Exponentiating with even elements of the Galois field yields the desired group elements, as well as the exponentiation of the odd Lie algebra elements

$$\begin{pmatrix} 0 & 1 & 0 \\ \overline{1} & 0 & 0 \\ 0 & 0 & 0 \end{pmatrix}, \quad \begin{pmatrix} 0 & 0 & 0 \\ 0 & 0 & 1 \\ 0 & \overline{1} & 0 \end{pmatrix}, \tag{9.51}$$

with the odd Galois elements.

Most are simple, except for $U_2(2) = \mathcal{S}_3$, $U_2(3) = \mathcal{A}_3$, $U_3(2) = (\mathcal{Z}_3 \times \mathcal{Z}_3) \rtimes \mathcal{Q}_8$

- ${}^2D_n(p^2) \equiv O_{2n}^-(q) = \mathcal{P}\Omega_{2n}^-(q)$. These are the finite classical orthogonal group defined for $n \geq 4$ ($n = 3$ reduces to $A_3(p^2)$), which leaves the quadratic form

$$(u, u)^- = \sum_{i=1, \neq k}^{2n} u_i^2 + v\, u_k^2,$$

invariant. All are simple, with notable isomorphisms, $O_4^-(q) = \mathcal{P}S\mathcal{L}_2(q^2)$, $O_6^-(q) = U_4(q)$.

- ${}^2E_6(p^2)$ are all simple.
- ${}^2D_4(p^3)$ are simple as well.

Suzuki and Ree groups

Their construction is much sneakier. It starts with the remark that the Dynkin diagrams of B_2, G_2 and F_4 with simple roots of different lengths also have a (more complicated) automorphism: flip the Dynkin diagram *and* interchange the short and long roots.

Looking at the root diagrams of B_2 and G_2, take the two simple roots and draw the line that bisects the angle between them, shown here for B_2.

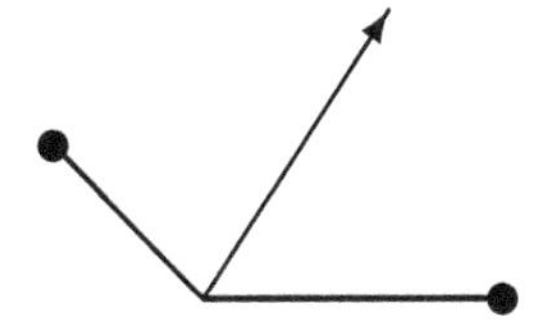

A reflection of the large root about the bisectrix (indicated by an arrow) yields a root of the same length, but in the direction of the shorter root, and vice versa. This reflection can be equally well applied to the other roots. The net result is simply a rotation of the Dynkin diagram about the origin by $\pi/4$ for B_2 and $\pi/6$ for G_2. Doing it again reproduces the original diagram, so this is an automorphism of order two. In the case of F_4, it is harder to draw but we can define a three-dimensional plane that bisects the four simple roots, and the reflections about this plane produce the same result.

As in the previous construction of twisted Chevalley procedure, this Dynkin symmetry must be matched with that of Galois fields which contain a specified Frobenius map. For B_2 and F_4, the Galois field must have characteristic two, and contain a map that is of the form the Frobenius map must satisfy

$$u \to u' = u^{\theta}, \qquad 2\theta^2 = 1.$$

The Galois field can only be $GF(2^{2l+1})$, for any integer l.

For G_2, the Galois field must be of characteristic three, and contain a similar map with $3\,\theta^2 = 1$, which restricts it to $GF(3^{2n+1})$. Only then are the group properties satisfied. Hence we have three more families

$$^2B_2\left(2^{2n+1}\right), \qquad ^2F_4\left(2^{2n+1}\right), \qquad ^2G_2\left(3^{2n+1}\right).$$

Most are simple with the exceptions

$$^2B_2(2) = \mathcal{Z}_5 \rtimes \mathcal{Z}_4, {}^2G_2(3) = \mathcal{PSL}_2(8) \rtimes \mathcal{Z}_3, {}^2F_4(2) = T \cdot 2,$$

where T is the Tits group, which is simple as well.

This almost completes the list of simple finite group. However, there are twenty-six more simple groups which defy all classification: the sporadic groups, to which we now turn.

9.3 A fleeting glimpse at the sporadic groups

Emile Léonard Mathieu discovered in 1860 (and 1873) five groups of multiply transitive permutations. Known today as Mathieu groups, they make up the first generation of sporadic groups, the twenty-six simple non-Abelian finite groups which are neither alternating nor Chevalley groups.

They are, in order of order:

- M_{11} group of $2^4 \cdot 3^2 \cdot 5 \cdot 11$ quadruply transitive permutations on eleven objects,
- M_{12} group of $2^6 \cdot 3^3 \cdot 5 \cdot 11$ quintuply transitive permutations on twelve objects,
- M_{22} with $2^7 \cdot 3^2 \cdot 5 \cdot 7 \cdot 11$ triply transitive permutations of twenty-two objects,
- M_{23} group of $2^7 \cdot 3^2 \cdot 5 \cdot 7 \cdot 11 \cdot 23$ quadruply transitive permutations on twenty-three objects, and
- M_{24} with $2^{10} \cdot 3^3 \cdot 5 \cdot 7 \cdot 11 \cdot 23$ quintuply transitive permutations on twenty-four objects.

They show up as automorphism groups of Steiner systems, as shown in the table below.

M_{11}	M_{12}	M_{22}	M_{23}	M_{24}
$S(4, 5, 11)$	$S(5, 6, 12)$	$S(3, 6, 22)$	$S(4, 7, 23)$	$S(5, 8, 24)$

These groups are quite extraordinary, but with the help of Galois fields, they are generated by relatively simple transformations. Complicated groups can be generated by simple operations; for instance the transformations

$$x \to x' = x + 1, \qquad x \to x' = -x^5 - 2x^2,$$

with $x \in GF(7)$, generate our old friend $\mathcal{PSL}_2(7)$.

One can adjoin to Galois fields one extra point, called the point at infinity, in order to define inversion as a group operation. Then M_{12} is generated by the three transformations

$$x \to x' = x + 1, \quad x \to x' = 4x^2 - 3x^7, \quad x \to x' = -1/x,$$

where $x \in GF(11)$. Its triply transitive subgroup M_{11} is generated by

$$x \to x' = x + 1, \quad x \to x' = 7x^4 - 6x^9, \quad x \to x' = -1/x.$$

In the same vein, M_{24}, the largest Mathieu group is also generated quite "simply" by

$$x \to x' = x + 1, \quad x \to x' = -3x^{15} + 4x^4, \quad x \to x' = -1/x,$$

where $x \in GF(23)$. It appears as the automorphism group of the perfect binary Golay code (for definitions and delightful reading, see T. M. Thompson, *From Error Correcting Codes Through Sphere Packings to Simple Groups*, Carus Mathematical Monographs, MAA, 1983).

It is well beyond the scope of this book to go beyond these.

Seven more sporadic groups make up the second generation; all are relevant in the context of the automorphism group of the Leech lattice, an infinite lattice of points in twenty-four dimensional Euclidean space. It yields the densest packing of spheres in twenty-four dimensions, where one sphere touches 196 560 others!

The Leech lattice can be defined in terms of the light-like vector in 26 $(1+25)$-dimensional discrete Minkowski space-time. The totally unique light-like vector with integer components

$$(70, 0, 1, 2, \ldots, 24),$$

serves to define the points of the Leech lattice by orthogonality. One cannot help but notice that this is the same (continuous) space where the quantum bosonic string roams.

The remaining fifteen sporadic groups make up the third generation. All are connected to the largest of them all, the Monster. It is beyond the scope of this book to go into any more details. These very unique mathematical objects have not yet proven very useful in physics. Yet, given Nature's attraction to unusual mathematical structures, it is easy to think they will some day assume physical relevance.

10

Beyond Lie algebras

10.1 Serre presentation

We have seen that Dynkin diagrams contain the information necessary to reconstruct the simple Lie algebras. Since that information is encoded in the Cartan matrix, there must be a way to build the algebra directly from it. This procedure, we alluded to earlier, is called the *Serre presentation* of the Lie algebra.

Let us review the properties of the Cartan matrices of the simple Lie algebras. The Cartan matrix for a rank r Lie algebra is a $(r \times r)$ matrix A with integer entries arranged such that:

- its diagonal elements are equal to 2;
- its off-diagonal elements are negative integers or zero;
- if one of its off-diagonal elements is zero, so is its transpose.

Our explicit constructions suggest three more properties:

- it can always be brought to symmetric form $\mathcal{S}$ by multiplication by a diagonal matrix $\mathcal{D}$ with positive entries: $\mathcal{S} = \mathcal{D}\,\mathcal{A}$;
- it cannot be written in block-diagonal form by rearranging its rows and columns;
- $\mathcal{A}$ has only positive eigenvalues.

For $SU(2)$, the Cartan matrix is just a number, with only one "diagonal" generator H_1, and one root $\boldsymbol{\alpha}_1$ with corresponding operators E_{α_1} and its conjugate $E_{-\alpha_1}$, with

$$[\, H_1,\; E_{\pm\alpha_1}\,] = \pm 2 E_{\pm\alpha_1},$$

defined in such a way that the numerical factor on the right-hand side is the Cartan matrix element. With this normalization,

$$[\, E_{\alpha_1},\; E_{-\alpha_1}\,] = H_1,$$

which completely defines the $SU(2)$ algebra.

Recall that for $SU(3)$, we have the (2×2) Cartan matrix

$$\mathcal{A}_{SU(3)} = \begin{pmatrix} 2 & -1 \\ -1 & 2 \end{pmatrix},$$

two diagonal generators H_i, and two simple roots, $\boldsymbol{\alpha}_i$ $i = 1, 2$, with corresponding generators $E_{\pm\alpha_i}$. With the normalization

$$[\, H_i,\ E_{\pm\alpha_j}\,] = \pm A_{ji} E_{\pm\alpha_j},$$

we obtain

$$[\, E_{\alpha_i},\ E_{-\alpha_j}\,] = \begin{cases} H_i & i = j, \\ 0 & i \neq j. \end{cases}$$

So far these yield two $SU(2)$ sub-algebras, which accounts for only six out of its eight elements. The extra generators are to be found in the commutator

$$[\, E_{\alpha_1},\ E_{\alpha_2}\,],$$

and its conjugate. This exhausts the number of $SU(3)$ generators. Hence, any other possible generator, written as multiple commutators between the E_{α_i}, must vanish, that is

$$[\, E_{\alpha_1},\ [\, E_{\alpha_1},\ E_{\alpha_2}\,]] = [\, E_{\alpha_2},\ [\, E_{\alpha_2},\ E_{\alpha_1}\,]] = 0.$$

Serre cleverly rewrites these relations in terms of the off-diagonal elements of the Cartan matrix as

$$\left[Ad(E_{\alpha_i}\right]^{1-A_{ji}} E_{\alpha_j} = 0, \quad i \neq j,$$

since $A_{12} = A_{21} = -1$, and where the adjoint operation $Ad(E)^k$ is just the k-fold commutator

$$[Ad(E)]^k\, X = \underbrace{[\, E,\ [\, E,\ \cdots [\, E,\ X\,] \cdots]]}_{k}.$$

A similar expression holds for the conjugate equation. This is called the *Serre presentation*, which can be shown to completely define the Lie algebra in terms of its Cartan matrix and its generators E_{α_i}. To simplify notation, we let $E_{\alpha_i} \equiv E_i$, $E_{-\alpha_i} \equiv \overline{E}_i$, $i = 1, 2, \cdots, r$.

To instill faith in this method, let us build the exceptional G_2 algebra from its Cartan matrix

$$\mathcal{A}_{G_2} = \begin{pmatrix} 2 & -3 \\ -1 & 2 \end{pmatrix}.$$

Like $SU(3)$, it has two generators E_i, $i = 1, 2$, but now we can form many more by commutation. Indeed, the Serre relations now read ($A_{21} = -1$)

$$[\, E_2, [\, E_2, E_1\,]] = 0,$$

and ($A_{12} = -3$)

$$[E_1, [E_1, [E_1, [E_1, E_2]]]] = 0.$$

These leave room to construct more generators by commutation. First we have

$$[E_2, E_1],$$

the next in line triple commutator being forbidden by the first Serre relation. From the second Serre relation, we can form

$$[E_1, [E_1, E_2]], \qquad [E_1, [E_1, [E_1, E_2]]].$$

Finally we have the four-fold commutator

$$[E_2, [E_1, [E_1, E_2]]].$$

Together with $E_{1,2}$, these account for six generators. Together with their conjugates and H_i, they form the fourteen-dimensional G_2 algebra, and the reader can verify, through the use of the Jacobi identities, that all other commutators vanish.

10.2 Affine Kac–Moody algebras

In our analysis of embeddings, we came across extended Dynkin diagrams, used to describe maximal embeddings without loss of rank. They are obtained by adding a node associated with the negative of the highest root to the original Dynkin diagram. Since the highest root is a linear combination of the simple roots, the corresponding extended Cartan matrix has vanishing determinant, and one zero eigenvalue, all other eigenvalues being positive.

Let us apply the Serre presentation to the singular (2×2) $SU(2)$ extended Cartan matrix

$$A^{(1)}_{SU(2)} = \begin{pmatrix} 2 & -2 \\ -2 & 2 \end{pmatrix},$$

and analyze in detail the algebraic structure it produces. By convention, we label the extra node with a zero subscript, associated with the root $\boldsymbol{\alpha}_0$.

The Serre-generated algebra starts with H_0, H_1, E_0, $\overline{E}_0$, E_1, and $\overline{E}_1$, and commutators

$$[H_0, H_1] = [E_1, \overline{E}_0] = 0,$$

$$[H_0, E_0] = 2E_0, \qquad [H_0, E_1] = -2E_1,$$

$$[H_1, E_0] = -2E_0, \qquad [H_1, E_1] = 2E_1,$$

$$[E_0, \overline{E}_0] = H_0, \qquad [E_1, \overline{E}_1] = H_1.$$

The Serre relations are now

$$[Ad(E_0)]^3\, E_1 = 0, \qquad [Ad(E_1)]^3\, E_0 = 0.$$

We first note that one linear combination of the Cartan sub-algebra commutes with all the members of the algebra since

$$[\, H_0 + H_1,\ E_i\,] = 0.$$

This is easily traced to the vanishing of the determinant of the extended Cartan matrix: these algebras have a non-trivial center, and this linear combination is just a c-number, written as

$$2k = H_0 + H_1.$$

Unlike Lie algebras, repeated commutations yield an *infinite* set of new elements. Start with

$$[\, E_1,\ E_0\,],$$

associated with the root

$$\boldsymbol{\delta} \equiv (\boldsymbol{\alpha}_0 + \boldsymbol{\alpha}_1).$$

It is a peculiar root of zero length,

$$(\boldsymbol{\delta},\ \boldsymbol{\delta}\,) = (\boldsymbol{\alpha}_0,\ \boldsymbol{\alpha}_0\,) + (\boldsymbol{\alpha}_0,\ \boldsymbol{\alpha}_1\,) + (\boldsymbol{\alpha}_1,\ \boldsymbol{\alpha}_0\,) + (\boldsymbol{\alpha}_1,\ \boldsymbol{\alpha}_1\,) = 0$$

and is orthogonal to the simple root of the algebra

$$(\boldsymbol{\delta},\ \boldsymbol{\alpha}_1\,) = (\boldsymbol{\alpha}_0,\ \boldsymbol{\alpha}_1\,) + (\boldsymbol{\alpha}_1,\ \boldsymbol{\alpha}_1\,) = 0.$$

The new elements

$$[\, E_0, [\, E_0,\ E_1\,]], \qquad [\, E_1, [\, E_1,\ E_0\,]],$$

are associated with the roots

$$(2\boldsymbol{\alpha}_0 + \boldsymbol{\alpha}_1) = \boldsymbol{\delta} - \boldsymbol{\alpha}_1, \qquad (2\boldsymbol{\alpha}_1 + \boldsymbol{\alpha}_0) = \boldsymbol{\delta} + \boldsymbol{\alpha}_1.$$

On the other hand, the vanishing of the commutator $[\, E_0,\ \overline{E}_1\,]$ tells us that $\boldsymbol{\alpha}_0 - \boldsymbol{\alpha}_1 = \boldsymbol{\delta} - 2\boldsymbol{\alpha}_1$ is not a root. However, $2\boldsymbol{\delta}$ is also a root since the element

$$[\, E_1, [\, E_0, [\, E_0,\ E_1\,]]],$$

does not vanish. We can then commute it with E_1 to obtain a new root $2\boldsymbol{\delta} + \boldsymbol{\alpha}_1$, and so on. In this way, we generate infinite towers of roots of the form

$$\pm\boldsymbol{\delta},\ \pm 2\boldsymbol{\delta},\ \cdots; \qquad (\pm\boldsymbol{\alpha}_1 \pm \boldsymbol{\delta}),\ (\pm\boldsymbol{\alpha}_1 \pm 2\boldsymbol{\delta}),\ \cdots,$$

yielding the infinite-dimensional *affine Kac–Moody algebra*.

This proliferation of roots is rather frightening, but it can be expressed diagrammatically in a two-dimensional root diagram: setting $\boldsymbol{\alpha}_1$ as a unit vector along the horizontal, and $\boldsymbol{\delta}$ along the vertical. We then have an infinite tower of roots as shown below.

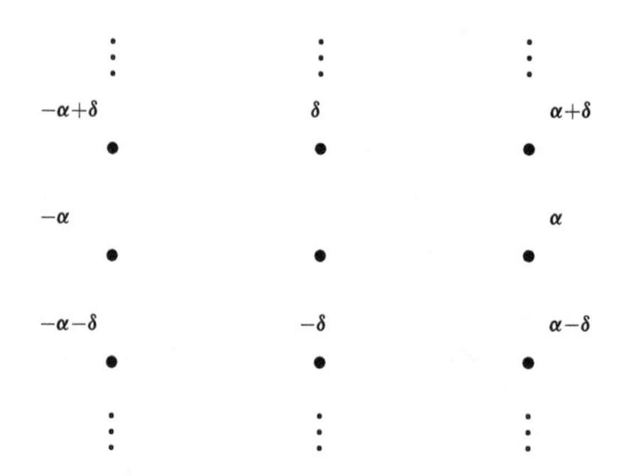

It is nothing but an infinite stack of $SU(2)$ weight diagrams, each differing by $\boldsymbol{\delta}$ in the vertical direction. Invariance of this weight diagram under vertical translations explains the mathematical term *affine*, which is associated with structures that reproduce under translations.

The Weyl group generalizes to the affine case in an obvious way: it is generated by the Weyl transformations on each line, that is $w(\boldsymbol{\alpha}_1) = -\boldsymbol{\alpha}_1$, *and* translations from one stack to the next, say $\boldsymbol{\alpha}_1 \rightarrow \boldsymbol{\alpha}_1 + \boldsymbol{\delta}$. The latter generate an infinite Abelian group.

This pattern generalizes to all affine Kac–Moody algebras: the same weight diagram as for the finite sub-algebra, translated *ad infinitum* by translations in the direction of the null root. The Weyl group has the corresponding structure.

The complicated-looking commutation relations can be cast elegantly by sticking close to those of the underlying $SU(2)$ algebra. We start from the familiar identification of its generators through

$$E_1 = \left(T_0^1 + iT_0^2\right) \equiv T_0^{(1+i2)}, \quad \overline{E}_1 = \left(T_0^1 - iT_0^2\right) \equiv T_0^{(1-i2)}, \quad \frac{H_1}{2} = T_0^3,$$

where we have added a zero subscript. They lead to the usual $SU(2)$ commutation relations

$$\left[T_0^a,\, T_0^b\right] = i\epsilon^{abc}\, T_0^c,$$

$a, b, c = 1, 2, 3$. We then organize the new elements in terms of their transformation properties with T_0^3. Since the new element associated with the special zero-length root $\boldsymbol{\delta}$,

$$[\,E_1,\, E_0\,],$$

commutes with T_0^3, we define it as

$$T_1^3 \equiv \frac{1}{2}[\, E_1,\, E_0\,], \qquad T_1^{3\,\dagger} \equiv T_{-1}^3.$$

This suggests the definitions

$$E_0 \; \equiv T_1^{(1-i2)}, \qquad \overline{E}_0 \; \equiv T_{-1}^{(1+i2)},$$

so as to satisfy

$$\left[T_0^{(1+i2)},\, T_1^{(1-i2)}\right] = 2T_1^3.$$

Note that the operators T_1^a and T_{-1}^a are no longer real but are hermitian conjugates of one another. It follows that

$$\left[T_{-1}^{(1+i2)},\, T_1^{(1-i2)}\right] = 2T_0^3 - 2k.$$

The new element

$$T_1^{(1+i2)} \equiv \frac{1}{2}[\, E_1, [\, E_1,\, E_0\,]]$$

yields, after a modest amount of Jacobi gymnastics,

$$\left[T_1^{(1+i2)},\, T_0^{(1-i2)}\right] = 2T_1^3.$$

More Jacobi acrobatics produces the suggestive commutator

$$\left[T_1^{(1+i2)},\, T_{-1}^{(1-i2)}\right] = 2T_0^3 + 2k$$

(note the change in sign of the c-number k), where

$$T_{-1}^{(1-i2)} \equiv \frac{1}{2}\left[\,\overline{E}_0, \left[\,\overline{E}_0,\, \overline{E}_1\right]\right],$$

is the hermitian conjugate. All other elements of the algebra are obtained by commutation. For instance define the element with root 2δ through

$$\left[T_1^{(1+i2)},\, T_1^{(1-i2)}\right] = \frac{1}{2}[\, E_0,\, [\, E_1, [\, E_0,\, E_1\,]]] \equiv 2T_2^3,$$

and so on. The full algebra is surprisingly simple to state in terms of $SU(2)$-like commutation relations,

$$\left[T_m^a,\, T_n^b\right] = i\epsilon^{abc}\, T_{m+n}^c + m\, k\delta^{ab}\delta_{m+n,0}, \quad m, n = 0, \pm 1, \pm 2, \ldots.$$

The undetermined constant k is at the heart of Kac–Moody algebras. It is useful to adjoin to this algebra an operator L_0 which counts the subscript, that is

$$[\, L_0,\, T_n^a\,] = n\, T_n^a.$$

Before leaving $SU(2)$, we note the existence of another Cartan-like matrix (negative integer off-diagonal entries and diagonal elements equal to 2) with vanishing determinant

$$A_1^{(2)} \equiv \begin{pmatrix} 2 & -1 \\ -4 & 2 \end{pmatrix}.$$

The Serre presentation applies equally well, and generates a new infinite-dimensional algebra, the *twisted affine Kac–Moody algebra*.

Coming back to the general case, the identification of the generators applies equally well to an arbitrary rank r Lie algebra. The elements belonging to the root δ are now the r-fold commutators involving one E_0, and we identify them with T_1^j where j runs over the r labels of the Cartan sub-algebra. One only needs to identify the generators with subscript equal to one in order to generate the full algebra.

The null root is defined in terms of the marks as

$$\delta = \sum_{i=0}^{r} c_i \, \boldsymbol{\alpha}_i,$$

so as to insure that it has zero length, and is orthogonal to all the other roots of the underlying Lie algebra,

$$(\delta, \boldsymbol{\alpha}_i) = 0, \qquad i = 1, 2, \ldots, r.$$

The procedure is of course more algebraically tedious, but the result is to produce the general commutation relations

$$\left[T^a_m, T^b_n\right] = i\, f^{abc}\, T^c_{m+n} + m\, k\delta^{ab}\delta_{m+n,0}, \quad m, n = 0, \pm 1, \pm 2, \ldots,$$

obtained by simply replacing ϵ^{abc} with a general structure function f^{abc}!

Representation theory

The representation theory of affine Kac–Moody algebras proceeds in much the same way as for Lie algebras, except that all are infinite-dimensional. This short section only skims the surface of this rich subject. The interested reader is referred to S. Kass, R. V. Moody, J. Patera, and R. Slansky [13], and Fuchs [5].

We use affine $SU(2)$ as an example, where the states are labeled by two integers $(a_0\, a_1)$. When both $a_0, a_1 \geq 0$ the state lies inside or on the walls of the Weyl chamber, and it can be used to label an irreducible representation. The rest of the states in the representation are obtained by subtracting the positive roots from the highest weight state. In the Dynkin basis, the positive roots are given by the row of the extended Cartan matrix,

$$\boldsymbol{\alpha}_0\!: \ (2, \, -2), \qquad \boldsymbol{\alpha}_1\!: \ (-2, \, 2).$$

We note that the linear combination with the co-marks

$$\check{c}_0\, a_0 + \check{c}_1\, a_1,$$

is invariant under subtraction by the roots because the co-marks are the components of the zero eigenvector of the Cartan matrix. It is called the *level* of the representation. In this case it is just $l = a_0 + a_1$, and it can serve as a partial label for the representation.

We can subtract $\boldsymbol{\alpha}_0$ a_0-times from the highest weight, and $\boldsymbol{\alpha}_1$ a_1-times. This results in a zig-zag pattern for the infinite number of states in the representation. The zig (zag) direction denotes subtraction by $\boldsymbol{\alpha}_0(\boldsymbol{\alpha}_1)$. To keep track of these states, weights are assigned a null weight which tells how many $\boldsymbol{\alpha}_0$ subtractions it is away from the highest weight. These weights assemble in an infinite number of $SU(2)$ representations.

Explicit examples of these representations are found in the physics literature.

When $k = 0$, this algebra is rather trivial to realize: introduce an infinite number of oscillators, each of which transform as $SU(2)$ doublets,

$$\left[a_m,\, a_n^\dagger\right] = \delta_{m,n}, \qquad m, n = 0, \pm 1, \pm 2, \ldots,$$

and define the generalized Schwinger representation

$$T^a_m = \frac{1}{2} \sum_{j=-\infty}^{\infty} a^\dagger_{j+m}\, \sigma^a\, a_j. \tag{10.1}$$

It is easily seen to satisfy the Kac–Moody algebra with $k = 0$. We also have

$$L_0 = \sum_{j=-\infty}^{\infty} j\, a_j^\dagger\, a_j. \tag{10.2}$$

Oscillator representations with $k \neq 0$ are less obvious, but c-numbers generically occur in commutators of quadratic combinations when normal ordering is invoked ($:\ :\rightarrow$ annihilation operators on the right),

$$\begin{aligned}\left[a_1^\dagger\, a_2^\dagger,\, a_1\, a_2\right] &= -a_1^\dagger\, a_1 - a_2\, a_2^\dagger = -a_1^\dagger\, a_1 - a_2^\dagger\, a_2 - 1,\\ &= -:\left(a_1^\dagger\, a_1 + a_2^\dagger\, a_2\right): -1.\end{aligned}$$

The Serre presentation is used to generate yet more general algebras with more complicated Cartan matrices. When $\mathcal{S}$ has one negative eigenvalue, the rest being positive, it leads to the *Hyperbolic Algebras*. Some have found application in the description of physics near a black hole singularity (M. Henneaux, D. Persson, and P. Spindel, *Living Rev. Rel.* **11**, 1 (2008))!

There are also connections with graph theory. The matrix elements of a matrix of the form $\mathcal{A} = \mathcal{D}\,\mathcal{S}$ must satisfy certain relations among themselves. For a (3×3) the matrix condition is

$$a_{12}a_{23}a_{31} = a_{13}a_{32}a_{21}.$$

This equation has a natural graphical explanation. Consider a graph with three points, with a_{ij} denoting the transformation from point i to point j. This equation translates into the equivalence of the transformation $1 \to 2 \to 3$ and its inverse $3 \to 2 \to 1$.

10.3 Super algebras

Super algebras are an extension of Lie algebras which contain two types of elements, even elements which satisfy commutation relations and contain Lie algebras, and odd elements which transform as representations of the Lie sub-algebras, and have the important property that their anticommutators close into the even elements. This last condition is necessary but not sufficient, as a suitably generalized Jacobi identity must also hold.

To see how this works, consider $SU(2)$. We know that its doublet representation satisfies the first criterion

$$\mathbf{2} \times \mathbf{2} = \mathbf{3},$$

and since this is a strict equality, the graded Jacobi identity must follow. Specifically, we denote the $SU(2)$ generators as T^A with $A = 1, 2, 3$, and the complex doublet by q_α, with $\alpha = 1, 2$. Clearly

$$[\, T^A,\, q_\alpha \,] = \frac{1}{2}\,(\sigma^A)_{\alpha\beta}\, q_\beta. \tag{10.3}$$

Rather than taking the complex conjugate, we form the combination

$$\widetilde{q}_\alpha = (\sigma_2)_{\alpha\beta}\, \bar{q}_\beta, \tag{10.4}$$

which also transforms as a doublet

$$[\, T^A,\, \widetilde{q}_\alpha \,] = \frac{1}{2}\,(\sigma^A)_{\alpha\beta}\, \widetilde{q}_\beta. \tag{10.5}$$

Next, we form the anticommutator

$$\{\, q_\alpha,\, q_\beta \,\} = (\sigma^A \sigma_2)_{\alpha\,\beta}\, T^A, \tag{10.6}$$

and we verify that the Jacobi identities

$$[\, T^A, \{\, q_\alpha,\, q_\beta \,\}\,] + \{\, q_\alpha, [\, q_\beta,\, T^A \,]\,\} + \{\, q_\beta, [\, q_\alpha,\, T^A \,]\,\} = 0, \tag{10.7}$$

and

$$[q_\alpha, \{q_\beta, q_\gamma\}] + [q_\beta, \{q_\gamma, q_\alpha\}] + [q_\gamma, \{q_\alpha, q_\beta\}] = 0, \tag{10.8}$$

hold. Similarly, the anticommutator

$$\{q_\alpha, \widetilde{q}_\beta\} = (\sigma^A \sigma_2)_{\alpha\beta} \, T^A, \tag{10.9}$$

also satisfies the graded Jacobi identities. Together with

$$[T^A, T^B] = i\epsilon^{ABC} \, T^C, \tag{10.10}$$

these define the smallest super algebra, made up of T^A, q_α and its conjugate. This is the simplest case of a symplectic super algebra.

All Lie super algebras can be constructed in the following manner: begin with a Lie algebra or direct products of Lie algebras with the property that its fundamental representation are such that their *symmetric* product, corresponding to the anticommutator, belong to the adjoint representation of the Lie algebra. Using our newly acquired knowledge of the properties of the low-lying representations of Lie algebras, it is straightforward to list the possibilities.

Since the symmetric product of the fundamental **2n** of $Sp(2n)$ is exactly its adjoint representation

$$\mathbf{2n} \times \mathbf{2n} = \mathbf{n(2n+1)},$$

and the graded Jacobi identities are automatically satisfied. This infinite family of super algebras is denoted by $OSp(2n)$, and the smallest one is $OSp(2)$, which we have just explicitly constructed.

The only other simple Lie algebra with the property that the symmetric product of its fundamental contains the adjoint is the exceptional group E_7. Its fundamental **56** is pseudoreal and satisfies

$$\mathbf{56} \times \mathbf{56} = \mathbf{133} + \mathbf{1463},$$

but the Jacobi identities are *not* satisfied: whenever the symmetric product contains other representations, the validity of the Jacobi identities is not automatic, and needs to be checked. With this caveat in mind, we simply list the simple super algebras, following the work of V. Kac.

Much of the machinery we have developed for Lie algebras applies to the super algebras, by a careful distinction of even and odd elements. Assign to any element of a Lie super algebra X an index x, which is equal to zero if X is an even element, and one if it is an odd element. This enables us to introduce the elegant notation for the *graded commutators*

$$\{X, Y] \equiv XY - (-1)^{xy} \, YX, \tag{10.11}$$

in terms of which the graded Jacobi identity becomes

$$\{X, \{Y, Z]] + \{Y, \{Z, X]] + \{Z, \{X, Y]] = 0. \tag{10.12}$$

The Lie super algebras are constructed from as many $OSp(2)$ as the rank, just as Lie algebras of rank r are generated by r copies of non-commuting $SU(2)$ algebras. The explicit construction is beyond the scope of this book, and we refer the reader to V. Kac's book [12] for detailed constructions.

The simple super algebras are as follows.

- $SU(n\,|\,m)$. The even elements form the Lie algebra $SU(n) \times SU(m) \times U(1)$. The odd elements transform as $(\mathbf{n},\, \overline{\mathbf{m}})$ and its complex conjugate. Hence its content

$$(\mathbf{1},\, \mathbf{1}) + \left(\mathbf{1},\, \mathbf{n^2 - 1}\right) + \left(\mathbf{n^2 - 1},\, \mathbf{1}\right) + (\mathbf{n},\, \overline{\mathbf{m}})_f + (\overline{\mathbf{n}},\, \mathbf{m})_f .$$

  The f subscript applies to the odd elements. The full super algebra can be conveniently written in terms of $(m+n) \times (n+m)$ hermitian matrices of the form

$$\begin{pmatrix} SU(n) & (\mathbf{n},\, \overline{\mathbf{m}})_f \\ (\overline{\mathbf{n}},\, \mathbf{m})_f & SU(m) \end{pmatrix} + \begin{pmatrix} n & 0 \\ 0 & m \end{pmatrix} .$$

  The diagonal matrix generates $U(1)$. The graded commutator of these matrices is such that the off-block-diagonal odd elements obey anticommutation relations while the block-diagonal elements satisfy commutation relations. These rules single out the *supertrace*, defined as the trace of the upper diagonal block *minus* the trace of the lower diagonal block. Thus when $n = m$, the supertrace of the diagonal matrix vanishes, and the $U(1)$ decouples from the algebra. The resulting super algebra, $SSU(n\,|\,n)$, contains only $SU(n) \times SU(n)$ in its Lie part. Its non-compact form $SSU(2, 2\,|\,4)$ is the symmetry group of the $N = 4$ super Yang–Mills theory.
- $OSp(n\,|\,2m)$. The Lie part is $SO(m) \times Sp(2m)$, and the odd elements transform as the real $(\mathbf{n},\, \mathbf{2m})_f$. This is the generalization of the smallest Lie super algebra discussed earlier. It is possible because orthogonal and symplectic have opposite symmetry properties: the antisymmetric (symmetric) product of the fundamental of the orthogonal (symplectic) vector representation contains the adjoint representation of the algebra.
- $A[n-1]$. The Lie part is $SU(n)$. The odd elements are the symmetric traceless $(n \times n)$ matrices which transform as the adjoint of $SU(n)$. Recall that $SU(n)$ algebras contain (for $n \geq 3$) the adjoint in both the symmetric (d-coupling) and antisymmetric (f-coupling since it is a Lie algebra) product of its adjoint representation.
- $P(n-1)$ and $\overline{P}(n-1)$. These super algebras also contain $SU(n)$ as their Lie part. The odd elements are different; for $P(n-1)$, they are the symmetric second rank tensor $S^{(ab)}$, and the second rank antisymmetric tensor $A_{[ab]}$. This is a *complex* super algebra, and its conjugate $\overline{P}(n-1)$ contains the symmetric $S_{(ab)}$, and the antisymmetric $A^{[ab]}$. These require some getting used to. For $n = 3$, the odd part is the sum of $\mathbf{6}$ and $\overline{\mathbf{3}}$. Hence many of the anticommutators vanish since only $\mathbf{6} \times \overline{\mathbf{3}}$ contains the adjoint $\mathbf{8}$.

- $F[4]$. Its even Lie part is $SO(7) \times SU(2)$. Its sixteen complex odd elements transform as $(\mathbf{8}, \mathbf{2})$. The $\mathbf{8}$ here is the real $SO(7)$ spinor representation. It has the same pseudoreality as the $SU(2)$ doublet. Now

$$(\mathbf{8}, \mathbf{2}) \times_s (\mathbf{8}, \mathbf{2}) = \left(\mathbf{8} \times_s \mathbf{8}, \mathbf{2} \times_s \mathbf{2}\right) + \left(\mathbf{8} \times_a \mathbf{8}, \mathbf{2} \times_a \mathbf{2}\right),$$

where $\times_{s,a}$ denotes the symmetric and antisymmetric products. Since $\mathbf{8} \times_{s,a} \mathbf{8}$ contain the $SO(7)$ singlet and adjoint $\mathbf{21}$ respectively, the above product contains the sum of the adjoints, as required.
- $G[3]$. The Lie part is $G_2 \times SU(2)$, under which the odd elements transform as $(\mathbf{7}, \mathbf{2})$. It has similar structure as $F[4]$, since $\mathbf{7} \times_{s,a} \mathbf{7}$ respectively contain the G_2 singlet and adjoint $\mathbf{14}$. It is the only super algebra with an exceptional group in its Lie part.
- $D(2\,|\,1\,;\,a)$. The even elements are $SU(2) \times SU(2) \times SU(2)$. The odd elements are doublets under each of these, that is they transform as $(\mathbf{2}, \mathbf{2}, \mathbf{2})$. In order to explain the parameter a, introduce spinor indices α, β, γ for each $SU(2)$. The odd elements are then represented as $\chi_{\alpha\,\beta\,\gamma}$, such that under each $SU(2)$, generated by R^A, S^A and T^A respectively,

$$\left[R^A, \chi_{\alpha\,\beta\,\gamma}\right] = \frac{1}{2}\left(\sigma^A\right)_{\alpha\alpha'} \chi_{\alpha'\,\beta\,\gamma},$$

$$\left[S^A, \chi_{\alpha\,\beta\,\gamma}\right] = \frac{1}{2}\left(\sigma^A\right)_{\beta\beta'} \chi_{\alpha\,\beta'\,\gamma},$$

$$\left[T^A, \chi_{\alpha\,\beta\,\gamma}\right] = \frac{1}{2}\left(\sigma^A\right)_{\gamma\gamma'} \chi_{\alpha\,\beta\,\gamma'}.$$

The anticommutator contains three terms, by symmetrizing on one $SU(2)$ and antisymmetrizing on the other two. Specifically

$$\begin{aligned}\left\{\chi_{\alpha\,\beta\,\gamma}, \chi_{\alpha'\,\beta'\,\gamma'}\right\} &= a_1\,(\sigma_2)_{\alpha\alpha'}(\sigma_2)_{\beta\beta'}\left(\sigma^A\sigma_2\right)_{\gamma\gamma'} T^A \\ &\quad + a_2\,(\sigma_2)_{\alpha\alpha'}(\sigma_2)_{\gamma\gamma'}\left(\sigma^A\sigma_2\right)_{\beta\beta'} S^A \\ &\quad + a_3\,(\sigma_2)_{\gamma\gamma'}(\sigma_2)_{\beta\beta'}\left(\sigma^A\sigma_2\right)_{\alpha\alpha'} R^A,\end{aligned}$$

with $a_1 + a_2 + a_3 = 0$, all from the graded Jacobi identity. We fix $a_1 = 1$ for normalization, which leaves us with $a_2 = a$ and $a_3 = -1 + a$, and one undetermined parameter. For the special choices $a = 1/2, 2, -1$, this algebra reduces to $OSp(4\,|\,2)$.

In addition to these super algebras, one finds the so-called hyperclassical simple super algebras. They also have even and odd elements, but their even elements do not form Lie algebras, with some exceptions. They are constructed out of Grassmann numbers θ_i and their derivatives ∂_i, $i, j = 1, 2, \ldots, n$, which satisfy anticommutation relations

$$\left\{\theta_i, \theta_j\right\} = \left\{\partial_i, \partial_j\right\} = 0, \tag{10.13}$$

with

$$\{\,\partial_i,\,\theta_j\,\} = \delta_{ij}. \tag{10.14}$$

There are three infinite families of hyperalgebras.

- $S(n)$ and $\overline{S}(n)$. The elements of $S(n)$ are of the form

$$S(n): \quad \partial_i \mathcal{P}_n \partial_j + (i \leftrightarrow j),$$

where $\mathcal{P}_n$ is the most general polynomial in the Grassmann variables; it terminates at $\theta_1\theta_2\cdots\theta_n$ because of the anticommutation relations. The even (odd) parts contain even (odd) products of Grassmann numbers and derivatives. It is easy to see that $S(3) = \overline{P}(2)$. However, for larger n we get new super algebras. For instance, $S(4)$ contains only representation of $SU(4)$,

$$S(4): \quad \mathbf{15} + \overline{\mathbf{10}} + (\mathbf{4} + \mathbf{20})_f,$$

and the even part is not a Lie algebra although it contains the $SU(4)$ adjoint.
- $W(n)$ and $\overline{W}(n)$. These contain elements of the form

$$W(n): \quad \mathcal{P}_n \partial_i.$$

We can see that while $W(2) = SU(2\,|\,1)$, we have

$$W(3): \quad \mathbf{8} + \overline{\mathbf{3}} + (\mathbf{3} + \overline{\mathbf{3}} + \overline{\mathbf{6}})_f,$$

which contains the $SU(3)$ adjoint and the complex $\overline{\mathbf{3}}$ in its even part.
- $H(n)$. Its elements are of the form

$$H(n) \quad \sum_{i=1}^{n} \partial_i \mathcal{P}_n \partial_i.$$

For low values of n, these are trivial. However, we have the isomorphism $H(4) = SSU(2\,|\,2)$. When $n > 4$, its even part contains the $SO(n)$ generators together with antisymmetric tensors with an even number of indices. Its odd elements are the antisymmetric tensors with an odd number of indices.

It is beyond the scope of this book to construct their representations.

11

The groups of the Standard Model

The symmetries of a physical system manifest themselves through its conservation laws; they are encoded in the Hamiltonian, the generator of time translations. In Quantum Mechanics, the hermitian operators which commute with the Hamiltonian generate the symmetries of the system.

Similar considerations apply to local Quantum Field Theory. There, the main ingredient is the Dirac–Feynman path integral taken over the exponential of the action functional. In local relativistic field theory, the action is the space-time integral of the Lagrange density, itself a function of fields, local in space-time, which represent the basic excitations of the system. In theories of fundamental interactions, they correspond to the elementary particles. Through E. Noether's theorem, the conservation laws are encoded in the symmetries of the action (or Lagrangian, assuming proper boundary conditions at infinity).

Physicists have identified four different forces in Nature. The force of gravity, the electromagnetic, weak, and strong forces. All four are described by actions which display stunningly similar mathematical structures, so much so that the weak and electromagnetic forces have been experimentally shown to stem from the same theory. Speculations of further syntheses abound, unifying all three forces except gravity into a *Grand Unified Theory*, or even including gravity in *Superstring or M Theories*!

In his remarkable 1939 James Scott lecture, Dirac speaks of the *mathematical quality of Nature* and even advocates a *principle of mathematical beauty* in seeking physical theories!

The so-called Standard Model of Elementary Particle Physics describes the fundamental interactions of Nature as we know it. It is not a complete theory as it only accounts for the gravitational force in the large, and fails in its quantum treatment. On the other hand, the other three forces are successfully described in the quantum realm.

The Standard Model symmetries can be split into two categories; the first type of symmetry acts directly on space-time; it is encapsulated into a set of transformations called the *Poincaré group*; the second type, which acts directly on the charges of the elementary particles, is called gauge symmetries; they are connected with the ubiquitous Lie algebras, $SU(3)$, $SU(2)$, and $U(1)$ through the Yang–Mills construction. Electric charge conservation is traced to a *local gauge* symmetry which acts on the fields with space-time dependent parameters. On the other hand, in the absence of gravity, the space-time transformations are *global*, the same for all space-time points. These invariances are still not sufficient to totally determine the multiplicity and interactions of the fields. There is a mysterious triplication of elementary particles into three families which awaits explanation, pointing perhaps to an hitherto unknown family symmetry.

For reasons that lie beyond the scope of this book, it turns out not to be possible to unify space-time and gauge symmetries under the aegis of one mathematical group, limiting their union to a direct product structure: the Standard Model Lagrangian is a polynomial of local fields, invariant under the direct product of the Poincaré group with the three gauge symmetries. Group theory comes in through the construction of invariants under all these symmetries.

11.1 Space-time symmetries

We first discuss symmetries which act on space and time. H. Lorentz had found Maxwell's equations to be invariant under six transformations, today called the *Lorentz group*, the non-compact group $SO(3, 1)$. Einstein boldly posited that the invariances of Maxwell's equations applied to Mechanics as well, thereby replacing Newton's laws with his own theory of Special Relativity. Theories which satisfy special relativity have a Lagrangian that is invariant under space-time rotations; invariance under space rotations translates into angular momentum conservation.

Noether's theorem associates time translation invariance with energy conservation, and invariance under space translations with momentum conservation. Poincaré obtained the ten-parameter *Poincaré group* by adding constant space and time translations, thus incorporating, *à la* Noether, conservation of momentum and energy.

Five years later, Bateman found the invariance group of the *free* Maxwell equations to be the larger fifteen-parameter *conformal group*, $SO(4, 2)$, which contains the Poincaré group as a subgroup. The conformal group is not a symmetry of Nature as we perceive it, since it contains no dimensionful parameters, such as the masses of elementary particles, the quark confinement scale, Newton's constant (Planck mass), to name a few. If it were to apply at all to the real world, it would be in the particular limit where *all* parameters with the dimension of mass are set to zero. Since much physics is to be gained from the study of Maxwell's equations, we discuss both the Poincaré and conformal groups.

11.1.1 The Lorentz and Poincaré groups

The Lorentz group $SO(3,1)$ contains six antisymmetric generators $M^{\mu\nu}$ which satisfy the commutation relations

$$[\, M^{\mu\nu},\, M^{\alpha\beta}\,] = i(\eta^{\mu\alpha}M^{\nu\beta} - \eta^{\nu\alpha}M^{\mu\beta} + \eta^{\nu\beta}M^{\mu\alpha} - \eta^{\mu\beta}M^{\nu\alpha}), \tag{11.1}$$

where the Greek indices run over $0, 1, 2, 3$, and with $\eta^{\mu\nu} = \eta_{\mu\nu} = \text{diag}\,(-1, 1, 1, 1)$. They act on the space-time coordinates x^μ as

$$x^\mu \to a^{\mu\nu}x^\nu, \tag{11.2}$$

where $a^{\mu\nu} = -a^{\nu\mu}$ are six infinitesimal constant rotation parameters which leave the quadratic form

$$\eta_{\mu\nu}x^\mu x^\nu = -(x^0)^2 + (x^1)^2 + (x^2)^2 + (x^3)^2,$$

invariant. We also have

$$[\, M^{\mu\nu},\, x^\sigma\,] = i(\eta^{\mu\sigma}x^\nu - \eta^{\nu\sigma}x^\mu). \tag{11.3}$$

Since the quadratic invariant is hyperbolic, these transformations cannot be represented with hermitian generators acting in a finite-dimensional space; all unitary representations of non-compact groups are infinite-dimensional.

In our study of compact Lie algebras, we have noted the isomorphism $SO(4) \sim SO(3) \times SO(3)$. It translates to the non-compact case by defining

$$M^{ij} = \epsilon^{ijk}\, J^k; \qquad M^{0i} = K^i, \tag{11.4}$$

where i, j, k run over the three space indices. Then we see that

$$[\, J^i,\, J^j\,] = i\epsilon^{ijk}\, J^k, \qquad [\, J^i,\, K^j\,] = i\epsilon^{ijk}\, K^k, \tag{11.5}$$

while

$$[\, K^i,\, K^j\,] = -i\epsilon^{ijk}\, J^k. \tag{11.6}$$

The all-important minus sign shows that $SO(3,1)$ splits into two sets of commuting generators,

$$\frac{1}{2}(J^i + iK^i), \qquad \frac{1}{2}(J^i - iK^i), \tag{11.7}$$

each of which satisfies the $SO(3)$ commutation relations. Neither is hermitian, and that is the difference with the compact case. Under the parity operation

$P: x^i \to -x^i$, we see that $J^i \to -J^i$, and $K^i \to -K^i$, so that the two $SU(2)$ go into one another; the same is true under complex conjugation $C: i \to -i$.

We can use this decomposition to obtain the representations of the Lorentz group. First represent the Lorentz generators in terms of (2×2) Pauli matrices as

$$J^i = \frac{1}{2}\sigma^i, \qquad K^i = \frac{i}{2}\sigma^i, \tag{11.8}$$

which act on a two-dimensional complex spinor field ψ_L. The simplest representation is then

$$\psi_L \to \psi'_L = e^{\frac{i}{2}\sigma^i(\omega^i + i\nu^i)}\,\psi_L, \tag{11.9}$$

where ω^i (ν^i) are the rotation (boost) parameters. It is called a (left-handed) Weyl spinor, and we describe it as $(\mathbf{2},\,\mathbf{1})$. The right-handed Weyl spinor $(\mathbf{1},\,\mathbf{2})$, transforms as

$$\psi_R \to \psi'_R = e^{\frac{i}{2}\sigma^i(\omega^i - i\nu^i)}\,\psi_R. \tag{11.10}$$

For those who like indices, spinors which transform as a left-handed Weyl spinor are denoted with a subscript α, while those transforming as right-handed spinors are given dotted indices, $\dot\alpha$

$$\psi_\alpha \sim (\mathbf{2},\,\mathbf{1}); \qquad \psi_{\dot\alpha} \sim (\mathbf{1},\,\mathbf{2}).$$

Clearly, while the space rotations are unitarily represented, the boosts are not. Left- and right-handed spinors can be related since the reader will verify that $\sigma^2\,\psi_R^*$ transforms as a left-handed spinor: all spinor representations can be written in terms of left-handed Weyl spinors. In this notation, real representations are of the form $(\mathbf{n},\,\mathbf{n})$: four-vectors are $(\mathbf{2},\,\mathbf{2})$, corresponding to the (2×2) matrix representation

$$x^\mu \to X = \begin{pmatrix} x^0 + x^3 & x^1 - ix^2 \\ x^1 + ix^2 & x^0 - x^3 \end{pmatrix}, \tag{11.11}$$

(with matrix elements written as $X_{\alpha\dot\alpha}$) whose determinant is the invariant quadratic form. The electric and magnetic fields assemble as $\vec{E} + i\vec{B} \sim (\mathbf{3},\,\mathbf{1})$, while $\vec{E} - i\vec{B} \sim (\mathbf{1},\,\mathbf{3})$. Since we are interested in the action of these generators on local fields, we introduce the operators conjugate to x^μ

$$[x^\mu,\, p^\nu] = i\eta^{\mu\nu}, \tag{11.12}$$

and separate the Lorentz generators into orbital and spin parts

$$M^{\mu\nu} = x^\mu p^\nu - x^\nu p^\mu + S^{\mu\nu}. \tag{11.13}$$

The spin parts $S^{\mu\nu}$ do not act on the space-time coordinates, but satisfy the same commutation relations as $M^{\mu\nu}$. For left-handed Weyl spinors, we would have $S^{ij} = \epsilon^{ijk}\sigma^k/2$, $S^{0i} = i\epsilon^{ijk}\sigma^k/2$.

The Poincaré group adds to the Lorentz group P^μ, the space-time translation generators (momenta) of

$$x^\mu \to x^\mu + a^\mu, \tag{11.14}$$

where a^μ is a constant. They transform as a four-vector,

$$[\, M^{\mu\nu},\ P^\sigma \,] = i(\eta^{\mu\sigma} P^\nu - \eta^{\nu\sigma} P^\mu), \tag{11.15}$$

and commute among themselves,

$$[\, P^\mu, P^\nu \,] \ = \ 0. \tag{11.16}$$

We can always take

$$P^\mu \ = -i\frac{\partial}{\partial x_\mu} \equiv p^\mu, \tag{11.17}$$

where $x_\mu = \eta_{\mu\nu} x^\nu$. We see that the Poincaré group is the semi-direct product of the Lorentz group with the Abelian translations. Its representations are characterized by the values of its two Casimir operators, $P_\mu P^\mu$, and $W^\mu W_\mu$, where W_μ is the Pauli–Lubanski vector

$$W_\mu \ = \frac{1}{2}\epsilon_{\mu\nu\rho\sigma} P^\nu M^{\rho\sigma}, \tag{11.18}$$

and where $\epsilon_{\mu\nu\rho\sigma}$ is the totally antisymmetric Levi–Civita symbol ($\epsilon_{0123} = 1$). They satisfy the commutation relations

$$[\, W_\mu, \ W_\nu \,] \ = -i\epsilon_{\mu\nu\rho\sigma}\ P^\rho W^\sigma, \tag{11.19}$$

and are orthogonal to the momenta

$$P^\mu W_\mu = 0.$$

The representations of the Poincaré group fall into three distinct types, since $P_\mu P^\mu$ can be negative, zero, or positive.

- $P^\mu P_\mu \ = -m^2$

With m real, the four-momentum is a *time-like* vector. Accordingly, we can by Lorentz transformations go to the rest frame, where

$$P^0 = m, \qquad P^i = 0.$$

Evaluation of the Pauli–Lubanski vector in the rest frame yields

$$W^0 = 0, \qquad W^i \ = -mS^i, \qquad S^i \ = \frac{1}{2}\epsilon^{ijk} S^{jk}, \qquad i, j, k = 1, 2, 3.$$

S^i are the three generators of $SO(3)$,

$$[\, S^i,\, S^j \,] = i\epsilon^{ijk}\, S^k,$$

whose quadratic Casimir operator is $s(s+1)$, with $s = 0, 1/2, 1, 3/2, \ldots$. These representations with

$$P^\mu P_\mu = -m^2, \qquad W^\mu W_\mu = m^2 s(s+1), \tag{11.20}$$

describe particles of mass m and spin s. We can further express the generators by singling out different ways in which the physical system evolves.

The most familiar one is the Newton–Wigner representation, to which we now turn our attention.

Newton–Wigner representation

The idea is to represent the generators on an initial surface in space-time. We identify x^0 with physical time, so that the initial surface is the slice of space-time $x^0 =$ constant. From the general representation eq. (11.13), the Pauli–Lubanski vector is given by

$$W_0 = \frac{1}{2}\epsilon_{ijk} p^i S^{jk}, \qquad W_i = -\epsilon_{ijk}\left(\frac{1}{2}p^0 S^{jk} - p^j S^{0k}\right). \tag{11.21}$$

The Lorentz generators split into space rotations M^{ij} which do not affect the initial surface, and the boosts M^{0i} which do; they will have a more complicated expression on the initial surface. To find it, we set

$$S^{0i} = A\,\epsilon^{ijk} p^j S^k, \tag{11.22}$$

and obtain, after a little bit of algebra ,

$$W^\mu W_\mu = [(p^0)^2 + (2Ap^0 + A^2 p^i p^i) p^j p^j] S^k S^k + (p^i S^i)^2 (2Ap^0 + A^2 p_j p_j + 1). \tag{11.23}$$

Comparison with eq. (11.20) requires

$$2Ap^0 + A^2 p^i p^i + 1 = 0,$$

that is,

$$A = \frac{1}{p^0 \pm m}, \tag{11.24}$$

which yields S^{0i} in terms of the other generators.

In these expressions, x^0 is to be identified with physical time, and the generators are expressed at an initial fixed value of x^0. This means that its conjugate variable $p^0 = -i\partial^0$ is no longer well defined: it must be expressed in terms of the remaining variables. This is achieved through the method of Lagrange multipliers. Following P. A. M. Dirac (*Rev. Mod. Phys.* **29**, 312 (1949)), we set

$$P^0 = p^0 + \lambda(p^\mu p_\mu + m^2), \tag{11.25}$$

and choose λ so as to eliminate p^0. The result is

$$P^0 = \pm\sqrt{p^i p^i + m^2}, \tag{11.26}$$

corresponding to positive and negative energy solutions. At $x^0 = 0$, we obtain the Newton–Wigner representation of the Poincaré group generators

$$P^0 = \pm\sqrt{m^2 - \partial^i\partial^i}, \qquad p^i = -i\partial^i, \tag{11.27}$$

$$M^{ij} = i(x^j\partial^i - x^i\partial^j) + \epsilon^{ijk}S^k, \quad M^{0i} = -\frac{1}{2}\{x^i, P^0\} - i\frac{\epsilon^{ijk}\partial^j S^k}{P^0 \pm m}, \tag{11.28}$$

accounting for the fact that x^i and P^0 no longer commute. The generators on the initial surface $x^0 = 0$, split into *kinematical* generators, p^i and M^{ij}, and *dynamical* generators, the Hamiltonian P^0 and the boosts M^{0i}, which move the system away from the initial surface. The Poincaré group therefore acts on Hilbert states labeled by kets of the form

$$\mid p^i;\ m,\ s,\ s_3\ >,$$

which are interpreted as particles of mass m, spin s, with three-momenta p^i and magnetic quantum number s_3. Their (positive or negative) energy is given by the mass-shell condition (11.26) in terms of their mass and three-momenta.

Light-cone representation

Dirac introduced the *front form*, a different way of expressing dynamical evolution, known today as the light-cone. The initial surface is light-like, $x^\mu x_\mu = 0$, labeled by constant values of

$$x^+ \equiv \frac{1}{\sqrt{2}}(x^0 + x^3). \tag{11.29}$$

For fixed x^+, its conjugate variable

$$p^- = \frac{1}{\sqrt{2}}(p^0 - p^3), \qquad [\,x^+,\ p^-\,] = -i, \tag{11.30}$$

is no longer well-defined, and must be expressed in terms of the remaining variables, that is x^-, p^+,

$$x^- = \frac{1}{\sqrt{2}}(x^0 - x^3), \qquad p^+ = \frac{1}{\sqrt{2}}(p^0 + p^3), \qquad [\,x^-,\ p^+\,] = -i, \tag{11.31}$$

and the transverse variables x^a, p^b, which satisfy

$$[\,x^a\,,\;p^b\,] = i\,\delta^{ab},\;\; a,b = 1,2. \tag{11.32}$$

There are seven kinematical Poincaré generators, p^+, p^a, M^{ab}, M^{+a}, and M^{+-}. The latter is kinematical only at $x^+ = 0$, since it contains x^+p^-.

There are only three dynamical Poincaré generators, the light-cone Hamiltonian P^- and the two boosts M^{-a}. Using Dirac's method of Lagrange multipliers, we find

$$P^- = \frac{p^a p^a + m^2}{2p^+}. \tag{11.33}$$

The remaining three translation generators P^+ and P^i are kinematical. We expect the kinematical generators to retain a simple form, for example

$$M^{ab} = x^a p^b - x^b p^a + S^{ab} \tag{11.34}$$

generates the transverse plane rotation. Finding the expression for S^{+a}, S^{+-}, and S^{-a} take a bit more doing.

Let us write the momentum vector in matrix form

$$P = \begin{pmatrix} p^0 + p^3 & p^1 - ip^2 \\ p^1 + ip^2 & p^0 - p^3 \end{pmatrix}, \tag{11.35}$$

with $\det P = m^2$. In the rest frame of the particle this matrix reduces to m times the unit matrix. It is easy to see that

$$P = \mathcal{U}\begin{pmatrix} m & 0 \\ 0 & m \end{pmatrix}\mathcal{U}^\dagger, \tag{11.36}$$

where

$$\mathcal{U} = \begin{pmatrix} \sqrt{\frac{\sqrt{2}p^+}{m}} & 0 \\ \frac{(p^1+ip^2)}{\sqrt{m\sqrt{2}p^+}} & \sqrt{\frac{m}{\sqrt{2}p^+}} \end{pmatrix} \tag{11.37}$$

is the matrix that transforms the rest frame into the light-cone frame. We can apply it to the Pauli–Lubanski vector as well,

$$W = \begin{pmatrix} W^0 + W^3 & W^1 - iW^2 \\ W^1 + iW^2 & W^0 - W^3 \end{pmatrix} = -m\,\mathcal{U}\begin{pmatrix} S^{12} & S^{23} - iS^{31} \\ S^{23} + iS^{31} & -S^{12} \end{pmatrix}\mathcal{U}^\dagger, \tag{11.38}$$

using its form in the rest frame. In light-cone coordinates, the components of the Pauli–Lubanski vector are

$$W^+ = -p^+ S^{12} + \epsilon^{ab} p^a S^{+b}, \qquad W^- = p^- S^{12} - \epsilon^{ab} p^a S^{-b}, \tag{11.39}$$

and

$$W^a = \epsilon^{ab}(p^- S^{+b} + p^b S^{-+} - p^+ S^{-b}), \tag{11.40}$$

where $\epsilon^{12} = -\epsilon^{21} = 1$. By comparing (11.38) with (11.39) for W^+, we find $S^{+a} = 0$, so that

$$M^{+a} = x^+ p^a - x^a p^+. \tag{11.41}$$

A similar procedure for the other components requires $S^{+-} = 0$, and thus

$$M^{+-} = -x^- p^+ + x^+ p^- = -x^- p^+, \tag{11.42}$$

at $x^+ = 0$. We can evaluate S^{-a} from comparing the entries in $W_1 + i W_2$ on both sides as well as W^-, yielding

$$S^{-a} = \frac{i}{p^+}(m S^a + i p^b S^{ab}). \tag{11.43}$$

We set $T^a = m S^a$, keeping the commutation relation,

$$[T^1, T^2] = i m^2 S^{12}. \tag{11.44}$$

This yields the light-cone representation of the Poincaré generators, first derived by Bacry and Chang (*Ann. Phys. (N.Y.)* **47**, 407 (1968)),

$$P^- = \frac{p^a p^a + m^2}{2p^+}, \qquad P^+ = p^+, \qquad P^a = p^a, \tag{11.45}$$

$$\begin{aligned} M^{+a} &= -x^a p^+, \quad M^{+-} = -x^- p^+, \quad M^{ab} = x^a p^b - x^b p^a + S^{ab}, \\ M^{-a} &= x^- p^a - \frac{1}{2}\{x^a, P^-\} + \frac{i}{p^+}(T^a + i p^b S^{ab}). \end{aligned} \tag{11.46}$$

In these coordinates, the Poincaré group acts Hilbert states labeled by the kets

$$| p^+, p^a; m, s, s_3 >,$$

representing massive states with light-cone momenta p^+, p^a, and spin s with p^- obeying the mass-shell condition (11.33).

This light-cone representation is very useful in other contexts; it readily generalizes to any number of transverse dimensions, for which we have the same representation, except that the indices a, b run over an arbitrary number of values, and thus

$$[T^a, T^b] = i m^2 S^{ab}. \tag{11.47}$$

It arises in the quantization of the Nambu–Goto string by J. Goldstone, P. Goddard, C. Rebbi and C. B. Thorn (*Nucl. Phys.* **B56**, 109 (1973)). For the string, the operators m^2 and S^{ab} are quadratic combinations of an infinite number of harmonic oscillators operators, while T^a are *cubic* polynomials in the same harmonic oscillators. Owing to normal ordering, this commutation relation is satisfied *only* when a, b range over twenty-four values: relativistic strings live only in a twenty-six dimensional space-time.

- $P^\mu P_\mu = 0$.

This case can be viewed in terms of the previous representation in the limit $m \to 0$. The Pauli–Lubanski vector is now light-like. Since both P^μ and W^μ are light-like and orthogonal to one another, they must be proportional. In the light frame where

$$P^- = P^1 = P^2 = 0, \quad W^- = W^1 = W^2 = 0, \quad W^+ = P^+ S^{12}, \tag{11.48}$$

so that

$$\frac{W^+}{P^+} = S^{12} \equiv \lambda, \tag{11.49}$$

where the value of the $SO(2)$ generator is the *helicity*. We can use the same representation for the operators, the difference being that we can no longer divide by m.

In the light-cone representation, the T^a operators transform as a transverse vector,

$$[\, S^{12},\, T^1\,] = T^2, \qquad [\, S^{12},\, T^2\,] = -T^1, \tag{11.50}$$

and commute with one another

$$[\, T^1,\, T^2\,] = 0. \tag{11.51}$$

This leads to two possibilities.

- $T^a = 0$. The ket labels are now just the momenta and the helicity λ, describing massless particles

$$|\, p^+,\, p^a; m = 0,\, \lambda\, >, \tag{11.52}$$

of fixed helicity. The values of the helicity are restricted, not by the algebra, but by the group. Exponentiation of the transverse rotations must be single-valued, which requires the helicity to be $\pm$ half-odd integers. This case represents a massless particle with a single value of helicity, $\lambda = 0, \pm 1/2, \pm 1, \ldots$.

From the point of view of group theory, massive representations can be thought of as a sum of massless representations; for instance the massive spin one vector representation consists of three states with magnetic component $-1, 0, 1$, each of which occurs as a massless representation.

- $T^a \neq 0$. In this case, T^a are simply the c-number components of a transverse vector. The states on which the Poincaré algebra is realized must now be labeled by the maximal number of commuting operators, that is the transverse vector ξ^a

$$T^a \mid p^+, p^a;\ \xi^a \ >= \xi^a \mid p^+, p^a;\ \xi^a \ > . \tag{11.53}$$

The representations are labeled by the length of the transverse vector ξ^a, since the Pauli–Lubanski vector is now space-like with

$$W^\mu W_\mu = \xi^a \xi^a. \tag{11.54}$$

Helicity is no longer a good quantum number as the light-cone boosts which contain T^i change the helicity by one unit. Hence the representation is made up of an infinite number of helicities, spaced by one unit. We have therefore two distinct representations, depending on whether the helicities are integer or half-odd integer. They are called "continuous spin representations" by Wigner in his original work. These have no simple interpretations in terms of particles, since they describe a massless object with an *infinite* number of integer-spaced helicities.

Finally, let us mention that the Poincaré group can be extended to a super algebra, the SuperPoincaré or Supersymmetry algebra. Yu. A. Golfand and E. P. Likthman (*JETP Lett.* **13**, 323 (1971)) included new generators that transform as spinors such that their *anticommutator* yields the translation generators

$$\{\, Q_\alpha,\ \bar{Q}_{\dot{\alpha}} \,\} = (\sigma^\mu)_{\alpha\dot{\alpha}}\, P_\mu, \tag{11.55}$$

and commute with the translations

$$[\, Q_\alpha,\ P^\mu \,] = [\, \bar{Q}_{\dot{\alpha}},\ P^\mu \,] = 0. \tag{11.56}$$

It follows that $P^\mu P_\mu$ is still an invariant and all of its representations can be discussed in terms of its value.

- Its $m = 0$ representation simply consists of two massless Poincaré algebra representations separated by half a unit of helicity, denoted as $(\lambda, \lambda + \frac{1}{2})$.
- The $m \neq 0$ representations are assembled out of massless representations, as in the Poincaré case. For instance a massive spin one-half particle is made up as $(\frac{1}{2}, 0)$, and $(0, -\frac{1}{2})$, so that it is accompanied by a complex scalar with the same mass. A massive spin one particle with three degrees of freedom belongs to $(1, \frac{1}{2}) + (\frac{1}{2}, -1) + (\frac{1}{2}, 0) + (0, -\frac{1}{2})$, and its supermultiplet contains two spinors and one complex scalar.
- The two continuous spin representations, one with integer helicity, the other with half-odd integer, neatly assemble into one massless supermultiplet.

11.1.2 The conformal group

The conformal group is the invariance group of the free Maxwell's equations. It is generated by a simple algebra, and is the simplest extension of the Poincaré group. While this group is of great theoretical interest, it is not a symmetry group of Nature as we know it. There are many ways to characterize it; for our purposes, we define it as the group of transformations which leaves the equation of a light ray invariant,

$$x^\mu \to x^{\mu\prime}; \qquad \eta_{\mu\nu}x^\mu x^\nu = \eta_{\mu\nu}x^{\mu\prime}x^{\nu\prime}, \tag{11.57}$$

It means that the light-cone analysis of the Poincaré algebra can be easily implemented.

The conformal algebra contains, in addition to the ten Poincaré generators, five new transformations, the dilatation

$$x^\mu \to s\, x^\mu, \tag{11.58}$$

and four conformal transformations

$$x^\mu \to \frac{x^\mu - c^\mu\,(x^\nu x_\nu)}{1 - 2c^\nu x_\nu + (c^\nu c_\nu)(x^\rho x_\rho)}, \tag{11.59}$$

with constant s and c^μ. The conformal transformations can be thought of as an inversion

$$x^\mu \to \frac{x^\mu}{(x^\nu x_\nu)}, \tag{11.60}$$

sandwiched between two translations. In space-time, the new generators are the dilatation

$$D = \frac{1}{2}(x^\mu p_\mu + p^\mu x_\mu) + d, \tag{11.61}$$

and the conformal transformations

$$K^\mu \;= -2x_\nu\, M^{\nu\mu} + (x^\nu x_\nu)p^\mu + 2(d - 2i)x^\mu + \kappa^\mu. \tag{11.62}$$

They do not act linearly on space-time. In the above, d and κ^μ are the non-orbital parts. Assuming that they satisfy the same commutators as the orbital parts, we find

$$[\,D,\, P^\mu\,] \;= i\,P^\mu, \qquad [\,D,\, K^\mu\,] \;= -i\,K^\mu, \tag{11.63}$$

and

$$[\,P^\mu,\, K^\nu\,] \;= 2i\; M^{\mu\nu} - 2i\;\eta^{\mu\nu}\;D, \qquad [\,K^\mu,\, K^\nu\,] = 0. \tag{11.64}$$

These commutation relation can be rewritten in a pleasing way by identifying

$$D = M^{56}, \quad K^\mu = (M^{5\mu} - M^{6\mu}), \quad P^\mu = (M^{5\mu} + M^{6\mu}), \tag{11.65}$$

which are seen to satisfy the $SO(4,2)$ Lie algebra,

$$[\,M^{AB},\, M^{CD}\,] = i(\eta^{AC}M^{BD} - \eta^{BC}M^{AD} + \eta^{BD}M^{AC} - \eta^{AD}M^{BC}), \tag{11.66}$$

with $A, B, \cdots = 0, 1, 2, 3, 5, 6$, and $\eta^{AB} = \eta_{AB}$ is a diagonal matrix with entries $(-1, 1, 1, 1, 1, -1)$.

The conformal group generators leave the quadratic form $\eta_{AB}\,\zeta^A\,\zeta^B$ invariant, where ζ^A are the six homogeneous coordinates on which they act linearly. It also has a cubic and a quartic Casimir operator, one more than the Poincaré group. On the light-cone

$$\eta_{AB}\,\zeta^A\,\zeta^B = 0, \tag{11.67}$$

the homogeneous coordinates are related to the space-time variables by stereographic projections

$$x^\mu = \frac{\sqrt{2}}{(\zeta^5 + \zeta^6)}\,\zeta^\mu, \qquad \eta_{\mu\nu}\,x^\mu\,x^\nu = 2\,\frac{(\zeta^6 - \zeta^5)}{(\zeta^6 + \zeta^5)}. \tag{11.68}$$

We note that $P^\mu P_\mu$ is no longer invariant; its eigenvalues are either zero or continuous. When it is zero, the particle interpretation of the Poincaré group is retained.

The $m = 0$ representations of the conformal group can be constructed using Dirac's front form since $x^2 = 0$ is conformally invariant, but there are other ways to do it: we note the especially elegant representation "witten" (*Commun. Math. Phys.* **252**, 189 (2004)) in terms of Penrose's *twistors*, the classical (boson-like) coordinates which transform as left- and/or right-handed Weyl spinors,

$$\lambda_\alpha \sim (\mathbf{2},\,\mathbf{1}), \qquad \tilde{\mu}_{\dot\alpha} \sim (\mathbf{1},\,\mathbf{2}). \tag{11.69}$$

In this language, indices can be raised by means of the antisymmetric symbols $\epsilon^{\alpha\beta}$ and $\epsilon^{\dot\alpha\dot\beta}$,

$$\lambda^\alpha = \epsilon^{\alpha\beta}\,\lambda_\beta, \qquad \tilde{\mu}^{\dot\alpha} = \epsilon^{\dot\alpha\dot\beta}\,\tilde{\mu}_{\dot\beta}. \tag{11.70}$$

We also introduce the derivative operators which satisfy

$$\frac{\partial}{\partial\lambda_\alpha}\,\lambda_\beta = \delta^\alpha_\beta, \qquad \frac{\partial}{\partial\tilde{\mu}_{\dot\alpha}}\,\tilde{\mu}_{\dot\beta} = \delta^{\dot\alpha}_{\dot\beta}. \tag{11.71}$$

The momenta are given in matrix form by

$$P_{\alpha\dot\alpha} = i\lambda_\alpha\,\frac{\partial}{\partial\tilde{\mu}^{\dot\alpha}}. \tag{11.72}$$

Its determinant clearly vanishes, and $m = 0$. The conformal generators are simply

$$K_{\alpha\dot\alpha} = i\tilde{\mu}_{\dot\alpha}\,\frac{\partial}{\partial\lambda^\alpha}, \tag{11.73}$$

with the dilatation in the form

$$D = \frac{i}{2}\left(\lambda_\alpha\,\frac{\partial}{\partial\lambda_\alpha} - \tilde{\mu}_{\dot\alpha}\,\frac{\partial}{\partial\tilde{\mu}_{\dot\alpha}}\right). \tag{11.74}$$

The Lorentz generators are written in symmetric bispinor form as

$$M_{\alpha\beta} = \frac{i}{2}\left(\lambda_\alpha \frac{\partial}{\partial \lambda^\beta} + \lambda_\beta \frac{\partial}{\partial \lambda^\alpha}\right), \quad M_{\dot\alpha\dot\beta} = \frac{i}{2}\left(\tilde\mu_{\dot\alpha} \frac{\partial}{\partial \tilde\mu^{\dot\beta}} + \tilde\mu_{\dot\beta} \frac{\partial}{\partial \tilde\mu^{\dot\alpha}}\right). \quad (11.75)$$

Thus while the conformal group acts non-linearly on space-time, its action on twistor space is linear.

A world with only massless particles, while in obvious contradiction with facts, may well be an interesting theoretical abstraction, because in the Standard Model masses result from spontaneous breaking of symmetries. For instance, the continuous spin representations of the Poincaré group are absent in both Nature and in the representations of the conformal group.

We close this section by determining the non-relativistic limit of the Poincaré and conformal groups. In that limit, the speed of light c is taken to infinity. It enters the generators through $x^0 = c\,t$, and the non-relativistic limit is obtained by building operators that survive it.

As $c \to \infty$, both M^{ij} and p^i retain their original form, while the Hamiltonian and boosts reduce to

$$c\,P^0 \to i\,\frac{\partial}{\partial t}, \qquad \frac{1}{c}\,M^{0i} \to -it\,\frac{\partial}{\partial x^i}. \quad (11.76)$$

Their action on $x^i(t)$ is then

$$x^i(t) \to x^i(t) + a^{ij}\,x^j(t) + a^i + v^i\,t, \quad (11.77)$$

where a^{ij}, a^i and v^i are constant. It is the group that leaves the equation

$$\frac{d^2x^i(t)}{d\,t^2} = 0, \quad (11.78)$$

invariant, in accordance with Newton's law for a free particle.

A similar analysis of conformal transformation shows that the conformal generators become

$$\frac{1}{c}\,K_0 \to i\,t^2\,\frac{\partial}{\partial t} - 2i\,t\,x^i\frac{\partial}{\partial x^i}, \qquad \frac{1}{c^2}\,K_i \to -i\,t^2\,\frac{\partial}{\partial x^i}. \quad (11.79)$$

The action of these generators on the trajectory $x^i(t)$ shows that it is quadratic in time. This means that conformal non-relativistic physics is invariant under

$$x^i(t) \to x^i(t) + a^i\,t^2, \quad (11.80)$$

a^i is a constant acceleration. The equation that replaces Newton's is of third order,

$$\frac{d^3x^i(t)}{d\,t^3} = 0. \quad (11.81)$$

In such a theory, one could not distinguish a particle at rest from one with a constant acceleration: protons uniformly accelerated around a ring would not radiate, much to the delight of accelerator physicists, but alas it is not so: conformal theories do not reproduce Nature.

11.2 Beyond space-time symmetries

It is natural to seek a unified description of space-time and internal symmetries, such as $SU(2)$-isospin and flavor $SU(3)$. We have already seen that by combining isospin and spin, Wigner and Stückelberg constructed the supermultiplet theory which could be used to catalog particles of different spin. With the advent of the flavor $SU(3)$ classification, F. Gürsey and L. Radicati (*Phys. Rev. Letters* **13**, 173 (1964)) and B. Sakita (*Phys. Rev.* **136**, B1756 (1964)) independently proposed to unify the eight-fold way with spin. This is akin to finding a group which naturally contains $SU(3)$ and $SU(2)$. The natural candidate is $SU(6)$ with embedding,

$$SU(6) \supset SU(3) \times SU(2), \qquad \mathbf{6} = (\mathbf{3},\ \mathbf{2}). \tag{11.82}$$

Hence the 35-dimensional adjoint deduced from $\mathbf{6} \times \bar{\mathbf{6}}$, yields

$$\mathbf{35} = (\mathbf{8},\ \mathbf{1}) + (\mathbf{1},\ \mathbf{3}) + (\mathbf{8},\ \mathbf{3}). \tag{11.83}$$

For physical identification, we are guided by the Fermi–Yang model, and recognize the pseudoscalar octet, and nine spin one vector mesons, which decompose under $SU(3)$ as an octet and a singlet. There are indeed two isoscalar vector mesons, the ω and ϕ particles, while the isotriplet vector is identified with the ρ-mesons, and the isodoublets correspond to the "K-star" vector resonances, and a linear combination of the ω and ϕ vector mesons in the octet, leading to the following.

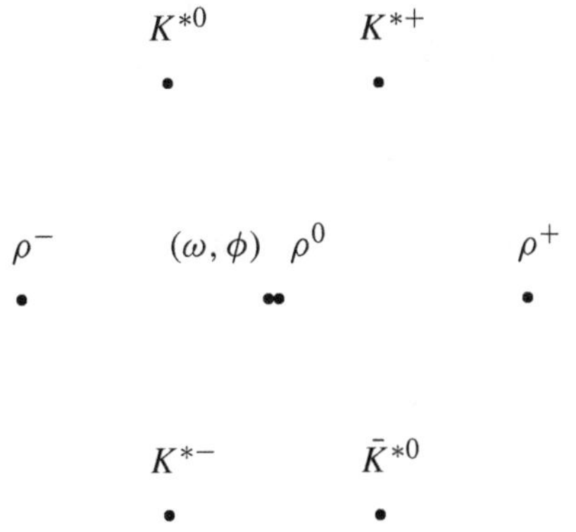

K^{*0} $\quad$ K^{*+}

ρ^- $\quad$ $(\omega, \phi)\ \rho^0$ $\quad$ ρ^+

K^{*-} $\quad$ $\bar{K}^{*0}$

So far so good. When this classification is generalized to the baryons, the spin one-half octet, and the spin three-half decuplet combine into one $SU(6)$ representation, the **56** with decomposition

$$\mathbf{56} = (\mathbf{8},\ \mathbf{2}) + (\mathbf{10},\ \mathbf{4}), \tag{11.84}$$

where we have labeled the spin representations by their dimensions. This unifies the baryon resonances in the octet and decuplet into one representation. The baryon representation is generated by the *symmetric* product of three fundamental sextets

$$\mathbf{56} = (\mathbf{6} \times \mathbf{6} \times \mathbf{6})_{sym}. \tag{11.85}$$

Symmetries can be used to relate the magnetic moments of the different particles in the baryon octet. Classically, the magnetic moment of a charge distribution is related to the electromagnetic current $\vec{J}$ by the formula

$$\vec{\mu} = \frac{1}{2c}\vec{r} \times \vec{J}. \tag{11.86}$$

In the quantum realm, this identification still applies, albeit altered by the spin of the particles. Hence the symmetry properties of the electromagnetic current imply physical consequences for the magnetic moments. Recall that the electric charge is given by the Gell-Mann–Nishijima formula (6.75)

$$Q = I_3 + \frac{Y}{2},$$

where I_3 is the third component of isospin and Y is the isospin singlet in $SU(3)$. This implies that the electromagnetic current is the sum of two parts; one that is an isospin singlet, and one that transforms as an isotriplet. The same applies to the magnetic moment. For $SU(2)$, this means, simply based on the symmetries, that (Wigner–Eckhardt theorem)

$$\mu = \langle\, I, I_3 \,|\, (a + b\,\hat{I}_3) \,|\, I, I_3 \,\rangle,$$

where the hat denotes the operator, and a, b are undetermined constants. Applied to the isotriplet of Σ^+, Σ^0 and Σ^-, this formula implies

$$\mu_{\Sigma^\pm} = a \pm \frac{b}{2}, \qquad \mu_{\Sigma^0} = a,$$

which yields one relation among physical quantities.

$$\mu_{\Sigma^+} + \mu_{\Sigma^-} = 2\,\mu_{\Sigma^0}. \tag{11.87}$$

To derive similar formulas for flavor $SU(3)$, we need to determine the transformation properties of the electric current. To that effect, we note that the electric charge commutes with an $SU(2)$ subgroup of $SU(3)$. Dubbed U-spin by

S. Meshkov, C. A. Levinson, and H. J. Lipkin (*Phys. Rev. Letters* **10**, 361 (1963)), its multiplets can be read off from the oblique lines in the octet diagram, which contains two U-spin doublets

$$(\Sigma^-, \Xi^-), \qquad (p, \Sigma^+),$$

and a U-spin triplet and singlet

$$(n, T \equiv \frac{1}{2}(\sqrt{3}\Lambda^0 - \Sigma^0), \Xi^0), \qquad S \equiv \frac{1}{2}(\Lambda^0 + \sqrt{3}\Sigma^0).$$

We already learned how to compute the matrix elements of tensor operators in the octet representation. In particular the part of the Hamiltonian which breaks $SU(3)$ is along the X^3_3 direction, and was given by eq. (6.88)

$$< X^3_3 >_{octet} = a + bY + c\left(\frac{1}{4}Y^2 - I(I+1)\right).$$

This formula can be taken over, by simply changing from hypercharge to charge and I-spin to U-spin, with

$$< X^1_1 >_{octet} = a + bQ + c\left(\frac{1}{4}Q^2 - U(U+1)\right).$$

The electromagnetic current is the traceless part since the electric charge is contained within $SU(3)$. The reader might verify that on the octet, we are left with

$$< J_{elec} >_{octet} = \beta\, Q + \gamma\left(1 + \frac{1}{4}Q^2 - U(U+1)\right), \tag{11.88}$$

written in terms of only two undetermined constants. Many relations among the baryon's magnetic moments follow from

$$\mu_* = \langle * \,|\, b\,\hat{Q} + c\left(1 + \frac{1}{4}\hat{Q}^2 - U(U+1)\right) |\, * \rangle. \tag{11.89}$$

As an example, we form

$$\mu_{\Sigma^0} = c\,\langle\, \Sigma^0 \,|\, 1 - U(U+1) \,|\, \Sigma^0 \,\rangle, \tag{11.90}$$

and by expressing the physical particles in terms of the U-triplet and U-singlet,

$$|\,\Sigma^0\,\rangle = \frac{1}{2}(\sqrt{3}|\,S\,\rangle - |\,T\,\rangle), \qquad |\,\Lambda^0\,\rangle = \frac{1}{2}(|\,S\,\rangle + \sqrt{3}|\,T\,\rangle),$$

we find

$$\mu_{\Sigma^0} = \frac{c}{2}. \tag{11.91}$$

A similar reasoning leads to

$$\mu_{\Lambda^0} = -\frac{c}{2}, \qquad \mu_{\Sigma^0 \to \Lambda^0} = -\frac{c\sqrt{3}}{8},$$

where the latter is the transition magnetic moment. The eight plus one (transition) magnetic moments of the octet baryons are determined in terms of the proton and the neutron magnetic moments, for instance

$$\mu_{\Sigma^-} = -\mu_p - \mu_n.$$

With $SU(6)$, these relations can be improved even further. The magnetic moment coupling of a fermion is given by

$$\Psi^\dagger \vec{\sigma}\, \Psi \cdot \vec{B}, \tag{11.92}$$

where $\vec{B}$ is the magnetic field, and Ψ is the Dirac field representing the fermion (its spinor indices are not shown).

In $SU(6)$, the baryons are represented by a Dirac field which is a symmetric third-rank tensor **56**, $\Psi_{(abc)}$, where $a, b, c = 1, 2 \ldots, 6$. Each $SU(6)$ index is really a pair of $SU(3)$ indices A, and spin indices α: $a = (A, \alpha), b = (B, \beta)$, etc., where $A, B, \ldots = 1, 2, 3$, and $\alpha, \beta, \ldots = 1, 2$. In index language, the decomposition (11.84) reads

$$\Psi_{(abc)} = \Psi_{((A,\alpha)(B,\beta)(C,\gamma))} \tag{11.93}$$

$$= \psi_{\alpha\beta\gamma} T_{ABC} + \sum_{sym} \left(\epsilon_{\alpha\beta}\psi_\gamma\, \epsilon_{ABD}\, T^D_C\right). \tag{11.94}$$

The first term contains $\psi_{\alpha\beta\gamma}$ the spin three-half Dirac field, and T_{ABC} a symmetric tensor that transforms as the **10** of $SU(3)$. The second term describes the baryon octet, with the spin one-half Dirac field ψ_α, and a tensor octet T^D_C. Its form is determined by the only invariant tensors in spin space $\epsilon_{\alpha\beta} = -\epsilon_{\beta\alpha}$, and the Levi–Civita tensor in triplet space. Finally, the sum is required to symmetrize over the three pairs of indices.

The magnetic moments of the baryons in the octet can thus be determined in terms of one of them. Indeed, after substituting this expression for the baryon octet in (11.92), M. A. B. Bég, B. W. Lee, and A. Pais (*Phys. Rev. Letters* **13**, 514 (1964)) find

$$2\mu_p + 3\mu_n = 0, \tag{11.95}$$

corresponding to $b = c$. This relation is in spectacular agreement with experiment! There are many other applications of $SU(6)$; to name a few: improving the $SU(3)$ mass relations by relating pseudoscalar to vector octet masses, in particular

$$m_K^2 - m_\pi^2 = m_{K^*}^2 - m_\rho^2,$$

as well as finding rules for the baryon decuplet masses, and also vector meson $\omega-\phi$ mixing. We leave to the reader the joy of rediscovering these remarkable relations.

11.2.1 Color and the quark model

Identifying the baryons with the totally symmetric **56** is at odds with the *quark model*. In the eight-fold way, the "elementary" particles belong to the $SU(3)$ octet and decuplet. Group-theoretically, these representations can be constructed as products of the fundamental representations. The octet appears in both

$$\mathbf{8} \in \mathbf{3} \times \bar{\mathbf{3}}, \qquad \mathbf{8} \in (\mathbf{3} \times \mathbf{3} \times \mathbf{3})_{mixed}, \tag{11.96}$$

while the decuplet is the three-fold symmetric product

$$\mathbf{10} = (\mathbf{3} \times \mathbf{3} \times \mathbf{3})_{sym}. \tag{11.97}$$

This led M. Gell-Mann and G. Zweig to speculate independently that the eight-fold way particles behave *as if* they are made up of some more elementary entities transforming as triplets of flavor $SU(3)$; they are the quarks of Gell-Mann (Zweig's aces), the nucleons as three-quark composites and the mesons as quark–antiquark composites. Scattering experiments later confirmed the quark hypothesis.

In the light of $SU(6)$, the quark model is problematical: quarks are fermions, since they belong to the **6**: flavor $SU(3)$ triplets with spin one-half. According to the Pauli exclusion principle (spin-statistics theorem in local quantum field theory), states containing several identical fermions must be antisymmetric under their interchange, and the three-quark combination should appear in the *antisymmetric* product of three $SU(6)$ sextets; yet it lies in the symmetric product.

The road to a solution for this mystery was found when W. Greenberg (*Phys. Rev. Letters* **13**, 598 (1964)) proposed a way to retain the spin-statistics connection: quarks have a hitherto unknown quantum number, which he couched in the language of *parastatistics of order three*, where the quark creation operator is the sum of three different creation operators, and the antisymmetry occurs over the three "colors," as they were later called. The rest of the triple product is symmetric, hence the nucleons and their resonances are in the **56**.

The next step was taken by M. Y. Han and Y. Nambu (*Phys. Rev.* **139**, B1006 (1965)), who boldly assumed that this new quantum number indicates that the strong force can be understood as a *gauge theory* based on color $SU(3)$, and that the color force was somehow caused by its eight vector particles. The observed particles are color singlets under this new $SU(3)$, but their hypothesis was short on dynamical details. This new theory is *quantum chromodynamics* (QCD). Color $SU(3)$ should not be confused with flavor $SU(3)$ of the eight-fold way.

11.3 Invariant Lagrangians

In field theory, Lagrange densities $\mathcal{L}$ are real polynomials built out of local fields. In fundamental theories of Nature, these polynomials are invariant under some symmetries, such as Poincaré and gauge symmetries. We can apply group theory to their construction.

If we use fields local in space-time, translation invariance is assured as long as the coordinates x^μ do not appear, except as arguments of local fields, but translation invariance does allow the derivative operator $\partial_\mu \sim (\mathbf{2},\ \mathbf{2})$ acting on the fields.

• Lagrangians constructed out of scalar fields $\varphi(x)$ are easiest to manufacture. Any function $V(\varphi(x))$ is Poincaré invariant. Use of an even number of derivative operators leads to an infinity of invariants, the simplest of which $\partial^\mu\varphi(x)\,\partial_\mu\varphi(x)$, has two derivatives; its variation yields a second-order differential equation, the massless Klein–Gordon equation. Invariants with more derivatives lead to higher-order equations of motions. A spinless particle is represented by a scalar field with Lagrangian

$$\mathcal{L} = \frac{1}{2}\,\partial^\mu\varphi(x)\,\partial_\mu\varphi(x) - V(\varphi(x)). \tag{11.98}$$

Since the action is dimensionless (in units of $\hbar$), φ has mass (length) dimension $1(-1)$. For a free particle of mass m, $V = m^2\varphi^2/2$.

The kinetic invariant of N scalar fields φ_a, $a = 1, 2\ldots N$,

$$\frac{1}{2}\sum_{a=1}^{N}\partial^\mu\varphi_a(x)\,\partial_\mu\varphi_a(x), \tag{11.99}$$

is a quadratic form, manifestly invariant under a global $SO(N)$ transformation

$$\delta\,\varphi_a(x) = \omega_{ab}\varphi_b(x), \qquad \omega_{ab} = -\omega_{ba}, \tag{11.100}$$

where ω_{ab} do not depend on x.

• Next we consider invariants built out of one Weyl spinor, represented by a complex two-component local Grassmann field $\psi_\alpha(x^\mu) \sim (\mathbf{2},\ \mathbf{1})$; its conjugate is $\bar{\psi}_{\dot\alpha} \sim (\mathbf{1},\ \mathbf{2})$. Group theory points to an immediate quadratic invariant in the antisymmetric product of the Weyl spinor with itself

$$(\mathbf{2},\ \mathbf{1}) \times (\mathbf{2},\ \mathbf{1}) = (\mathbf{1},\ \mathbf{1})_a + (\mathbf{3},\ \mathbf{1})_s. \tag{11.101}$$

In terms of the fields, $\epsilon^{\alpha\beta}\psi_\alpha(x)\psi_\beta(x)$, and its conjugate $\epsilon^{\dot\alpha\dot\beta}\bar{\psi}_{\dot\alpha}(x)\bar{\psi}_{\dot\beta}(x)$. In field theory, the spin-statistics connection is assured by taking fermion fields to be Grassmann, so this invariant does not vanish; when added to its conjugate to make it real, it is interpreted as the *Majorana mass* of the Weyl spinor.

The product of the Weyl fermion with its conjugate

$$(\mathbf{2},\ \mathbf{1}) \times (\mathbf{1},\ \mathbf{2}) = (\mathbf{2},\ \mathbf{2}), \tag{11.102}$$

transforms as a Lorentz vector, but we can make it Lorentz-invariant by using the derivative operator, obtaining

$$\psi^{\dagger}_{\dot{\alpha}}(x)\,(\sigma^{\mu})_{\dot{\alpha}\alpha}\,\partial_{\mu}\,\psi_{\alpha}(x), \tag{11.103}$$

with the matrix $\sigma^{\mu} = (1, \sigma^{i})$ written in terms of Pauli matrices. There are two other invariants of this kind; one, with the derivative operator acting on both spinors, is a surface term which does not affect the equations of motion, the other with the derivative operator acting on the conjugate spinor is needed for a real invariant. The equation derived from this invariant is the massless Dirac equation for a left-handed Weyl spinor. From the kinetic term, we note that the mass dimension of ψ is $3/2$; thus the Majorana term must be multiplied with a parameter with dimension of mass; hence its name.

We note that since ψ is a complex field, the kinetic term is invariant under the global $U(1)$ phase transformation

$$\delta\,\psi(x) = i\omega\,\psi(x), \qquad \delta\,\bar{\psi}(x) = -i\omega\,\bar{\psi}(x), \tag{11.104}$$

where ω is independent of x.

The Dirac kinetic term of N Weyl fermions $\psi_a(x)$ is the quadratic sum

$$\sum_{a=1}^{N} \psi^{\dagger}_{a\,\dot{\alpha}}(x)\,(\sigma^{\mu})_{\dot{\alpha}\alpha}\,\partial_{\mu}\,\psi_{a\,\alpha}(x), \tag{11.105}$$

manifestly invariant under the same global $U(1)$ phase transformation on all the fields, and global $SU(N)$ transformations

$$\delta\,\psi_a(x) = \frac{i}{2}\omega^{A}\,(\boldsymbol{\lambda}^{A})_a^{\ b}\,\psi_b(x), \tag{11.106}$$

where $\boldsymbol{\lambda}^{A}$ are the Gell-Mann matrices, and the $(N^2 - 1)$ parameters ω^{A} do not depend on x. The Majorana mass terms assemble into

$$m_{ab}\,\epsilon^{\alpha\beta}\,\psi_{a\,\alpha}(x)\psi_{b\,\beta}(x), \tag{11.107}$$

where m_{ab} is a *symmetric* mass matrix. This term is no longer invariant under $SU(N) \times U(1)$.

If the theory contains scalar fields, $\varphi_n(x)$, the Majorana invariant can be turned into interaction terms, called *Yukawa couplings*,

$$Y_{n\,ab}\,\varphi_n(x)\,\psi_{a\,\alpha}(x)\,\epsilon^{\alpha\beta}\,\psi_{b\,\beta}(x), \tag{11.108}$$

where $Y_{n\,ab}$ are dimensionless coupling parameters.

• The treatment of theories with local four-vector fields $A_\mu(x)$ which transform as $(\mathbf{2},\mathbf{2})$ is more complicated. Naively, such a field contains four degrees of freedom, one for each component. Yet, the group theory of the Poincaré group indicates that a massless spin one particle contains one helicity state. However, this state is complex, and when we add its conjugate, we obtain two states of opposite helicity ± 1. This is consistent with two types of photons, corresponding to left- and right-circularly polarized states.

The description of a spin one massless particle in terms of a four-vector field must account for this overcounting. The answer is to be found in Maxwell's theory of electromagnetism, where the overcounting riddle is solved by *gauge invariance*. Maxwell's theory is a gauge theory associated with a $U(1)$ phase transformation.

The vector field is introduced by generalizing the derivative to a *covariant* derivative, constructed as to ensure that the global symmetry of the kinetic term is upgraded to a *local* symmetry where the transformation parameters depend on the space-time coordinates x^μ. Taking as an example the kinetic term for a Weyl spinor, we generalize it to

$$\psi^\dagger_{\dot\alpha}(x)\,(\sigma^\mu)_{\dot\alpha\alpha}\,\mathcal{D}_\mu\,\psi_\alpha(x), \tag{11.109}$$

where

$$\mathcal{D}_\mu = \partial_\mu + i\,A_\mu(x), \tag{11.110}$$

is the covariant derivative. In order to preserve the symmetry (11.104), it must satisfy under the gauge transformation

$$\mathcal{D}_\mu \to e^{i\omega(x)}\,\mathcal{D}_\mu\,e^{-i\omega(x)}, \qquad \psi(x) \to e^{i\omega(x)}\,\psi(x). \tag{11.111}$$

For this to happen, the vector field must transform inhomogeneously as

$$A_\mu(x) \to e^{i\omega(x)}\,A_\mu(x)\,e^{-i\omega(x)} - i\,e^{i\omega(x)}\left(\partial_\mu\,e^{-i\omega(x)}\right). \tag{11.112}$$

The covariant derivative clearly induces a gauge coupling of the vector field with the fermion bilinear *current*. From the covariant derivative, we construct the field strength

$$F_{\mu\nu} = i\,[\mathcal{D}_\mu,\,\mathcal{D}_\nu\,], \tag{11.113}$$

which is invariant under the gauge transformation. The Maxwell kinetic term ensues

$$\frac{1}{4g^2}\,F_{\mu\nu}\,F^{\mu\nu}, \tag{11.114}$$

where g is the coupling strength. In Maxwell's theory, the electron is described by two Weyl spinors $e(x)$ and $\bar e(x)$, with equal gauge couplings to the vector field, resulting in the QED Lagrangian

$$\mathcal{L}_{QED} = \frac{1}{4g^2} F_{\mu\nu} F^{\mu\nu} + e^{\dagger}(x)\,\sigma^{\mu}\,\mathcal{D}_{\mu}\,e(x) + \bar{e}^{\dagger}(x)\,\sigma^{\mu}\,\mathcal{D}_{\mu}\,\bar{e}(x) + m\bar{e}(x)\,e(x) + c.c., \tag{11.115}$$

suppressing all spinor indices. It is invariant under the local $U(1)$, and we have added the Dirac mass term, accounting for the electron mass.

11.4 Non-Abelian gauge theories

C.N. Yang and R.L. Mills (*Phys. Rev.* **96**, 191 (1954)) generalized the QED construction to gauge transformation generated by arbitrary Lie algebras. Our discussion is limited to the basic ideas, and the interested reader is urged to consult the many excellent physics and mathematics treatises dedicated to this rich subject.

Start with the kinetic term for N Weyl fermions, assembled into one Ψ. We have seen that it is invariant under a global $SU(N)$ transformation

$$\Psi(x) = \mathcal{U}\,\Psi(x), \tag{11.116}$$

and we wish to extend the invariance when $\mathcal{U} = \mathcal{U}(x)$ depends on x, and $\mathcal{U}(x)\mathcal{U}^{\dagger}(x) = 1$. Following the Abelian case, we construct a covariant derivative such that

$$\Psi(x) \rightarrow \mathcal{U}(x)\,\Psi(x), \qquad \mathcal{D}_{\mu}\,\Psi(x) \rightarrow \mathcal{U}(x)\,\mathcal{D}_{\mu}\,\Psi(x). \tag{11.117}$$

This requires the covariant derivative to be a matrix with the transformation

$$\mathcal{D}_{\mu} \rightarrow \mathcal{U}(x)\,\mathcal{D}_{\mu}\,\mathcal{U}^{\dagger}(x). \tag{11.118}$$

We introduce $(N^2 - 1)$ vector fields through

$$\left(\mathcal{D}_{\mu}\right)^{b}_{b} = \partial_{\mu}\,\delta^{b}_{a} + \frac{i}{2}\,A^{C}(x)\,(\boldsymbol{\lambda}^{C})^{b}_{a}. \tag{11.119}$$

In terms of

$$\mathbb{A}_{\mu}(x) \equiv \frac{1}{2}\,A^{C}(x)\,(\boldsymbol{\lambda}^{C})^{b}_{a}, \tag{11.120}$$

the gauge transformation reads

$$\mathbb{A}_{\mu}(x) \rightarrow \mathcal{U}(x)\,\mathbb{A}_{\mu}(x)\,\mathcal{U}^{\dagger}(x) - i\,\mathcal{U}(x)\left(\partial_{\mu}\,\mathcal{U}^{\dagger}(x)\right). \tag{11.121}$$

The field strength

$$\mathbb{F}_{\mu\nu} = -i\,[\,\mathcal{D}_{\mu},\,\mathcal{D}_{\nu}\,], \tag{11.122}$$

$$= \partial_{\mu}\,\mathbb{A}_{\nu} - \partial_{\nu}\,\mathbb{A}_{\mu} + i[\,\mathbb{A}_{\nu},\,\mathbb{A}_{\nu}\,] \tag{11.123}$$

is a matrix that transforms covariantly

$$\mathbb{F}_{\mu\nu} \rightarrow \mathcal{U}(x)\,\mathbb{F}_{\mu\nu}\,\mathcal{U}^{\dagger}(x). \tag{11.124}$$

The Yang–Mills Lagrangian for the vector fields is then

$$\mathcal{L}_{YM} = \frac{1}{4g^2} \operatorname{Tr}\left(\mathbb{F}_{\mu\nu}\mathbb{F}^{\mu\nu}\right), \tag{11.125}$$

tracing over the Lie algebra. Although our construction was for $SU(N)$ transformations, it applies to any other Lie algebra. We can abstract from this construction that Yang–Mills theories contain the following.

- Self-interacting massless spin one *gauge bosons*, one for each parameter of the Lie algebra. Gauge bosons can also interact with matter, if present. All gauge interactions depends on one coupling "constant." Matter consists of the following.
- Spin one-half particles, which transform as some representation of the Lie algebra, and
- Spin zero particles, also transforming as some representation of the Lie algebra.

Before describing the Standard Model in group-theoretical terms, two remarks are in order. One is that the restriction of matter to spins no higher than one-half comes from the additional requirement of renormalizabilty, not from any group-theoretical considerations.

Secondly, specifying a gauge theory and its matter content does not fully determine the theory. Fermions and scalars can have interactions outside the realm of gauge interactions: Yukawa couplings and self-couplings of the scalars. Renormalizability of the theory requires the mass dimension of these interactions to be at most four.

In fact, the Standard Model must have Yukawa couplings as necessary to account for the masses of the elementary particles, and weak decays, as well as scalar self-interactions to generate masses for some of its vector bosons.

11.5 The Standard Model

The Standard Model of particle physics of strong, weak and electromagnetic forces is described by three gauge theories $SU(3) \times SU(2) \times U(1)$, linked through their matter content.

Quantum Chromodynamics (QCD) is the color $SU(3)$ gauge theory with eight massless vector bosons called *gluons*. Color matter consists of six *flavors* of quarks, three with charge $2/3$, the up quark u, the charmed quark c, and the top quark t, together with three charge $-1/3$ quarks, the down quark d, the strange quark s, and the bottom quark b. The gluons coupling to the quarks conserves parity.

The weak and electromagnetic forces are described by two gauge theories associated with $SU(2)$ and $U(1)$.

In our notation, all fermions are left-handed Weyl spinors under the Lorentz group. This means that it takes two Weyl spinors to define a quark or a charged lepton. For instance the left-handed up quark is described by $\mathbf{u}$, which transforms as a color triplet $\mathbf{3^c}$, while $\bar{\mathbf{u}}$, transforming as $\bar{\mathbf{3}}^{\mathbf{c}}$ of color, is the conjugate right-handed up quark.

It is convenient to label all fermions by their transformation properties under the three gauge groups $SU(2) \times SU(3)^c \times U(1)^Y$ with their hypercharge Yas a subscript, determined from the Gell-Mann–Nishijima relation for the electric charge

$$Q = I_3 + \frac{Y}{2}.$$

All spin one-half elementary particles split into two types: leptons, which are color singlets,

$$\text{leptons:} \quad \begin{pmatrix} \nu_e \\ e \end{pmatrix} \sim (\mathbf{2},\ \mathbf{1^c})_{-1}; \qquad \bar{e} \sim (\mathbf{1},\ \mathbf{1^c})_2,$$

and quarks, which have color,

$$\text{quarks:} \quad \begin{pmatrix} u \\ d \end{pmatrix} \sim (\mathbf{2},\ \mathbf{3^c})_{\frac{1}{3}}; \quad \bar{u} \sim (\mathbf{1},\ \bar{\mathbf{3}}^{\mathbf{c}})_{-\frac{4}{3}}; \quad \bar{d} \sim (\mathbf{1},\ \bar{\mathbf{3}}^{\mathbf{c}})_{\frac{1}{3}}.$$

Stable matter is made up of these particles.

This pattern is triplicated with two more exact copies, with the same gauge properties. There are two heavier charged leptons, the muon and tau leptons with their associated neutrinos,

$$\begin{pmatrix} \nu_\mu \\ \mu \end{pmatrix}, \begin{pmatrix} \nu_\tau \\ \tau \end{pmatrix} \sim (\mathbf{2},\ \mathbf{1^c})_{-1}; \qquad \bar{\mu}, \bar{\tau} \sim (\mathbf{1},\ \mathbf{1^c})_2,$$

as well as the heavier charm, strange, top and bottom quarks

$$\begin{pmatrix} c \\ s \end{pmatrix}, \begin{pmatrix} t \\ b \end{pmatrix} \sim (\mathbf{2},\ \mathbf{3^c})_{\frac{1}{3}}; \quad \bar{c}, \bar{t} \sim (\mathbf{1},\ \bar{\mathbf{3}}^{\mathbf{c}})_{-\frac{4}{3}}; \quad \bar{s}, \bar{b} \sim (\mathbf{1},\ \bar{\mathbf{3}}^{\mathbf{c}})_{\frac{1}{3}}.$$

These elementary constituents have similar group properties, although with widely different masses. The top quark weighs by itself almost as much as a hafnium atom! The fermions of the heavier families decay into the lightest one, and were present in Boltzmann abundance in the early Universe. It is beyond the scope of this book to go into any more details.

The Standard Model also contains a spinless weak doublet and color singlet $\sim (\mathbf{2},\ \mathbf{1^c})_{-1}$. Its self-interactions are engineered by hand to produce the *spontaneous breakdown* of the $SU(2) \times U(1)$ gauge symmetry to Maxwell's $U(1)$. The $SU(3) \times U(1)$ gauge symmetries are unbroken, with nine massless gauge bosons, the photon and the eight gluons. The other gauge bosons are massive, to account for the finite range of the weak interactions.

In theories with spontaneous breaking, the Lagrangian is invariant under the gauge symmetries, but its lowest energy configuration is not. This is done by adding to the theory spinless fields; the lowest energy configuration of the theory is at the minimum of their potential. Careful crafting of this potential is necessary to ensure breaking of the symmetries. The symmetry breaking generates a mass for the gauge bosons associated with the broken symmetry generators, the Higgs mechanism.

While fixing the potential in such a way appears to be an unsatisfactory *ad hoc* feature of the Standard Model, it also generates the masses for the fermions. The Higgs mechanism leaves behind a massive scalar particle, the Higgs, whose coupling to matter is proportional to the mass of the particles with which it couples. The Higgs still awaits discovery, but since it cannot be arbitrarily heavy, its existence provides an experimental test of the validity of this idea.

11.6 Grand Unification

As its name indicates, the strong force between quarks is strong, so much so that quarks are forever prisoners inside nucleons. No free quarks have ever been observed. However, the QCD coupling "constant" varies with the scale at which it is measured, in such a way that at short distances, the strong interaction between quarks becomes weaker and weaker, a phenomenon called *asymptotic freedom*. Thus, it could even be that, at some very short distance, quarks may not be too dissimilar from leptons. Jogesh C. Pati and Abdus Salam (*Phys. Rev.* **D8**, 1240 (1973)) were the first to suggest that quarks and leptons could be unified: quark color stems from a broken $SU(4)$, whose quartet representation contains one quark and one lepton, through the embedding

$$SU(4) \supset SU(3)^c \times U(1); \qquad \mathbf{4} = \mathbf{3} + \mathbf{1}, \tag{11.126}$$

where the $U(1)$ charge is identified with baryon number ($1/3$ for quarks) minus lepton number.

$SU(5)$

In a different realization of this idea, H. Georgi and S. L. Glashow (*Phys. Rev. Letters* **32**, 438 (1974)) gather *all* gauge symmetries of the Standard Model in one gauge group, $SU(5)$,

$$SU(5) \supset SU(2) \times SU(3)^c \times U(1), \tag{11.127}$$

with embedding defined by the decomposition of its fundamental quintet

$$\mathbf{5} = (\mathbf{2},\, \mathbf{1^c})_1 + (\mathbf{1},\, \mathbf{3^c})_{-\frac{1}{3}}, \qquad \bar{\mathbf{5}} = (\mathbf{2},\, \mathbf{1^c})_{-1} + (\mathbf{1},\, \bar{\mathbf{3}}^{\mathbf{c}})_{\frac{1}{3}}, \tag{11.128}$$

as the $SU(2)$ doublet is pseudoreal, and the subscript denotes the $U(1)$ charge. The other fundamental **10** and the adjoint **24** representations are obtained from Kronecker products (see Appendix 2 for more details)

$$\mathbf{5} \times \mathbf{5} = \mathbf{10}_{antisym} + \mathbf{15}_{sym}, \qquad \mathbf{5} \times \bar{\mathbf{5}} = \mathbf{24} + \mathbf{1}. \tag{11.129}$$

Multiplying out the representations in (11.128), we obtain

$$\mathbf{10} = (\mathbf{2},\ \mathbf{3}^c)_{\frac{1}{3}} + (\mathbf{1},\ \bar{\mathbf{3}}^c)_{-\frac{4}{3}} + (\mathbf{1},\ \mathbf{1}^c)_2, \tag{11.130}$$

as well as the adjoint representation

$$\mathbf{24} = (\mathbf{3},\ \mathbf{1}^c)_0 + (\mathbf{1},\ \mathbf{8}^c)_0 + (\mathbf{1},\ \mathbf{1}^c)_0 + (\mathbf{2},\ \mathbf{3}^c)_{\frac{2}{3}} + (\mathbf{2},\ \bar{\mathbf{3}}^c)_{-\frac{2}{3}}. \tag{11.131}$$

We can now interpret these group-theoretical results.

The elementary particles of each family fit in two representations

$$\bar{\mathbf{5}}: \quad \left[\begin{pmatrix} \nu_e \\ e \end{pmatrix},\ \bar{d} \right]; \qquad \mathbf{10}: \quad \left[\begin{pmatrix} u \\ d \end{pmatrix},\ \bar{u},\ \bar{e} \right]. \tag{11.132}$$

Remarkably, the hypercharge is in $SU(5)$, from which the third integral charges of the quarks arise naturally (a similar relation was determined in the Standard Model by requiring that it be free from anomalies). This pattern is repeated for the other two families.

The gauge bosons of an $SU(5)$ Yang–Mills theory reside in its adjoint representation. Comparing with (11.131), we recognize the eight gluons $(\mathbf{1},\ \mathbf{8}^c)$, the hypercharge $(\mathbf{1},\ \mathbf{1}^c)$, and the weak bosons $(\mathbf{3},\ \mathbf{1}^c)$ of the Standard Model.

$SU(5)$ predicts new forces beyond the Standard Model, mediated by the other twelve gauge bosons. As their charges indicate, they cause transitions between leptons and the anti-down quark when acting on the $\bar{\mathbf{5}}$, as well as quark–antiquark transitions when acting on the **10**. They are called *leptoquark* gauge bosons. If massless, they would cause immediate proton decay. The experimental limit on the proton lifetime ($\gtrsim 10^{33}$ years) requires these gauge bosons to be very massive, so that the $SU(5)$ symmetry, while extraordinary for taxonomy, must be broken at distances very much smaller than a nuclear radius.

This conclusion is quantitatively in accord with the strength of the couplings. For leptons and quarks to unify, the strength of their couplings must be similar. Using the experimentally determined values of the three gauge couplings of the Standard Model, as boundary condition, their renormalization group running implies a possible merging at scales which are also extraordinarily small.

The Achilles' heel of grand unified theories is that the symmetries they imply must be broken. In analogy with the Standard Model, the symmetry breaking is assumed to be spontaneous, that it is a property of the lowest energy state. This requires an elaborate machinery. To this day, there are no credible ways to achieve such breakings of symmetry. On the other hand, the elementary fermions fit beautifully into representations of these gauge group, and the evolution of the couplings towards a single value at shorter scales gives credence to the idea of a single gauge group.

$SO(10)$

The Pati–Salam $SU(4)$ can be extended to $SO(10)$ (H. Fritzsch and P. Minkowski, *Ann. Phys.* **93**, 193 (1975) and independently H. Georgi in *Proceedings of the American Institute of Physics* 23 ed. C. E. Carlson, 575 (1974)) through the embedding

$$SO(10) \supset SU(4) \times SU(2)_L \times SU(2)_R. \tag{11.133}$$

The first $SU(2)_L$ is the one in the Standard Model, while the second describes three gauge bosons which cause transitions among right-handed particles. Under this embedding (see Appendix 2), the $SO(10)$ spinor representation is given by

$$\mathbf{16} = (\,\mathbf{4},\ \mathbf{2},\ \mathbf{1}\,) + (\,\bar{\mathbf{4}},\ \mathbf{1},\ \mathbf{2}\,). \tag{11.134}$$

After interpreting leptons as the fourth color, we recognize the particles of the Standard Model, and a new lepton. It is the right-handed partner of the neutrino. This is easiest seen through the embedding

$$SO(10) \ \subset \ SU(5) \times U(1), \tag{11.135}$$

with

$$\mathbf{16} = \bar{\mathbf{5}}_3 + \mathbf{10}_{-1} + \mathbf{1}_{-5}. \tag{11.136}$$

In $SU(5)$, the electric charge is within, and the new particle has no color, no weak nor electric charge; it can rightly be interpreted as the right-handed partner of the neutrino in each family. Its Majorana mass bilinear transforms under $SU(4) \times SU(2) \times SU(2)$ as

$$(\,\mathbf{10},\ \mathbf{1},\ \mathbf{3}\,) + (\,\mathbf{6},\ \mathbf{1},\ \mathbf{1}\,), \tag{11.137}$$

and is commensurate with the scale of the breaking of Pati–Salam's $SU(4)$, which is expected to be very large. Such a large Majorana mass suggests a natural mechanism, the *see-saw mechanism*, for generating tiny neutrino masses. These lead to neutrino oscillations of the type observed in the recent past. It provides yet another circumstantial evidence for the idea of grand unification.

Its adjoint representation

$$\mathbf{45} = \mathbf{24}_0 + \mathbf{1}_0 + \mathbf{10}_4 + \overline{\mathbf{10}}_{-4}, \tag{11.138}$$

shows more gauge bosons than in $SU(5)$, all of which cause interactions not yet observed in Nature. Breaking the larger $SO(10)$ symmetry presents a challenge; the mechanism by which it occurs remains a mystery to be solved by future generations.

E_6

The next rung on the ladder of gauge symmetries

$$SU(3) \times U(1) \subset SU(3) \times SU(2) \times U(1) \subset SU(5) \subset SO(10) \subset ?,$$

is the exceptional group E_6. Its gauge theory (F. Gürsey, P. Ramond, and P. Sikivie, *Phys. Lett.* **B60**, 177 (1976)) uses the embedding

$$E_6 \supset SO(10) \times U(1), \tag{11.139}$$

with each family described by

$$\mathbf{27} = \mathbf{16}_1 + \mathbf{10}_{-2} + \mathbf{1}_4. \tag{11.140}$$

Here the **10** is the vector representation of $SO(10)$, which breaks under $SU(5)$ as the vector-like pair $\mathbf{5} + \bar{\mathbf{5}}$, with masses expected to be out of present experimental range. This sequence is familiar to mathematicians, in terms of plucking dots out of Dynkin diagrams

$$SU(3) \times U(1) \subset SU(3) \times SU(2) \times U(1) \subset SU(5) \subset SO(10) \subset E_6 \subset E_7 \subset E_8$$

Amazingly, $E_8 \times E_8$ appears as the gauge symmetry of the Heterotic String model. Nature does seem to like exceptional structures!

11.7 Possible family symmetries

The patterns of elementary fermions appear in triplicate, pointing to a three-fold symmetry. It is as mysterious today as it was when the first heavy lepton, the muon, was discovered (e.g. I. I. Rabi's famous remark "who ordered that?").

The impulse has been to assemble the three families in a *family group*, with two- and three-dimensional representations. This group may be continuous or finite. If continuous, it means either $SU(2)$ or $SU(3)$, as Lie algebra with such representations; if finite, the family group has to be a finite subgroup of $SU(2)$, $SO(3)$, or $SU(3)$. The recent discovery of large neutrino mixing angles has prompted interest in exploring finite family groups. In the following, we list the finite groups with such representations. While this connection is pure speculation, it allows for the introduction of a number of finite groups that may be of physical interest.

11.7.1 Finite $SU(2)$ and $SO(3)$ subgroups

All finite groups with two-dimensional representations were identified as early as 1876 by Felix Klein, and summarized by G. A. Miller, H. F. Blichfeldt, and L. E. Dickson [16]. Consider the transformations

$$Z = \begin{pmatrix} z_1 \\ z_2 \end{pmatrix} \rightarrow \mathcal{M}\, Z, \tag{11.141}$$

where $\mathcal{M}$ is a unitary matrix with unit determinant. Klein constructed three quadratic forms,

$$\vec{r} = Z^{\dagger} \vec{\sigma}\, Z, \tag{11.142}$$

where $\vec{\sigma}$ are the Pauli matrices. These transformations act on the vector $\vec{r}$ as rotations about the origin. These are five types of such rotations:

- rotations of order n about one axis that generate the finite Abelian group $\mathcal{Z}_n$;
- rotations with two rotation axes that generate the dihedral group $\mathcal{D}_n$;
- rotations that map the tetrahedron into itself form $\mathcal{T} = \mathcal{A}_4$, the tetrahedral group;
- rotations that map the cube into itself form $\mathcal{O} = \mathcal{S}_4$, the octahedral group;
- rotations that map the icosahedron (dodecahedron) into itself form $\mathcal{I} = \mathcal{A}_5$, the icosahedral group.

Recall that the dihedral groups have only singlet and *real* doublet representations ($n > 2$). $\mathcal{T}$, $\mathcal{O}$ and $\mathcal{I}$ are the symmetry groups of the platonic solids. Since they are $SO(3)$ subgroups, all have real triplet representations, but no *real* two-dimensional representations.

Each can be thought of as the quotient group of its *binary*-equivalent group with at least one two-dimensional "spinor" representation and twice the number of elements (double cover). In ATLAS notation, these are

$$2 \cdot \mathcal{D}_n, \qquad 2 \cdot \mathcal{A}_4, \qquad 2 \cdot \mathcal{S}_4, \qquad 2 \cdot \mathcal{A}_5. \tag{11.143}$$

The first is just $\mathcal{Q}_{2n}$, the binary dihedral or dicyclic group introduced earlier.

Since $\mathcal{A}_4 = \mathcal{PSL}_2(3)$, the binary tetrahedral group is $\mathcal{SL}_2(3)$, obtained by dropping the projective restriction, with 24 elements. Its presentation is closely related to that of $\mathcal{A}_4$

$$\begin{aligned} \mathcal{PSL}_2(3): &\quad < a, b, \,|\, a^2 = b^3 = (ab)^3 = e >, \\ \mathcal{SL}_2(3): &\quad < a, b, \,|\, a^2 = b^3 = (ab)^3 >. \end{aligned}$$

In the binary group, a^2 is **not** the identity, but the negative of the identity, and it is thus of order four: $\mathcal{A}_4$, with no element of order four, is not a subgroup, rather it is its largest *quotient* group

$$\mathcal{A}_4 = \mathcal{SL}_2(3)/\mathcal{Z}_2. \tag{11.144}$$

The same reasoning applies to $\mathcal{I} = \mathcal{A}_5 = \mathcal{PSL}_2(5)$, and the binary icosahedral group is $\mathcal{SL}_2(5)$ with 120 elements, dropping the projective requirement. No longer simple, it has a normal subgroup of order two.

Its presentation is of the same form as for $\mathcal{A}_5$

$$\mathcal{PSL}_2(5): \quad < a, b \,|\, a^2 = b^3 = (ab)^5 = e >,$$
$$\mathcal{SL}_2(5): \quad < a, b \,|\, a^2 = b^3 = (ab)^5 > .$$

This generates (at least) one two-dimensional *spinor* representation to those of $\mathcal{A}_5$. In that representation, the generators are written in terms of η, $\eta^5 = 1$, as

$$ab = \begin{pmatrix} -\eta^2 & 0 \\ 0 & -\eta^3 \end{pmatrix}, \qquad b = \frac{1}{\sqrt{5}} \begin{pmatrix} \eta^4 - \eta^2 & \eta - 1 \\ 1 - \eta^4 & \eta - \eta^3 \end{pmatrix}. \tag{11.145}$$

$\mathcal{SL}_2(5)$ has in fact four spinor-like representations, $\mathbf{2_s}$, $\mathbf{2'_s}$, $\mathbf{4_s}$, and $\mathbf{6_s}$. The character table is given by the following, where $\varphi = (1+\sqrt{5})/2$, $\tilde{\varphi} = (-1+\sqrt{5})/2$.

$\mathcal{SL}_2(5)$	C_1	$12C_2^{[5]}$	$12C_3^{[5]}$	$1C_4^{[2]}$	$12C_5^{[10]}$	$12C_6^{[10]}$	$20C_7^{[3]}$	$20C_8^{[6]}$	$30C_9^{[4]}$
$\chi^{[1]}$	1	1	1	1	1	1	1	1	1
$\chi^{[3_1]}$	3	φ	$-\tilde{\varphi}$	3	$-\tilde{\varphi}$	φ	0	0	-1
$\chi^{[3_2]}$	3	$-\tilde{\varphi}$	φ	3	φ	$-\tilde{\varphi}$	0	0	-1
$\chi^{[4]}$	4	-1	-1	4	-1	-1	1	1	0
$\chi^{[5]}$	5	0	0	5	0	0	-1	-1	0
$\chi^{[2_s]}$	2	$-\varphi$	$\tilde{\varphi}$	-2	$-\tilde{\varphi}$	φ	-1	1	0
$\chi^{[2'_s]}$	2	$\tilde{\varphi}$	$-\varphi$	-2	φ	$-\tilde{\varphi}$	-1	1	0
$\chi^{[4_s]}$	4	-1	-1	-4	-1	-1	1	0	-1
$\chi^{[6_s]}$	6	1	1	-6	-1	-1	0	0	0

The Kronecker products of its representations follows an intriguing pattern, known as the *McKay correspondence*. Start from the singlet irrep and multiply it with the spinor irrep $\mathbf{2_s}$, to get of course

$$\mathbf{1} \times \mathbf{2_s} = \mathbf{2_s}.$$

We associate one dot with $\mathbf{1}$, and one for $\mathbf{2_s}$, and connect them by a line. Next we generate new representations by continuing to take products with the spinor:

$$\mathbf{2_s} \times \mathbf{2_s} = \mathbf{1} + \mathbf{3_1}, \qquad \mathbf{2_s} \times \mathbf{3_1} = \mathbf{2_s} + \mathbf{4_s},$$

and

$$\mathbf{2_s} \times \mathbf{4_s} = \mathbf{5} + \mathbf{3_1}, \qquad \mathbf{2_s} \times \mathbf{5} = \mathbf{6_s} + \mathbf{4_s},$$

creating four more dots for $\mathbf{3_1}$, $\mathbf{4_s}$, $\mathbf{5}$, and $\mathbf{6_s}$, connecting the new irrep with a horizontal line as we go. The next product contains three irreps

$$\mathbf{2_s} \times \mathbf{6_s} = \mathbf{5} + \mathbf{3_2} + \mathbf{4},$$

so we branch out two lines to the new representations. Now we have

$$\mathbf{2_s} \times \mathbf{4} = \mathbf{6_s} + \mathbf{2'_s},$$

generating one new representation. However, the two chains now terminate because further multiplication does not generate any new irrep

$$\mathbf{2_s} \times \mathbf{2'_s} = \mathbf{4}, \qquad \mathbf{2_s} \times \mathbf{3_2} = \mathbf{6_s}.$$

Remarkably, McKay's graphical rules generate the following extended E_8 Dynkin diagram with each of its dots associated with an irrep of the binary icosahedral group.

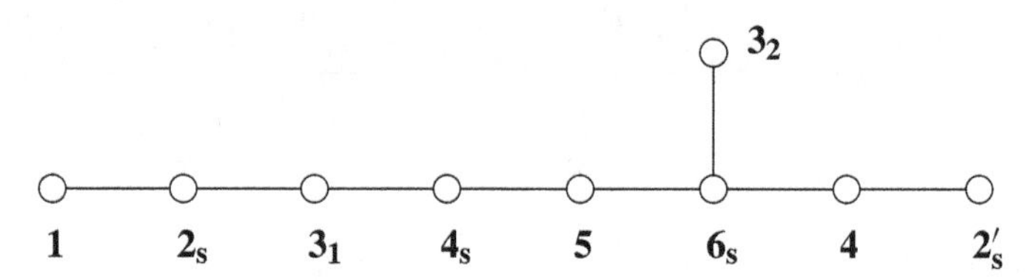

This pattern holds true for the binary extensions of $\mathcal{Z}_n$, $\mathcal{D}_n$, $\mathcal{T}$, $\mathcal{O}$, in McKay correspondence with the extended Dynkin diagrams of $SU(n+1)$, $SO(2n)$, E_6 and E_7, respectively.

11.7.2 Finite SU(3) subgroups

The finite subgroups of $SU(3)$ are described in Miller *et al.* [16], and discussed extensively by W. M. Fairbanks, T. Fulton, and W. H. Klink (*J. Math. Phys.* **5**, 1038 (1964)).

While not all finite subgroups of $SU(3)$ have complex triplet representations, they find two infinite families and two simple groups with triplet representations:

- $\Delta(3n^2) = (\mathcal{Z}_n \times \mathcal{Z}_n) \rtimes \mathcal{Z}_3$
- $\Delta(6n^2) = (\mathcal{Z}_n \times \mathcal{Z}_n) \rtimes \mathcal{S}_3$
- $\mathcal{A}_5$ (see Appendix 1)
- $\mathcal{PSL}_2(7)$ of order 168 (see Appendix 1).

Four $SU(3)$ subgroups have no triplet representations:

- Hessian group: $\Sigma(216) = (\mathcal{Z}_3 \times \mathcal{Z}_3) \rtimes \mathcal{SL}_2(3)$
- $\Sigma(72) = (\mathcal{Z}_3 \times \mathcal{Z}_3) \rtimes \mathcal{Q} \subset \Sigma(216)$

- $\Sigma(36) = (\mathcal{Z}_3 \times \mathcal{Z}_3) \rtimes \mathcal{Z}_4 \subset \Sigma(72)$
- $\mathcal{A}_6$, the order 360 alternating group.

In the $SU(2)$ case, we used $\mathcal{Z}_2$, the center of $SO(3)$, to construct the double cover of its subgroups without spinor representations. Here, the center of $SU(3)$ is $\mathcal{Z}_3$, allowing us to construct the triple cover of its subgroups without triplets. This generates four groups with complex triplet representations, and triple the number of elements:

$$3 \cdot \Sigma(216), \qquad 3 \cdot \Sigma(72), \qquad 3 \cdot \Sigma(366), \qquad 3 \cdot \mathcal{A}_6,$$

shown here in ATLAS language.

12

Exceptional structures

Nature (and humans) seems to like those special mathematical structures which have unusual properties. For instance, the platonic solids have held a special place in both human imagination and group theory, where they are associated with exceptional groups. For that reason, it is interesting to study these groups, using their unique underlying algebraic structures.

12.1 Hurwitz algebras

Consider a complex number and its conjugate

$$z = x + iy, \qquad \bar{z} = x - iy,$$

where x and y are real numbers. Its norm is defined as

$$N(z) \;\equiv\; \sqrt{\bar{z}\,z} = \sqrt{x^2 + y^2}.$$

The square of two complex numbers z, w is also a complex number, whose norm we can define. Trivially,

$$N(z\,w) = \sqrt{z\,w\bar{z}\bar{w}} = N(z)\,N(w), \tag{12.1}$$

and the norm of the product is the product of the norms. This is also true of real numbers for which the norm is the number itself. Can we generalize to numbers with more imaginary units? It turns out that there are only two more algebras which share this property. They are called *Hurwitz algebras*: the real numbers R, the complex numbers C, the *quaternions* Q with three imaginary units, and the *octonions* (Cayley numbers) O, with seven imaginary units. All have norms such that the norm of the product of any two members is the product of their norms.

Quaternions

We have already encountered quaternions in the context of finite groups. A quaternion q and its conjugate $\bar{q}$ are defined as

$$q \equiv x_0 + e_1 x_1 + e_2 x_2 + e_3 x_3, \qquad \overline{q} = x_0 - e_1 x_1 - e_2 x_2 - e_3 x_3, \qquad (12.2)$$

where x_i, $i = 0, 1, 2, 3$ are real numbers, and e_a, $a = 1, 2, 3$ are three imaginary units

$$e_1 e_1 = e_2 e_2 = e_3 e_3 = -1. \qquad (12.3)$$

They must satisfy

$$e_a e_b = \epsilon_{abc} e_c, \quad a \neq b = 1, 2, 3, \qquad (12.4)$$

in order to share the same norm property as complex numbers for any two quaternions q and q',

$$N(q\, q') = N(q)\, N(q'), \qquad (12.5)$$

where the norm is defined in the usual manner

$$N(q) \equiv \sqrt{\overline{q}\, q} = \sqrt{x_0^2 + x_1^2 + x_2^2 + x_3^2}. \qquad (12.6)$$

These are Hamilton's *quaternions*, the natural generalization of complex numbers, and a unit quaternion represents a point on the unit four-sphere.

Let η, χ be pure imaginary quaternions, $\eta = -\bar{\eta}$, $\chi = -\bar{\chi}$, and consider the transformation

$$q \rightarrow q' = e^{\eta}\, q\, e^{\chi}, \qquad (12.7)$$

and take its conjugate (reversing the order of operators)

$$\overline{q} \rightarrow \overline{q}' = e^{-\chi}\, \overline{q}\, e^{-\eta}.$$

The norm is clearly left untouched by these six-parameter transformations, the norm group $SO(4)$. By taking $\chi = -\eta$, we specialize to the $SU(2)$ subgroup

$$q \rightarrow q' = e^{\eta}\, q\, e^{-\eta}, \qquad (12.8)$$

or infinitesimally,

$$\delta q = [\, \eta,\, q\,]. \qquad (12.9)$$

Setting $\eta = e_m$, apply this transformation to the quaternion multiplication law, finding

$$[\, e_m,\, e_i\, e_j - \epsilon_{ijk}\, e_k\,] = 0. \qquad (12.10)$$

The transformations (12.8) leave quaternion multiplication invariant: $SU(2)$ is the *automorphism group* of the quaternion algebra. To physicists, quaternions are simply i times the Pauli matrices, and as such satisfy an associative but non-commutative algebra.

Octonions

Octonions are numbers with seven imaginary units e_α,

$$\omega = a_0 + \sum_1^7 a_\alpha\, e_\alpha; \qquad \overline{\omega} = a_0 - \sum_1^7 a_\alpha\, e_\alpha, \tag{12.11}$$

where a_0, a_α are real numbers. They satisfy the Cayley algebra

$$e_\alpha\, e_\beta = -\delta_{\alpha\beta} + \psi_{\alpha\beta\gamma}\, e_\gamma, \tag{12.12}$$

where $\psi_{\alpha\beta\gamma}$ are totally antisymmetric octonion structure functions, whose only non-zero elements are

$$\psi_{123} = \psi_{246} = \psi_{435} = \psi_{651} = \psi_{572} = \psi_{714} = \psi_{367} = 1. \tag{12.13}$$

We define the octonion norm as

$$N(\omega) = \sqrt{\omega\bar{\omega}} = \sqrt{(a_0)^2 + \sum_{\alpha=1}^7 a_\alpha^2}. \tag{12.14}$$

The $\psi_{\alpha\beta\gamma}$ are such that the magical norm property

$$N(\omega\omega') = N(\omega)N(\omega'), \tag{12.15}$$

holds for any two octonions ω, ω'. Like the quaternions, the Cayley algebra is non-commutative, but a little computation shows that it is not associative as well.

The associator of three octonions $\omega_{1,2,3}$ is defined as

$$(\,\omega_1,\, \omega_2,\, \omega_3\,) \equiv (\omega_1\, \omega_2)\, \omega_3 - \omega_1\, (\omega_2\, \omega_3). \tag{12.16}$$

Using the algebra, we find

$$(\,e_\alpha,\, e_\beta,\, e_\gamma\,) \;= -2\widetilde{\psi}_{\alpha\beta\gamma\delta}\, e_\delta, \tag{12.17}$$

where

$$\widetilde{\psi}_{\alpha_1\alpha_2\alpha_3\alpha_4} = \frac{1}{3!}\epsilon_{\alpha_1\alpha_2\alpha_3\alpha_4\alpha_5\alpha_6\alpha_7}\psi_{\alpha_5\alpha_6\alpha_7} \tag{12.18}$$

is the dual of the octonion structure functions. The associator is completely antisymmetric under the interchange of the three octonions: octonions form an *alternative* algebra. Octonions also satisfy the even less intuitive Moufang identities (with many more to be found in Gürsey and Tze [8]),

$$(\omega_1\,\omega_2)\,(\omega_3\,\omega_1) = \omega_1\,(\omega_2\,\omega_3)\,\omega_1,$$
$$(\omega_1\,\omega_2\,\omega_1)\omega_3 = \omega_1\,(\omega_2(\omega_1\,\omega_3)),$$
$$\omega_1(\omega_2\,\omega_3\,\omega_2) = ((\omega_1\,\omega_2)\omega_3)\,\omega_2.$$

Since octonions are described by eight real numbers, the group that leaves the quadratic norm invariant is $SO(8)$. In the quaternion case, the norm group was $SO(4)$, and we used a transverse quaternion to reduce it to $SO(3)$, which turned out to be the automorphism group of the quaternion algebra. We follow a similar procedure for octonions.

Imaginary octonions are given in terms of seven parameters, so we decompose $SO(8)$ in sevens, and find the sequence

$$SO(8) \supset SO(7) \supset G_2. \tag{12.19}$$

Written in terms of the imaginary octonion α, the seven coset generators of $SO(8)/SO(7)$ which map the real part ω_0 into the imaginary units ω_α, are simply

$$SO(8)/SO(7){:}\quad \omega \;\rightarrow\; \omega' = e^{\alpha}\,\omega\,e^{\alpha}, \qquad \delta\,\omega = \alpha\,\omega + \omega\alpha. \tag{12.20}$$

Seven of the twenty-one $SO(7)$ transformations that map the seven imaginary components into themselves are given by

$$SO(7)/G_2{:}\quad \omega \;\rightarrow\; \omega' = e^{\beta}\,\omega\,e^{-\beta}, \qquad \delta\,\omega = \beta\,\omega - \omega\beta, \tag{12.21}$$

where β is an imaginary octonion. The remaining fourteen transformations generate the exceptional group G_2. Its action on octonions is not trivial, reflecting the fact that while group operations are associative, the octonion algebra is not. G_2 turns out to be the automorphism group of the octonion algebra. Its infinitesimal action can be written in terms of two imaginary octonions ζ and η, and involves the associator and a double commutator

$$G_2{:}\quad \delta\,\omega = 3(\,\zeta,\,\eta,\,\omega\,) - [\,[\,\zeta,\,\eta\,],\,\omega\,], \tag{12.22}$$

and it is not exactly straightforward to prove that this transformation respects octonion multiplication.

12.2 Matrices over Hurwitz algebras

In this section, we consider antihermitian matrices whose elements are members of the Hurwirtz algebras. For the associative Hurwitz algebras, the commutator of these matrices are expected to close and form Lie algebras. On the other hand, for matrices whose elements are non-associative octonions, one expects that the Jacobi identity will fail. Yet in a few special cases, Lie algebras can be generated this way, the four remaining exceptional groups, F_4, E_6, E_7, and E_8.

Antisymmetric matrices over real numbers

The commutator of two antisymmetric matrices is itself antisymmetric. These $(2n + 1) \times (2n + 1)$ and $(2n \times 2n)$ antisymmetric matrices generate the Lie algebras, $SO(2n + 1)$ and $SO(2n)$, respectively. Rotations from the i to the j axis are represented by matrices whose elements are zero except for the ij entry equal to one and ji entry equal to minus one.

Antihermitian matrices over complex numbers

The commutator of two such matrices is antihermitian and traceless. The Lie algebras A_{n-1}, are represented in their simplest form by antihermitian $n \times n$ matrices over the complex numbers

$$\begin{pmatrix} ia_1 & z_{12} & \cdots & z_{1n} \\ -\overline{z}_{12} & ia_2 & \ldots & z_{2n} \\ \cdot & \cdot & \cdot & \cdot \\ -\overline{z}_{1n} & -\overline{z}_{2n} & \ldots & ia_n \end{pmatrix},$$

where z_{ij} are complex numbers and a_i are real. The commutator of two such matrices yields an antihermitian traceless matrix, which compels us to impose the trace condition

$$\sum_i^n a_i = 0.$$

The $n(n - 1)/2$ complex off-diagonal elements together with the $(n - 1)$ real diagonal elements account for the $n^2 - 1$ parameters of $SU(n)$.

Antihermitian matrices over quaternions

What matrices over quaternions? Antihermitian matrices over quaternions are of the form

$$M(\mathbf{Q}) = \begin{pmatrix} p_1 & q_{12} & \cdots & q_{1n} \\ -\overline{q}_{12} & p_2 & \ldots & q_{2n} \\ \cdot & \cdot & \ldots & \cdot \\ -\overline{q}_{1n} & -\overline{q}_{2n} & \ldots & p_n \end{pmatrix},$$

where q_{ij} are quaternions and the p_i are imaginary quaternions,

$$\overline{p}_i = -p_i.$$

In analogy, let us impose the trace condition

$$\sum_i^n p_i = 0.$$

The number of parameters in this matrix is

$$4 \times \frac{n(n-1)}{2} + 3 \times (n-1) = \frac{2n(2n+1)}{2} - 3,$$

parameters, since imaginary quaternions have three real parameters. This is three short of the number of parameters of the $Sp(2n)$ algebras.

Since quaternions are not commutative, the commutator of two quaternionic matrices is not automatically traceless, as it is for complex numbers. It is straightforward to see that the trace of the commutator of two quaternionic matrices is the sum of quaternion commutators which, as per eq. (12.9), can be recast as elements of the $SU(2)$ automorphism group.

The action of $Sp(2n)$ splits as $(n \times n)$ traceless antihermitian quaternionic matrices, together with the quaternion automorphism group acting diagonally.

12.3 The Magic Square

The natural next step is to consider antihermitian matrices over octonions. The analogy with the quaternions fails because their trace is a transverse octonion, which transforms as the **7** of their automorphism group G_2. Still, let us focus on traceless matrices, and consider the automorphism group separately. For $(n \times n)$ traceless octonionic matrices, we find

$$8 \times \frac{n(n-1)}{2} + 7 \times (n-1) = 4n^2 + 3n - 7,$$

parameters, since an imaginary octonion has seven parameters.

But here we encounter an apparently fatal roadblock: octonions are non-associative, and thus octonionic matrices will not satisfy associativity (Jacobi identity). This is consistent since matrices over the real, complex and quaternions account for all four infinite families of Lie algebras, leaving out the five exceptional groups. The first G_2 is accounted for as the octonions' automorphism group. What about the 52-parameter F_4?

We note that the (3×3), octonionic traceless antihermitian matrix contains 36 parameters, which together with the 14 in G_2 yield 52 parameters, the same as F_4. This is strong incentive to consider (3×3) matrices over octonions!

The story begins, as can be expected, with $SO(8)$ triality. The first point to make is that the action of the octonion's norm group $SO(8)$, can be decomposed with respect to G_2 into *three* different transformations acting on an octonion ω

$$t = d + l_\alpha + r_\beta, \tag{12.23}$$

where d is an element of G_2,

$$d: \quad \omega \to \omega + \delta\omega = \omega + 3(\zeta, \eta, \omega) - [[\zeta, \eta], \omega], \tag{12.24}$$

and

$$l_\alpha: \quad \omega \to \omega + \alpha\,\omega,$$
$$r_\beta: \quad \omega \to \omega + \omega\,\beta, \tag{12.25}$$

where ζ, η, α and β are imaginary octonions. This decomposition can be shown to be *unique*. To see this, let $t = 0$, so that

$$d = -l_\alpha - r_\beta, \tag{12.26}$$

meaning that $l_\alpha + r_\beta$ is an element of G_2. The infinitesimal form of its transformation, eq. (12.22), requires $\alpha = -\beta$. Hence

$$l_\alpha - r_\alpha: \quad \omega\omega' \to (\omega\omega') + [\alpha(\omega\omega') - (\omega\omega')\alpha],$$

but, it can also be interpreted as an element of G_2,

$$l_\alpha - r_\alpha: \quad \omega\omega' \to \omega\omega' + (\alpha\omega - \omega\alpha)\omega' + \omega(\alpha\omega' - \omega'\alpha).$$

Equating the two forms and the use of alternativity, yields,

$$(\alpha, \omega, \omega') = 0,$$

which means that α associates with *any* octonion; it must be zero. We have shown that $t = 0$ implies $d = \alpha = \beta = 0$. Since dim $SO(8)$ – dim $G_2 = 28 - 14 = 14$, the generality of this decomposition ensues.

Let t_1 be an infinitesimal element of $SO(8)$, with action on an octonion ω,

$$\omega \to \omega + t_1(\omega).$$

The *principle of triality* states that we can associate uniquely to t_1 two other elements of $SO(8)$, t_2 and t_3, with the property,

$$t_1(\overline{\overline{\omega\omega'}}) = t_2(\omega)\omega' + \omega\, t_3(\omega'), \tag{12.27}$$

for any two octonions ω and ω'. They are uniquely determined according to the following.

t_1	t_2	t_3
d	d	d
l_α	r_α	$-l_\alpha - r_\alpha$
r_α	$-l_\alpha - r_\alpha$	l_α

The uniqueness can be proven by considering the three cases $t_1 = d$, $t_1 = l$, and $t_1 = r$, separately, and by clever use of the octonion algebra.

Triality enables us to establish a one-to-one correspondence between any $SO(8)$ element and the set of triples $\{ t_1, t_2, t_3 \}$. The $SO(8)$ algebra is then realized in the form

$$[\{t_1, t_2, t_3\}, \{u_1, u_2, u_3\}] = \{[t_1, u_1], [t_2, u_2], [t_3, u_3]\}. \tag{12.28}$$

Consider (3×3) antihermitian octonionic matrices with only off-diagonal elements; there are $3 \times 8 = 24$ such matrices. Together with the 28 $SO(8)$ transformations, they assemble into the 52-parameter exceptional group F_4.

Write the twenty-four matrices with no diagonal elements

$$M(\omega_i) = \begin{pmatrix} 0 & \omega_1 & \bar{\omega}_2 \\ -\bar{\omega}_1 & 0 & \omega_3 \\ -\omega_2 & -\bar{\omega}_3 & 0 \end{pmatrix}, \tag{12.29}$$

for any three octonions $\omega_1, \omega_2, \omega_3$.

The infinitesimal action on this matrix of the $SO(8)$ triple $\{ t_1, t_2, t_3 \}$, is *defined* to be

$$\delta M(\omega_i) \equiv \begin{pmatrix} 0 & t_3(\omega_1) & \bar{t}_2(\omega_2) \\ -\bar{t}_3(\omega_1) & 0 & t_1(\omega_3) \\ -t_2(\omega_2) & -\bar{t}_1(\omega_3) & 0 \end{pmatrix}, \tag{12.30}$$

where the t_i satisfy eq. (12.27).

To satisfy algebraic closure, we need to determine the commutator of two matrices in such a way that it yields off-diagonal matrices, and $SO(8)$ triples. The commutator of $M(\omega_i)$ with $M(\omega_i')$ contains a matrix of the same kind, but also a diagonal matrix, which needs to be interpreted as an $SO(8)$ transformation.

The naive commutator of two matrices of the same type

$$X_a = \begin{pmatrix} 0 & \omega_a & 0 \\ -\bar{\omega}_a & 0 & 0 \\ 0 & 0 & 0 \end{pmatrix}, \tag{12.31}$$

where ω_a are arbitrary octonions, is diagonal. According to our intuition, it must be interpreted as an $SO(8)$ triple,

$$[X_1, X_2] = \{t_{1\,(\omega_1\,\omega_2)}, t_{2\,(\omega_1\,\omega_2)}, t_{3\,(\omega_1\,\omega_2)}\}. \tag{12.32}$$

The form of these transformations is determined from their action on the matrices according to eq. (12.27). They are given by

$$\begin{aligned} t_{1\,(\omega_1\,\omega_2)}: &\quad \omega \rightarrow \omega + \frac{1}{4}\bar{\omega}_2(\omega_1\omega) - (1 \leftrightarrow 2), \\ t_{2\,(\omega_1\,\omega_2)}: &\quad \omega \rightarrow \omega + \frac{1}{4}(\omega\omega_1)\bar{\omega}_2 - (1 \leftrightarrow 2), \\ t_{3\,(\omega_1\,\omega_2)}: &\quad \omega \rightarrow \omega + \frac{1}{2}(\omega\bar{\omega}_1 - \omega_1\bar{\omega})\omega_2 - (1 \leftrightarrow 2). \end{aligned} \tag{12.33}$$

In deriving these formulæ, much care is to be exercised because of non-associativity. As an application, we have the double "commutator,"

$$[\,[\,X_a,\,X_b\,],\,X_c\,] = \begin{pmatrix} 0 & t_{3\,(\omega_a\,\omega_b)}(\omega_c) & 0 \\ -\bar{t}_{3\,(\omega_a\,\omega_b)}(\omega_c) & 0 & 0 \\ 0 & 0 & 0 \end{pmatrix}. \tag{12.34}$$

We can verify the Jacobi identity by taking the cyclic sum over a, b, c, since

$$\sum_{cycl} t_{3\,(\omega_a\,\omega_b)}(\omega_c) = 0, \tag{12.35}$$

using eq. (12.33).

In general, the "commutator" between two matrices is given by another matrix, and a set of triples

$$\begin{aligned} [\,M(\omega_i),\, M(\omega_i')\,] = M(\omega_i'') & \\ + (\,t_{1\,(\omega_3\,\omega_3')},\, t_{2\,(\omega_3\,\omega_3')},\, t_{3\,(\omega_3\,\omega_3')}\,) & \\ + (\,t_{2\,(\omega_2\,\omega_2')},\, t_{3\,(\omega_2\,\omega_2')},\, t_{1\,(\omega_2\,\omega_2')}\,) & \\ + (\,t_{3\,(\omega_1\,\omega_1')},\, t_{1\,(\omega_1\,\omega_1')},\, t_{2\,(\omega_1\,\omega_1')}\,), & \end{aligned} \tag{12.36}$$

where

$$\omega_1'' = \bar{\omega}_2'\bar{\omega}_1 - \omega_2\bar{\omega}_1', \quad \omega_2'' = \bar{\omega}_1'\bar{\omega}_3 - \bar{\omega}_1\bar{\omega}_3', \quad \omega_1'' = \bar{\omega}_3'\bar{\omega}_2 - \bar{\omega}_3\omega_2'. \tag{12.37}$$

Further checks show that the twenty-four M matrices, together with the twenty-eight similarity triples, close on the exceptional group F_4. This remarkable construction of a *composition algebra* is due to J. Tits.

The same procedure, applied to real numbers, complex numbers, and quaternions, yields three classical groups $SO(3)$, $SU(3)$, and $Sp(6)$, respectively. Their dimensions are given by

$$3n_H + 2(n_H - 1) + d_H, \tag{12.38}$$

where n_H is the number of parameters in the four Hurwitz algebra $(1, 2, 4, 8)$, and d_H is the dimension of the automorphism group of the algebra, $0, 0, 3, 14$. The embeddings

$$F_4 \supset SO(3) \times G_2, \quad F_4 \supset Sp(6) \times SU(2) \quad F_4 \supset SU(3) \times SU(3),$$

easily follows from these constructions.

The F_4 construction can be thought of as involving two Hurwitz algebras, the reals and octonions. In fact, it generalizes to two arbitrary Hurwitz algebras, and produces the remaining exceptional groups.

While triality is of course trivial for real and complex numbers, it can still be defined for quaternions but the decomposition is not unique. With $SO(8)$ replaced by $SO(4)$, it is written in the form

$$t = d + r_q, \tag{12.39}$$

where d is an element of the automorphism group $SU(2)$, and $\bar{q} = -q$ is a pure imaginary quaternion. The similarity triples are as follows.

t_1	t_2	t_3
d	d	d
d	l_q	$-r_q$
l_q	0	$-r_q$
0	r_q	$-l_q$

Starting with two independent Hurwitz algebras, we build antihermitian matrices with no diagonal elements, whose matrix elements are products of the two algebras, H and H':

$$M(\omega_i, \rho_i) = \begin{pmatrix} 0 & \omega_1\rho_1 & \bar{\omega}_2\bar{\rho}_2 \\ -\bar{\omega}_1\bar{\rho}_1 & 0 & \omega_3\rho_3 \\ -\omega_2\rho_2 & -\bar{\omega}_3\bar{\rho}_3 & 0 \end{pmatrix},$$

where $\omega_i(\rho_i)$ belong to $H(H')$, respectively. The diagonal part of their commutator is interpreted as elements of the similarity triples of H and H'.

These matrices and the two similarity triples generate new composition algebras. If H represents real numbers, this is exactly our construction of F_4.

When both H and H' are octonions, the composition algebra contains two similarity triples, which generate the 56 parameters of $SO(8) \times SO(8)$. Each matrix element contains 8^2 possibilities, so that there are $3(64) = 192$ matrices. All together they generate the 248 parameter exceptional group E_8.

Since the construction is symmetric, there are ten composition algebras, as shown in the following *Magic Square*.

	R	C	Q	$\mathcal{O}$
R	$SO(3)$	$SU(3)$	$Sp(6)$	F_4
Z	$SU(3)$	$SU(3) \times SU(3)$	$SU(6)$	E_6
Q	$Sp(6)$	$SU(6)$	$SO(12)$	E_7
$\mathcal{O}$	F_4	E_6	E_7	E_8

We can read off interesting *algebraic embeddings* along its rows or columns. For instance, the algebras in the last column contains those in the third column times $SU(2)$, and the second column times $SU(3)$, and the first column times G_2. For example, the last row yields

$$E_8 \supset E_7 \times SU(2), \quad E_8 \supset E_6 \times SU(3), \quad E_8 \supset F_4 \times G_2,$$

and the same applies to the other three rows.

Appendix 1

Properties of some finite groups

$\mathcal{D}_3 \sim \mathcal{S}_3 \sim \mathcal{Z}_3 \rtimes \mathcal{Z}_2$

Order: 6, permutation/dihedral group

Presentation: $<a, b \mid a^3 = e, b^2 = e, bab^{-1} = a^{-1}>$

Classes: $C_2^{[2]}(b, ab, a^2b),\ C_3^{[3]}(a, a^2)$

Permutations: (1 2), (1 2 3)

$\mathcal{S}_3$	C_1	$3C_2^{[2]}$	$2C_3^{[3]}$
$\chi^{[\mathbf{1}]}$	1	1	1
$\chi^{[\mathbf{1}_1]}$	1	−1	1
$\chi^{[\mathbf{2}]}$	2	0	−1

Irrep:

$$\mathbf{2}:\ a = \begin{pmatrix} e^{2\pi i/3} & 0 \\ 0 & e^{-2\pi i/3} \end{pmatrix}; \qquad b = \begin{pmatrix} 0 & 1 \\ 1 & 0 \end{pmatrix}$$

Kronecker products:

$$\mathbf{1}_1 \times \mathbf{1}_1 = \mathbf{1}$$

$$\mathbf{1}_1 \times \mathbf{2} = \mathbf{2}$$

$$\mathbf{2} \times \mathbf{2} = \mathbf{1} + \mathbf{1}_1 + \mathbf{2}$$

$\mathcal{D}_4 \sim \mathcal{Z}_4 \rtimes \mathcal{Z}_2$

Order: 8, dihedral/octic group

Presentation: $<a, b \mid a^4 = e, b^2 = e, bab^{-1} = a^{-1}>$

Classes: $C_2^{[2]}(a^2),\ C_3^{[2]}(b, a^2b),\ C_4^{[2]}(ab, a^3b),\ C_5^{[4]}(a, a^3)$

Permutations: (1 2 3 4), (1 2)

$\mathcal{D}_4$	C_1	$C_2^{[2]}$	$2C_3^{[2]}$	$2C_4^{[2]}$	$2C_5^{[4]}$
$\chi^{[1]}$	1	1	1	1	1
$\chi^{[1_1]}$	1	1	−1	1	−1
$\chi^{[1_2]}$	1	1	−1	−1	1
$\chi^{[1_3]}$	1	1	1	−1	−1
$\chi^{[2]}$	2	−2	0	0	0

Irrep:

$$\mathbf{2}:\ a = \begin{pmatrix} i & 0 \\ 0 & -i \end{pmatrix}; \qquad b = \begin{pmatrix} 0 & 1 \\ 1 & 0 \end{pmatrix}$$

Kronecker products:

$$\mathbf{1_1} \times \mathbf{1_1} = \mathbf{1_2} \times \mathbf{1_2} = \mathbf{1_3} \times \mathbf{1_3} = \mathbf{1}$$

$$\mathbf{1_i} \times \mathbf{1_j} = \mathbf{1_k}, \quad i \neq j \neq k,$$

$$\mathbf{1_i} \times \mathbf{2} = \mathbf{2}$$

$$\mathbf{2} \times \mathbf{2} = \mathbf{1} + \mathbf{1_1} + \mathbf{1_2} + \mathbf{1_3}$$

$\mathcal{Q} \sim \mathcal{Q}_4$

Order: 8, quaternion, dicyclic

Presentation: $<a, b \,|\, a^4 = e, a^2 = b^2, bab^{-1} = a^{-1}>$

Classes: $C_2^{[4]}(a, a^3)$, $C_3^{[2]}(a^2)$, $C_4^{[4]}(b, a^2b)$, $C_5^{[4]}(ab, a^3b)$

Permutations: (1 2 3 4)(5 6 7 8), (1 6 3 8)(2 5 4 7)

$\mathcal{Q}$	C_1	$2C_2^{[4]}$	$C_3^{[2]}$	$2C_4^{[4]}$	$2C_5^{[4]}$
$\chi^{[\mathbf{1}]}$	1	1	1	1	1
$\chi^{[\mathbf{1_1}]}$	1	−1	1	1	−1
$\chi^{[\mathbf{1_2}]}$	1	−1	1	−1	1
$\chi^{[\mathbf{1_3}]}$	1	1	1	−1	−1
$\chi^{[\mathbf{2}]}$	2	0	−2	0	0

Irrep:

$$\mathbf{2}: \quad a = \begin{pmatrix} i & 0 \\ 0 & -i \end{pmatrix}; \qquad b = \begin{pmatrix} 0 & 1 \\ -1 & 0 \end{pmatrix}$$

Kronecker products:

$$\mathbf{1_1} \times \mathbf{1_1} = \mathbf{1_2} \times \mathbf{1_2} = \mathbf{1_3} \times \mathbf{1_3} = \mathbf{1}$$

$$\mathbf{1_i} \times \mathbf{1_j} = \mathbf{1_k}, \quad i \neq j \neq k,$$

$$\mathbf{1_i} \times \mathbf{2} = \mathbf{2}$$

$$\mathbf{2} \times \mathbf{2} = \mathbf{1} + \mathbf{1_1} + \mathbf{1_2} + \mathbf{1_3}$$

$\mathcal{D}_5 \sim \mathcal{Z}_5 \rtimes \mathcal{Z}_2$

Order: 10, dihedral group

Presentation: $<a, b \,|\, a^5 = e, b^2 = e, ab = ba^{-1}>$

Classes: $C_2^{[2]}(b, ab, a^2b, a^3b, a^4b)$, $C_3^{[5]}(a, a^4)$, $C_4^{[5]}(a^2, a^3)$

Permutations: (1 2 3 4 5), (2 5)(3 4)

$\mathcal{D}_5$	C_1	$5C_2^{[2]}$	$2C_3^{[5]}$	$2C_4^{[5]}$
$\chi^{[\mathbf{1}]}$	1	1	1	1
$\chi^{[\mathbf{1_1}]}$	1	-1	1	1
$\chi^{[\mathbf{2_1}]}$	2	0	$2\cos(2\pi/5)$	$2\cos(4\pi/5)$
$\chi^{[\mathbf{2_2}]}$	2	0	$2\cos(4\pi/5)$	$2\cos(2\pi/5)$

Irrep:

$$\mathbf{2_k}\colon\; a = \begin{pmatrix} e^{2k\pi i/5} & 0 \\ 0 & e^{-2k\pi i/5} \end{pmatrix}, \quad k = 1, 2; \qquad b = \begin{pmatrix} 0 & 1 \\ 1 & 0 \end{pmatrix}$$

Kronecker products:

$$\mathbf{1_1} \times \mathbf{1_1} = \mathbf{1}$$

$$\mathbf{1_1} \times \mathbf{2_k}, = \mathbf{2_k}, \quad k = 1, 2,$$

$$\mathbf{2_1} \times \mathbf{2_2} = \mathbf{2_1} + \mathbf{2_2}$$

$$\mathbf{2_1} \times \mathbf{2_1} = \mathbf{1} + \mathbf{1_1} + \mathbf{2_2}, \qquad \mathbf{2_2} \times \mathbf{2_2} = \mathbf{1} + \mathbf{1_1} + \mathbf{2_1}$$

$\Gamma\ (\mathcal{Q}_6) \sim \mathcal{Z}_3 \rtimes \mathcal{Z}_4$

Order: 12, dicyclic group

Presentation: $<a, b \mid a^6 = e, b^2 = a^3, bab^{-1} = a^{-1}>$

Classes: $C_2^{[6]}(a, a^5),\ C_3^{[3]}(a^2, a^4),\ C_4^{[2]}(a^3),\ C_5^{[4]}(b, a^2b, a^4b),\ C_6^{[4]}(ab, a^3b, a^5b),$

Permutations: (1 2 3), (1 2)(4 5 6 7)

$\mathcal{Q}_6$	C_1	$2C_2^{[6]}$	$2C_3^{[3]}$	$C_4^{[2]}$	$3C_5^{[4]}$	$3C_6^{[4]}$
$\chi^{[1]}$	1	1	1	1	1	1
$\chi^{[1_1]}$	1	−1	1	−1	i	$-i$
$\chi^{[1_2]}$	1	1	1	1	−1	−1
$\chi^{[\bar{1}_1]}$	1	−1	1	−1	$-i$	i
$\chi^{[2_1]}$	2	1	−1	−2	0	0
$\chi^{[2_2]}$	2	−1	−1	2	0	0

Irrep:

$$\mathbf{2_k}:\ a = \begin{pmatrix} e^{k\pi i/3} & 0 \\ 0 & e^{-k\pi i/3} \end{pmatrix}; \qquad b = \begin{pmatrix} 0 & 1 \\ (-1)^k & 0 \end{pmatrix}, \quad k = 1, 2.$$

Kronecker products:

$$\mathbf{1_1} \times \bar{\mathbf{1}}_1 = \mathbf{1_2} \times \mathbf{1_2} = \mathbf{1}$$

$$\mathbf{1_1} \times \mathbf{1_2} = \bar{\mathbf{1}}_1, \qquad \bar{\mathbf{1}}_1 \times \mathbf{1_2} = \mathbf{1_1},$$

$$\mathbf{1_1} \times \mathbf{2_1} = \bar{\mathbf{1}}_1 \times \mathbf{2_1} = \mathbf{2_2}$$

$$\mathbf{1_1} \times \mathbf{2_2} = \bar{\mathbf{1}}_1 \times \mathbf{2_2} = \mathbf{2_1}$$

$$\mathbf{1_2} \times \mathbf{2_k} = \mathbf{2_k}, \quad k = 1, 2,$$

$$\mathbf{2_1} \times \mathbf{2_1} = \mathbf{1} + \mathbf{1_2} + \mathbf{2_2}, \qquad \mathbf{2_2} \times \mathbf{2_2} = \mathbf{1} + \mathbf{1_2} + \mathbf{2_2},$$

$$\mathbf{2_1} \times \mathbf{2_2} = \mathbf{1_1} + \bar{\mathbf{1}}_1 + \mathbf{2_1}$$

$\mathcal{D}_6$

Order: 12, dihedral group

Presentation: $<a, b \mid a^6 = e, b^2 = e, ab = ba^{-1}>$

Classes: $C_2^{[2]}(a^3)$, $C_3^{[2]}(b, a^2b, a^4b)$, $C_4^{[4]}(ab, a^3b, a^5b)$, $C_5^{[3]}(a^2, a^4)$, $C_6^{[6]}(a, a^5)$

Permutations: (1 2 3)(4 5), (2 3)

$\mathcal{D}_6$	C_1	$C_2^{[2]}$	$3C_3^{[2]}$	$3C_4^{[2]}$	$2C_5^{[3]}$	$2C_6^{[6]}$
$\chi^{[\mathbf{1}]}$	1	1	1	1	1	1
$\chi^{[\mathbf{1_1}]}$	1	−1	−1	1	1	−1
$\chi^{[\mathbf{1_2}]}$	1	−1	1	−1	1	−1
$\chi^{[\mathbf{1_3}]}$	1	1	−1	−1	1	1
$\chi^{[\mathbf{2_1}]}$	2	−2	0	0	−1	1
$\chi^{[\mathbf{2_2}]}$	2	2	0	0	−1	−1

Irrep:

$$\mathbf{2_k}: \; a = \begin{pmatrix} e^{k\pi i/3} & 0 \\ 0 & e^{-k\pi i/3} \end{pmatrix}, \quad k = 1, 2; \qquad b = \begin{pmatrix} 0 & 1 \\ 1 & 0 \end{pmatrix}$$

Kronecker products:

$$\mathbf{1_1} \times \mathbf{1_1} = \mathbf{1_2} \times \mathbf{1_2} = \mathbf{1_3} \times \mathbf{1_3} = \mathbf{1}$$
$$\mathbf{1_i} \times \mathbf{1_j} = \mathbf{1_k}, \quad i \neq j \neq k,$$
$$\mathbf{1_1} \times \mathbf{2_1} = \mathbf{2_2}, \qquad \mathbf{1_1} \times \mathbf{2_2} = \mathbf{2_1},$$
$$\mathbf{1_2} \times \mathbf{2_1} = \mathbf{2_2}, \qquad \mathbf{1_2} \times \mathbf{2_2} = \mathbf{2_1},$$
$$\mathbf{1_3} \times \mathbf{2_1} = \mathbf{2_1}, \qquad \mathbf{1_3} \times \mathbf{2_2} = \mathbf{2_2},$$
$$\mathbf{2_1} \times \mathbf{2_1} = \mathbf{2_2} \times \mathbf{2_2} = \mathbf{1} + \mathbf{1_3} + \mathbf{2_2}$$
$$\mathbf{2_1} \times \mathbf{2_2} = \mathbf{1_1} + \mathbf{1_2} + \mathbf{2_1}$$

$Hol(\mathcal{Z}_5) \sim (\mathcal{Z}_2 \times \mathcal{Z}_2) \rtimes \mathcal{Z}_3$

Order: 20, holomorph of $\mathcal{Z}_5$

Presentation: $< a, b \,|\, a^5 = e, b^4 = e, b^{-1}ab = a^3 >$

Classes: $C_2^{[2]}(b^2, ab^2, a^2b^2, a^3b^2, a^4b^2)$, $C_3^{[4]}(b, ab, a^2b, a^3b, a^4b)$, $C_4^{[4]}(b^3, ab^3, a^2b^3, a^3b^3, a^4b^3)$, $C_5^{[5]}(a, a^2, a^3, a^4)$

Permutations: (1 2 3 4 5), (2 4 5 3)

$Hol(\mathcal{Z}_5)$	C_1	$5C_2^{[2]}$	$5C_3^{[4]}$	$5C_4^{[4]}$	$4C_5^{[5]}$
$\chi^{[\mathbf{1}]}$	1	1	1	1	1
$\chi^{[\mathbf{1}_1]}$	1	-1	i	$-i$	1
$\chi^{[\mathbf{1}_2]}$	1	1	-1	-1	1
$\chi^{[\mathbf{1}_3]}$	1	-1	$-i$	i	1
$\chi^{[\mathbf{4}]}$	4	0	0	0	-1

Irrep:

$$\mathbf{4}: a = \begin{pmatrix} e^{2\pi i/5} & 0 & 0 & 0 \\ 0 & e^{4\pi i/5} & 0 & 0 \\ 0 & 0 & e^{8\pi i/5} & 0 \\ 0 & 0 & 0 & e^{6\pi i/5} \end{pmatrix}; \quad b = \begin{pmatrix} 0 & 1 & 0 & 0 \\ 0 & 0 & 1 & 0 \\ 0 & 0 & 0 & 1 \\ 1 & 0 & 0 & 0 \end{pmatrix}$$

Kronecker products:

$$\mathbf{1}_1 \times \bar{\mathbf{1}}_1 = \mathbf{1}_2 \times \mathbf{1}_2 = \mathbf{1}$$

$$\mathbf{1}_1 \times \mathbf{1}_2 = \bar{\mathbf{1}}_1, \qquad \bar{\mathbf{1}}_1 \times \mathbf{1}_2 = \mathbf{1}_1$$

$$\mathbf{1}_1 \times \mathbf{4} = \bar{\mathbf{1}}_1 \times \mathbf{4} = \mathbf{1}_2 \times \mathbf{4} = \mathbf{4},$$

$$\mathbf{4} \times \mathbf{4} = \mathbf{1} + \mathbf{1}_1 + \bar{\mathbf{1}}_1 + \mathbf{1}_2 + \mathbf{4} + \mathbf{4} + \mathbf{4}$$

$\mathcal{A}_4 \sim \mathcal{T}$

Order: 12, alternating group, tetrahedral group

Presentation: $<a, b \,|\, a^2 = b^3 = (ab)^3 = e>$

Classes: $C_2^{[3]}(b, ab, aba, ba)$, $C_3^{[3]}(b^2, ab^2, b^2a, ab^2a)$, $C_4^{[2]}(a, b^2ab, bab^2)$

Permutations: $a = (1\,2)(3\,4)$, $b = (1\,2\,3)$

$\mathcal{A}_4$	C_1	$4C_2^{[3]}$	$4C_3^{[3]}$	$3C_4^{[2]}$
$\chi^{[\mathbf{1}]}$	1	1	1	1
$\chi^{[\mathbf{1}_1]}$	1	$e^{2\pi i/3}$	$e^{4\pi i/3}$	1
$\chi^{[\bar{\mathbf{1}}_1]}$	1	$e^{4\pi i/3}$	$e^{2\pi i/3}$	1
$\chi^{[\mathbf{3}]}$	3	0	0	-1

Irrep:

$$\mathbf{3}:\ a = \begin{pmatrix} -1 & 0 & 0 \\ 0 & 1 & 0 \\ 0 & 0 & -1 \end{pmatrix}; \quad b = \begin{pmatrix} 0 & 1 & 0 \\ 0 & 0 & 1 \\ 1 & 0 & 0 \end{pmatrix}$$

Kronecker products:

$$\mathbf{1}_1 \times \mathbf{1}_1 = \bar{\mathbf{1}}_1; \qquad \bar{\mathbf{1}}_1 \times \bar{\mathbf{1}}_1 = \mathbf{1}_1; \qquad \mathbf{1}_1 \times \bar{\mathbf{1}}_1 = \mathbf{1}$$

$$\mathbf{1}_1 \times \mathbf{3} = \bar{\mathbf{1}}_1 \times \mathbf{3} = \mathbf{3},$$

$$\mathbf{2} \times \mathbf{3}_1 = \mathbf{2} \times \mathbf{3}_2 = \mathbf{3}_1 + \mathbf{3}_2,$$

$$\mathbf{3} \times \mathbf{3} = \mathbf{1} + \mathbf{1}_1 + \bar{\mathbf{1}}_1 + \mathbf{3} + \mathbf{3}$$

$\mathcal{S}_4$

Order: 24, permutations group, symmetries of the cube

Presentation: $<a, b \mid a^4 = e, b^2 = e, (ab)^3 = e>$

Classes: $C_2^{[2]}(b, a^2ba^2, ba^2ba, aba^2b, a^3ba, aba^3)$, $C_3^{[2]}(a^2, ba^2b, a^2ba^2b)$, $C_4^{[3]}(ab, a^3ba^2, a^3b, aba^2, ba^3, ba, a^2ba^3, a^2ba)$, $C_5^{[4]}(a, a^3, aba, ba^2, a^2b)$

Permutations: (1 2 3 4)

$\mathcal{S}_4$	C_1	$6C_2^{[2]}$	$3C_3^{[2]}$	$8C_4^{[3]}$	$6C_5^{[4]}$
$\chi^{[\mathbf{1}]}$	1	1	1	1	1
$\chi^{[\mathbf{1}_1]}$	1	−1	1	1	−1
$\chi^{[\mathbf{2}]}$	2	0	2	−1	0
$\chi^{[\mathbf{3}_1]}$	3	1	−1	0	−1
$\chi^{[\mathbf{3}_2]}$	3	−1	−1	0	1

Irreps:

$$\mathbf{2}\text{: } a = \begin{pmatrix} 0 & e^{2\pi i/3} \\ e^{4\pi i/3} & 0 \end{pmatrix}; \quad b = \begin{pmatrix} 0 & 1 \\ 1 & 0 \end{pmatrix}$$

$$\mathbf{3}_1\text{: } a = \begin{pmatrix} 0 & 0 & -1 \\ 0 & -1 & 0 \\ -1 & 0 & 0 \end{pmatrix}; \quad b = \begin{pmatrix} 1 & 0 & 0 \\ 0 & 0 & 1 \\ 0 & 1 & 0 \end{pmatrix}$$

$$\mathbf{3}_2\text{: } a = \begin{pmatrix} 0 & 0 & 1 \\ 0 & 1 & 0 \\ 1 & 0 & 0 \end{pmatrix}; \quad b = \begin{pmatrix} -1 & 0 & 0 \\ 0 & 0 & 1 \\ 0 & 1 & 0 \end{pmatrix}$$

Kronecker products:

$$\mathbf{1}_1 \times \mathbf{2} = \mathbf{2}; \qquad \mathbf{1}_1 \times \mathbf{3}_1 = \mathbf{3}_2; \qquad \mathbf{1}_1 \times \mathbf{3}_2 = \mathbf{3}_1$$

$$\mathbf{2} \times \mathbf{2} = \mathbf{1} + \mathbf{1}_1 + \mathbf{2},$$

$$\mathbf{2} \times \mathbf{3}_1 = \mathbf{2} \times \mathbf{3}_2 = \mathbf{3}_1 + \mathbf{3}_2,$$

$$\mathbf{3}_1 \times \mathbf{3}_1 = \mathbf{3}_2 \times \mathbf{3}_2 = \mathbf{1} + \mathbf{2} + \mathbf{3}_1 + \mathbf{3}_2$$

$$\mathbf{3}_1 \times \mathbf{3}_2 = \mathbf{1}_1 + \mathbf{2} + \mathbf{3}_1 + \mathbf{3}_2$$

$\mathcal{Z}_7 \rtimes \mathcal{Z}_3$

Order: 21, Frobenius group

Presentation: $<a, b \mid a^7 = e, b^3 = e, b^{-1}ab = a^4>$

Classes: $C_2(b, ba, ba^2, ba^3, ba^4, ba^5, ba^6)$,
$C_3(b^2, b^2a, b^2a^2, b^2a^3, b^2a^4, b^2a^5, b^2a^6)$, $C_4(a, a^2, a^4)$, $C_5(a^3, a^5, a^6)$

Permutations: (1 2 3 4 5 6 7), (2 5 3)(4 6 7)

$\mathcal{Z}_7 \rtimes \mathcal{Z}_3$	C_1	$7C_2^{[3]}$	$7C_3^{[3]}$	$3C_4^{[7]}$	$3C_5^{[7]}$
$\chi^{[\mathbf{1}]}$	1	1	1	1	1
$\chi^{[\mathbf{1}']}$	1	$e^{2\pi i/3}$	$e^{4\pi i/3}$	1	1
$\chi^{[\bar{\mathbf{1}}']}$	1	$e^{4\pi i/3}$	$e^{2\pi i/3}$	1	1
$\chi^{[\mathbf{3}]}$	3	0	0	$\frac{1}{2}(-1+i\sqrt{7})$	$\frac{1}{2}(-1-i\sqrt{7})$
$\chi^{[\bar{\mathbf{3}}]}$	3	0	0	$\frac{1}{2}(-1-i\sqrt{7})$	$\frac{1}{2}(-1+i\sqrt{7})$

$$\mathbf{3}:\ a = \begin{pmatrix} e^{2\pi i/7} & 0 & 0 \\ 0 & e^{4\pi i/7} & 0 \\ 0 & 0 & e^{\pi i/7} \end{pmatrix}; \qquad b = \begin{pmatrix} 0 & 1 & 0 \\ 0 & 0 & 1 \\ 1 & 0 & 0 \end{pmatrix}$$

$$\bar{\mathbf{3}}:\ a = \begin{pmatrix} e^{5\pi i/7} & 0 & 0 \\ 0 & e^{10\pi i/7} & 0 \\ 0 & 0 & e^{6\pi i/7} \end{pmatrix}; \qquad b = \begin{pmatrix} 0 & 1 & 0 \\ 0 & 0 & 1 \\ 1 & 0 & 0 \end{pmatrix}$$

Kronecker products:

$$\mathbf{1}' \times \mathbf{1}' = \bar{\mathbf{1}}'; \qquad \mathbf{1}' \times \bar{\mathbf{1}}' = \mathbf{1}; \qquad \mathbf{3} \times \mathbf{1}' = \mathbf{3}$$

$$\mathbf{3} \times \bar{\mathbf{1}}' = \mathbf{3}; \qquad \mathbf{3} \times \mathbf{3} = (\mathbf{3} + \bar{\mathbf{3}})_s + \bar{\mathbf{3}}_a$$

$$\mathbf{3} \times \bar{\mathbf{3}} = \mathbf{1} + \mathbf{1}' + \bar{\mathbf{1}}' + \mathbf{3} + \bar{\mathbf{3}}$$

$\mathcal{A}_5$

Order: 60, alternating group, icosahedral group

Presentation: $<a, b \mid a^2 = b^3 = (ab)^5 = e>$

Classes: $C_2(ab),\ C_3((ab)^2),\ C_4(a),\ C_5(b)$

Permutations: $a = (12)(34)\ ,\ b = (135)$

$\mathcal{A}_5$	C_1	$12C_2^{[5]}$	$12C_3^{[5]}$	$15C_4^{[2]}$	$20C_5^{[3]}$
$\chi^{[\mathbf{1}]}$	1	1	1	1	1
$\chi^{[\mathbf{3}_1]}$	3	φ	$-\widetilde{\varphi}$	-1	0
$\chi^{[\mathbf{3}_2]}$	3	$-\widetilde{\varphi}$	φ	-1	0
$\chi^{[\mathbf{4}]}$	4	-1	-1	0	1
$\chi^{[\mathbf{5}]}$	5	0	0	1	-1

$\varphi = \frac{1}{2}(1+\sqrt{5}), \qquad \widetilde{\varphi} = \frac{1}{2}(-1+\sqrt{5})$

$$\mathbf{3}_1\colon\ a = \begin{pmatrix} -1 & 0 & 0 \\ 0 & -1 & 0 \\ \varphi & \varphi & 1 \end{pmatrix};\quad b = \begin{pmatrix} 0 & 1 & 0 \\ 0 & 0 & 1 \\ 1 & 0 & 0 \end{pmatrix}$$

$$\mathbf{3}_2\colon\ a = \begin{pmatrix} -1 & 0 & 0 \\ 0 & -1 & 0 \\ -\widetilde{\varphi} & -\widetilde{\varphi} & 1 \end{pmatrix};\quad b = \begin{pmatrix} 0 & 1 & 0 \\ 0 & 0 & 1 \\ 1 & 0 & 0 \end{pmatrix}$$

$$\mathbf{4}\colon\ a = \begin{pmatrix} 1 & 0 & 0 & 0 \\ 0 & 1 & 0 & 0 \\ 0 & 0 & -1 & 0 \\ 0 & 0 & 0 & -1 \end{pmatrix};\quad b = \frac{1}{4}\begin{pmatrix} -1 & -\sqrt{5} & -\sqrt{5} & -\sqrt{5} \\ \sqrt{5} & 1 & 1 & -3 \\ -\sqrt{5} & -1 & 3 & -1 \\ \sqrt{5} & -3 & 1 & 1 \end{pmatrix}$$

$$\mathbf{5}\colon\ a = \begin{pmatrix} 1 & 0 & 0 & 0 & 0 \\ 0 & -1 & 0 & 0 & 0 \\ 0 & 0 & -1 & 0 & 0 \\ 0 & 0 & 0 & 1 & 0 \\ 0 & 0 & 0 & 0 & 1 \end{pmatrix};\quad b = -\frac{1}{2}\begin{pmatrix} 1 & 0 & 1 & -\omega^2 & -\omega \\ 0 & 1 & 1 & -1 & -1 \\ -1 & -1 & 0 & \omega & \omega^2 \\ \omega & 1 & \omega^2 & 0 & -\omega \\ \omega^2 & 1 & \omega & -\omega^2 & 0 \end{pmatrix};\quad \omega^3=1$$

$\mathcal{A}_5$ Kronecker products:

$\mathbf{3}_1 \times \mathbf{3}_1 = \mathbf{3}_{1a} + (\mathbf{1}+\mathbf{5})_s;\quad \mathbf{3}_2 \times \mathbf{3}_2 = \mathbf{3}_{2a} + (\mathbf{1}+\mathbf{5})_s;\quad \mathbf{3}_1 \times \mathbf{3}_2 = \mathbf{4}+\mathbf{5}$

$\mathbf{3}_1 \times \mathbf{4} = \mathbf{3}_2 + \mathbf{4} + \mathbf{5};\quad \mathbf{3}_2 \times \mathbf{4} = \mathbf{3}_1 + \mathbf{4} + \mathbf{5}$

$\mathbf{3}_1 \times \mathbf{5} = \mathbf{3}_1 + \mathbf{3}_2 + \mathbf{4} + \mathbf{5};\quad \mathbf{3}_2 \times \mathbf{5} = \mathbf{3}_1 + \mathbf{3}_2 + \mathbf{4} + \mathbf{5}$

$\mathbf{4} \times \mathbf{4} = (\mathbf{3}_1 + \mathbf{3}_2)_a + (\mathbf{1} + \mathbf{4} + \mathbf{5})_s$

$\mathbf{5} \times \mathbf{5} = (\mathbf{3}_1 + \mathbf{3}_2 + \mathbf{4})_a + (\mathbf{1} + \mathbf{4} + \mathbf{5} + \mathbf{5})_s;\quad \mathbf{4} \times \mathbf{5} = \mathbf{3}_1 + \mathbf{3}_2 + \mathbf{4} + \mathbf{5} + \mathbf{5}$

$\mathcal{PSL}_2(7)$

Order: 168

Presentation: $<a, b \mid a^2 = b^3 = (ab)^7 = [a, b]^4 = e>$

Classes: $C_1^{[1]}$, $C_2^{[2]}(a)$, $C_3^{[3]}(b)$, $C_4^{[4]}([a, b])$, $C_5^{[7]}(ab)$, $C_6^{[7]}(ab^2)$

$\mathcal{PSL}_2(7)$	$C_1^{[1]}$	$21C_2^{[2]}(a)$	$56C_3^{[3]}(b)$	$42C_4^{[4]}([a, b])$	$24C_5^{[7]}(ab)$	$24C_6^{[7]}(ab^2)$
$\chi^{[1]}$	1	1	1	1	1	1
$\chi^{[3]}$	3	−1	0	1	b_7	$\bar{b}_7$
$\chi^{[\bar{3}]}$	3	−1	0	1	$\bar{b}_7$	b_7
$\chi^{[6]}$	6	2	0	0	−1	−1
$\chi^{[7]}$	7	−1	1	−1	0	0
$\chi^{[8]}$	8	0	−1	0	1	1

$b_7 = \frac{1}{2}(-1 + i\sqrt{7})$, $\quad \bar{b}_7 = \frac{1}{2}(-1 - i\sqrt{7})$; $\quad \eta^7 = 1$

$$\mathbf{3}: \quad a = \frac{i}{\sqrt{7}} \begin{pmatrix} \eta^2 - \eta^5 & \eta - \eta^6 & \eta^4 - \eta^3 \\ \eta - \eta^6 & \eta^4 - \eta^3 & \eta^2 - \eta^5 \\ \eta^4 - \eta^3 & \eta^2 - \eta^5 & \eta - \eta^6 \end{pmatrix}; \quad b = \frac{i}{\sqrt{7}} \begin{pmatrix} \eta^3 - \eta^6 & \eta^3 - \eta & \eta - 1 \\ \eta^2 - 1 & \eta^6 - \eta^5 & \eta^6 - \eta^2 \\ \eta^5 - \eta^4 & \eta^4 - 1 & \eta^5 - \eta^3 \end{pmatrix}$$

$\mathcal{PSL}_2(7)$ **Kronecker products**:

$\mathbf{3} \times \mathbf{3} = \bar{\mathbf{3}}_a + \mathbf{6}_s$; $\quad \mathbf{3} \times \bar{\mathbf{3}} = \mathbf{1} + \mathbf{8}$;

$\mathbf{3} \times \mathbf{6} = \bar{\mathbf{3}} + \mathbf{7} + \mathbf{8}$; $\quad \bar{\mathbf{3}} \times \mathbf{6} = \mathbf{3} + \mathbf{7} + \mathbf{8}$;

$\mathbf{3} \times \mathbf{7} = \mathbf{6} + \mathbf{7} + \mathbf{8}$; $\quad \bar{\mathbf{3}} \times \mathbf{7} = \mathbf{6} + \mathbf{7} + \mathbf{8}$;

$\mathbf{3} \times \mathbf{8} = \mathbf{3} + \mathbf{6} + \mathbf{7} + \mathbf{8}$; $\quad \bar{\mathbf{3}} \times \mathbf{8} = \bar{\mathbf{3}} + \mathbf{6} + \mathbf{7} + \mathbf{8}$;

$\mathbf{6} \times \mathbf{6} = (\mathbf{1} + \mathbf{6} + \mathbf{6} + \mathbf{8})_s + (\mathbf{7} + \mathbf{8})_a$; $\quad \mathbf{6} \times \mathbf{7} = \mathbf{3} + \bar{\mathbf{3}} + \mathbf{6} + \mathbf{7} + \mathbf{7} + \mathbf{8} + \mathbf{8}$;

$\mathbf{6} \times \mathbf{8} = \mathbf{3} + \bar{\mathbf{3}} + \mathbf{6} + \mathbf{6} + \mathbf{7} + \mathbf{7} + \mathbf{8} + \mathbf{8}$;

$\mathbf{7} \times \mathbf{7} = (\mathbf{1} + \mathbf{6} + \mathbf{6} + \mathbf{7} + \mathbf{8})_s + (\mathbf{3} + \bar{\mathbf{3}} + \mathbf{7} + \mathbf{8})_a$; $\quad \mathbf{7} \times \mathbf{8} = \mathbf{3} + \bar{\mathbf{3}} + \mathbf{6} + \mathbf{6} + \mathbf{7} + \mathbf{7} + \mathbf{8} + \mathbf{8} + \mathbf{8}$;

$\mathbf{8} \times \mathbf{8} = (\mathbf{1} + \mathbf{6} + \mathbf{6} + \mathbf{7} + \mathbf{8} + \mathbf{8})_s + (\mathbf{3} + \bar{\mathbf{3}} + \mathbf{7} + \mathbf{7} + \mathbf{8})_a$

Appendix 2

Properties of selected Lie algebras

We list the basic properties of the Lie algebras that occur frequently in physics. Their Dynkin diagrams are obtained by taking out the node ⊗ with number (0) from the extended Dynkin diagrams. The numbering of the nodes is shown, together with the dimensions of the fundamental representations. Representations are labeled by their dimensions in boldface type with their Dynkin label as a subscript whenever needed to avoid confusion. The adjoint representation (Adjt) is identified with a supersript. Useful properties of the algebra are listed, in particular the congruence properties of the representations, the marks, co-marks, the Coxeter number, and the level vector.

Useful embeddings are shown for each algebra; for those which contain a $U(1)$ factor, the $U(1)$ charge appears as a subscript. Much of this information can be found in R. Slansky, *Phys. Rep.* **79**, 1–128 (1981), as well as in the excellent tables of W. G. McKay and J. Patera, *Tables of Dimensions, Indices, and Branching Rules for Representations of Simple Lie Algebras*, Marcel Dekker, Inc., New York (1981), as well as W. G. McKay, J. Patera, and D. W. Rand, *Tables of Representations of Simple Lie Algebras*, Volume I Exceptional Simple Lie Algebras, Publications CRM, Montréal (1990). For the reader who finds these appendices insufficient, we recommend *Schur*, the open source software, from the late B. Wybourne. It can be found at http://sourceforge.net/projects/schur/

$SU(3)$

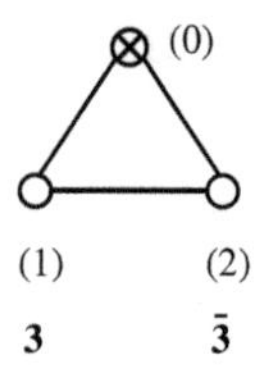

Representations & Dynkin indices

Reps	$I^{(2)}$	$I^{(3)}$	$I^{(4)}$
$\mathbf{3}_{(10)}$	1	1	4
$\bar{\mathbf{3}}_{(01)}$	1	−1	4
$\mathbf{6}_{(20)}$	5	7	68
$\bar{\mathbf{6}}_{(02)}$	5	−7	68
$\mathbf{8}_{(11)}$	6	0	72
$\mathbf{10}_{(30)}$	15	27	396
$\overline{\mathbf{10}}_{(03)}$	15	−27	396
$\mathbf{15}_{(21)}$	20	14	464
$\mathbf{15}_{(40)}$	35	77	1484
$\mathbf{27}_{(22)}$	15	0	1944

Kronecker products

$$\mathbf{3}_{(10)} \times \mathbf{3}_{(10)} = \mathbf{6}^{s}_{(20)} + \bar{\mathbf{3}}^{a}_{(01)}$$

$$\mathbf{3}_{(10)} \times \bar{\mathbf{3}}_{(01)} = \mathbf{8}^{\text{Adjt}}_{(11)} + \mathbf{1}_{(00)}$$

$$\mathbf{3}_{(10)} \times \mathbf{6}_{(20)} = \mathbf{10}_{(30)} + \mathbf{8}^{\text{Adjt}}_{(11)}$$

$$\bar{\mathbf{3}}_{(01)} \times \mathbf{6}_{(20)} = \mathbf{15}_{(21)} + \mathbf{3}_{(10)}$$

$$\mathbf{6}_{(20)} \times \mathbf{6}_{(20)} = \mathbf{15}^{s}_{(40)} + \bar{\mathbf{6}}^{s}_{(02)} + \mathbf{15}^{a}_{(21)}$$

$$\mathbf{6}_{(20)} \times \bar{\mathbf{6}}_{(02)} = \mathbf{27}_{(22)} + \mathbf{8}^{\text{Adjt}}_{(11)} + \mathbf{1}_{(00)}$$

$$\mathbf{3}_{(10)} \times \mathbf{8}^{\text{Adjt}}_{(11)} = \mathbf{15}_{(21)} + \bar{\mathbf{6}}_{(02)} + \mathbf{3}_{(10)}$$

$$\mathbf{8}^{\text{Adjt}}_{(11)} \times \mathbf{8}^{\text{Adjt}}_{(11)} = \mathbf{27}^{s}_{(22)} + \mathbf{8}^{\text{Adjt},\,s}_{(11)} + \mathbf{1}^{s}_{(00)} + \mathbf{10}^{a}_{(30)} + \overline{\mathbf{10}}^{a}_{(03)} + \mathbf{8}^{\text{Adjt},\,a}_{(11)}$$

Embeddings

$SU(\mathbf{3}) \supset SU(\mathbf{2}) \times U(\mathbf{1})$

$\mathbf{3} = \mathbf{2}_{-1} + \mathbf{1}_2; \qquad \mathbf{8} = \mathbf{3}_0 + \mathbf{1}_0 + \mathbf{2}_3 + \mathbf{2}_{-3}$

$SU(\mathbf{3}) \supset SO(\mathbf{3})$

$\mathbf{3} = \mathbf{3}; \qquad \mathbf{8} = \mathbf{3} + \mathbf{5}$

Representation congruence: $\mathcal{Z}_3$; **Coxeter number:** 3

Marks: (1, 1, 1); **Level vector:** (2, 2)

$SU(4) \sim SO(6)$

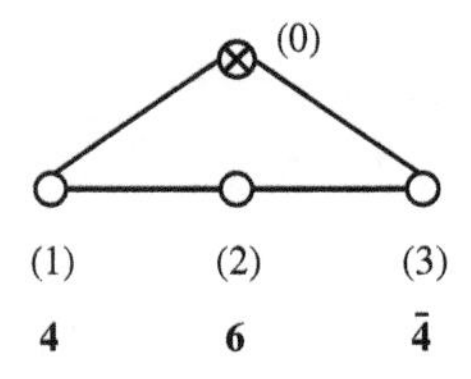

Representations & Dynkin indices

	$I^{(2)}$	$I^{(3)}$	$I^{(4)}$
$\mathbf{4}_{(100)}$	1	1	9
$\bar{\mathbf{4}}_{(001)}$	1	−1	9
$\mathbf{6}_{(010)}$	2	0	24
$\mathbf{10}_{(200)}$	6	8	168
$\mathbf{15}_{(101)}$	8	0	192
$\mathbf{20}_{(011)}$	13	−7	381
$\mathbf{20}_{(020)}$	16	0	576
$\mathbf{20}_{(003)}$	21	29	1101
$\mathbf{36}_{(201)}$	33	21	1473
$\mathbf{45}_{(210)}$	48	48	2496
$\mathbf{64}_{(111)}$	64	0	3072
$\mathbf{84}_{(202)}$	112	0	7488

Kronecker products

$$\mathbf{4}_{(100)} \times \mathbf{4}_{(100)} = \mathbf{10}^{s}_{(200)} + \mathbf{6}^{a}_{(010)}$$

$$\mathbf{4}_{(100)} \times \bar{\mathbf{4}}_{(001)} = \mathbf{15}^{\text{Adjt}}_{(101)} + \mathbf{1}_{(000)}$$

$$\mathbf{4}_{(100)} \times \mathbf{6}_{(010)} = \overline{\mathbf{20}}_{(110)} + \bar{\mathbf{4}}_{(001)}$$

$$\mathbf{6}_{(010)} \times \mathbf{6}_{(010)} = \mathbf{20}^{s}_{(020)} + \mathbf{1}^{s}_{(000)} + \mathbf{15}^{a}_{(101)}$$

$$\mathbf{4}_{(100)} \times \mathbf{15}^{\text{Adjt}}_{(101)} = \mathbf{4}_{(001)} + \mathbf{20}_{(011)} + \mathbf{36}_{(201)}$$

$$\mathbf{6}_{(010)} \times \mathbf{15}^{\text{Adjt}}_{(101)} = \mathbf{4}_{(001)} + \overline{\mathbf{10}}_{(002)} + \mathbf{10}_{(200)} + \mathbf{64}_{(111)}$$

$$\mathbf{15}^{\text{Adjt}}_{(101)} \times \mathbf{15}^{\text{Adjt}}_{(101)} = \mathbf{1}^{s}_{(000)} + \mathbf{15}^{\text{Adjt},\,s}_{(101)} + \mathbf{20}^{s}_{(020)} + \mathbf{84}^{s}_{(202)} + \mathbf{15}^{\text{Adjt},\,a}_{(101)} + \mathbf{45}^{a}_{(210)} + \overline{\mathbf{45}}^{a}_{(012)}$$

Embeddings

$SU(\mathbf{4}) \supset SU(\mathbf{3}) \times U(\mathbf{1})$

$$\mathbf{4} = \mathbf{3}_{-1} + \mathbf{1}_{3}; \qquad \mathbf{15} = \mathbf{3}_{-4} + \bar{\mathbf{3}}_{4} + \mathbf{1}_{0} + \mathbf{8}_{0}$$

$SU(\mathbf{4}) \supset SU(\mathbf{2}) \times SU(\mathbf{2})$

$$\mathbf{4} = (\mathbf{2}, \mathbf{2}); \qquad \mathbf{15} = (\mathbf{3}, \mathbf{1}) + (\mathbf{1}, \mathbf{3}) + (\mathbf{3}, \mathbf{3})$$

$SU(\mathbf{4}) \supset SO(\mathbf{5})$

$$\mathbf{4} = \mathbf{4}; \qquad \mathbf{15} = \mathbf{5} + \mathbf{10}$$

Representation congruence: $\mathcal{Z}_4$; **Coxeter number:** 4

Marks: (1, 1, 1, 1); **Level vector:** (3, 4, 3)

$SU(5)$

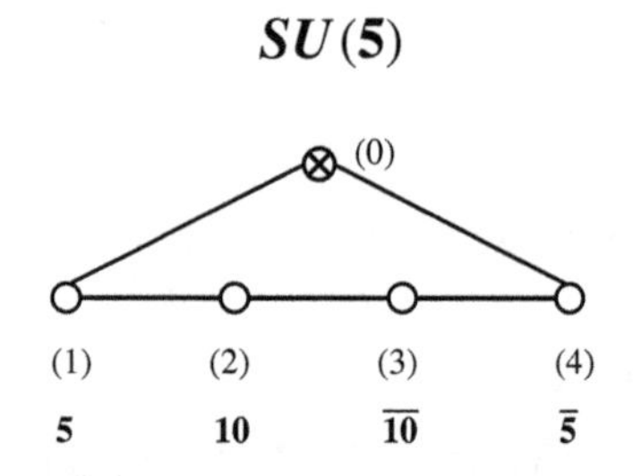

Representations & Dynkin indices

	$I^{(2)}$	$I^{(3)}$	$I^{(4)}$
$\mathbf{5}_{(1000)}$	1	1	16
$\bar{\mathbf{5}}_{(0001)}$	1	−1	16
$\mathbf{10}_{(0100)}$	3	1	72
$\mathbf{15}_{(2000)}$	7	9	328
$\mathbf{24}_{(1001)}$	10	0	400
$\mathbf{35}_{(0003)}$	28	−44	2392
$\mathbf{40}_{(0011)}$	22	−16	1168
$\mathbf{45}_{(0101)}$	24	−6	1224
$\mathbf{50}_{(0020)}$	35	−15	2360
$\mathbf{70}_{(2001)}$	49	29	3544
$\mathbf{70}_{(0004)}$	84	−156	11064
$\mathbf{75}_{(0110)}$	50	0	3200
$\mathbf{105}_{(0012)}$	210	−99	8176
$\mathbf{126}_{(2010)}$	105	75	9000
$\mathbf{175}_{(1101)}$	140	30	11360
$\mathbf{200}_{(2002)}$	200	0	21200

Kronecker products

$$\mathbf{5}_{(1000)} \times \mathbf{5}_{(1000)} = \mathbf{15}^{s}_{(2000)} + \mathbf{10}^{a}_{(0100)}$$

$$\mathbf{5}_{(1000)} \times \bar{\mathbf{5}}_{(0001)} = \mathbf{24}^{\mathrm{Adjt}}_{(1001)} + \mathbf{1}_{(0000)}$$

$$\mathbf{5}_{(1000)} \times \mathbf{10}_{(0100)} = \overline{\mathbf{40}}_{(1100)} + \overline{\mathbf{10}}_{(0010)}$$

$$\bar{\mathbf{5}}_{(0001)} \times \mathbf{10}_{(0100)} = \mathbf{45}_{(0101)} + \mathbf{5}_{(1000)}$$

$$\mathbf{10}_{(0100)} \times \mathbf{10}_{(0100)} = \overline{\mathbf{50}}^{s}_{(0200)} + \bar{\mathbf{5}}^{s}_{(1000)} + \overline{\mathbf{45}}^{a}_{(1010)}$$

$$\mathbf{10}_{(0100)} \times \overline{\mathbf{10}}_{(0010)} = \mathbf{1}_{(0000)} + \mathbf{24}^{\mathrm{Adjt}}_{(1001)} + \mathbf{75}_{(0110)}$$

$$\mathbf{5}_{(1000)} \times \mathbf{24}^{\mathrm{Adjt}}_{(1001)} = \mathbf{5}_{(1000)} + \mathbf{45}_{(0101)} + \mathbf{70}_{(2001)}$$

$$\mathbf{10}_{(1000)} \times \mathbf{24}^{\mathrm{Adjt}}_{(1001)} = \mathbf{10}_{(0100)} + \mathbf{15}_{(2000)} + \mathbf{40}_{(0011)} + \mathbf{175}_{(1101)}$$

$$\mathbf{24}^{\mathrm{Adjt}}_{(1001)} \times \mathbf{24}^{\mathrm{Adjt}}_{(1001)} = \mathbf{1}^{s}_{(0000)} + \mathbf{24}^{\mathrm{Adjt},\,s}_{(1001)} + \mathbf{75}^{s}_{(0110)} + \mathbf{200}^{s}_{(2002)} + \mathbf{24}^{\mathrm{Adjt},\,a}_{(1001)} + \mathbf{126}^{a}_{(2010)} + \overline{\mathbf{126}}^{a}_{(0102)}$$

Embeddings

$SU(\mathbf{5}) \supset SU(\mathbf{4}) \times U(\mathbf{1})$

$$\mathbf{5} = \mathbf{4}_{-1} + \mathbf{1}_{4}; \qquad \mathbf{24} = \mathbf{4}_{-5} + \bar{\mathbf{4}}_{5} + \mathbf{1}_{0} + \mathbf{15}_{0}$$

$SU(\mathbf{5}) \supset SU(\mathbf{3}) \times SU(\mathbf{2}) \times U(\mathbf{1})$

$$\mathbf{5} = (\mathbf{3}, \mathbf{1})_{-2} + (\mathbf{1}, \mathbf{2})_{3};$$

$$\mathbf{24} = (\mathbf{8}, \mathbf{1})_{0} + (\mathbf{1}, \mathbf{3})_{0} + (\mathbf{1}, \mathbf{1})_{0} + (\mathbf{3}, \mathbf{2})_{-5} + (\bar{\mathbf{3}}, \mathbf{2})_{5}$$

$SU(\mathbf{5}) \supset SO(\mathbf{5})$

$$\mathbf{5} = \mathbf{5}; \qquad \mathbf{24} = \mathbf{10} + \mathbf{14}$$

Representation congruence: $\mathcal{Z}_5$; **Coxeter number:** 5;

Marks: (1, 1, 1, 1); **Level vector:** (4, 6, 6, 4)

$SU(6)$

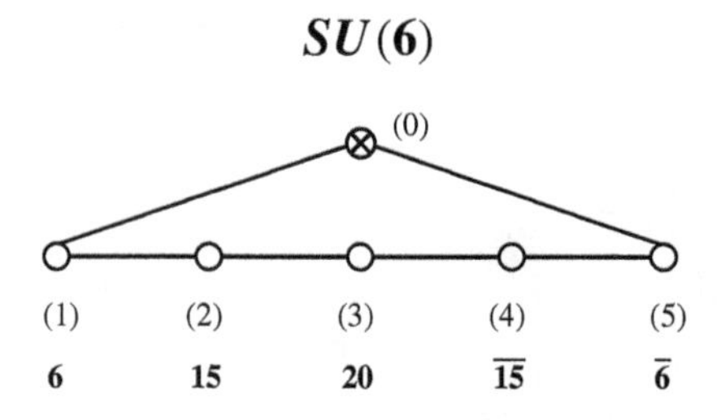

Representations & Dynkin indices

	$I^{(2)}$	$I^{(3)}$	$I^{(4)}$
$\mathbf{6}_{(10000)}$	1	1	25
$\bar{\mathbf{6}}_{(00001)}$	1	−1	25
$\mathbf{15}_{(01000)}$	4	2	160
$\mathbf{20}_{(00100)}$	6	0	270
$\mathbf{21}_{(20000)}$	8	10	560
$\mathbf{35}_{(10001)}$	12	0	720
$\mathbf{56}_{(30000)}$	36	54	4500
$\mathbf{70}_{(11000)}$	33	27	2745
$\mathbf{84}_{(01001)}$	38	−4	2990
$\mathbf{105}_{(00101)}$	52	−22	4480
$\mathbf{105}_{(00020)}$	64	−40	6880
$\mathbf{120}_{(20001)}$	68	38	7220
$\mathbf{126}_{(00004)}$	120	−210	22800
$\mathbf{175}_{(00200)}$	120	0	14400
$\mathbf{189}_{(01010)}$	108	0	10800
$\mathbf{210}_{(00110)}$	131	−37	14315
$\mathbf{210}_{(00012)}$	152	−170	20720
$\mathbf{252}_{(00005)}$	330	−726	87450
$\mathbf{280}_{(20010)}$	192	108	24480
$\mathbf{315}_{(30001)}$	264	258	42960
$\mathbf{336}_{(00102)}$	248	−196	34040

$\mathbf{384}_{(11000)}$	256	80	31360
$\mathbf{405}_{(20002)}$	324	0	49680
$\mathbf{420}_{(00021)}$	358	−308	56470

Kronecker products

$$\mathbf{6}_{(10000)} \times \mathbf{6}_{(10000)} = \mathbf{21}^{s}_{(20000)} + \mathbf{15}^{a}_{(01000)}$$

$$\mathbf{6}_{(10000)} \times \bar{\mathbf{6}}_{(00001)} = \mathbf{35}^{\text{Adjt}}_{(10001)} + \mathbf{1}_{(00000)}$$

$$\mathbf{6}_{(10000)} \times \mathbf{15}_{(01000)} = \overline{\mathbf{70}}_{(11000)} + \mathbf{20}_{(00100)}$$

$$\mathbf{6}_{(10000)} \times \overline{\mathbf{15}}_{(01000)} = \bar{\mathbf{6}}_{(00001)} + \overline{\mathbf{84}}_{(10010)}$$

$$\mathbf{6}_{(10000)} \times \mathbf{20}_{(00100)} = \overline{\mathbf{15}}_{(10010)} + \overline{\mathbf{105}}_{(10100)}$$

$$\mathbf{15}_{(01000)} \times \mathbf{15}_{(01000)} = \overline{\mathbf{105}}^{s}_{(02000)} + \overline{\mathbf{105}}^{a}_{(10100)} + \overline{\mathbf{15}}^{s}_{(00010)}$$

$$\mathbf{15}_{(01000)} \times \overline{\mathbf{15}}_{(00010)} = \mathbf{1}_{(00000)} + \mathbf{189}_{(01010)} + \mathbf{35}^{\text{Adjt}}_{(10001)}$$

$$\mathbf{20}_{(00100)} \times \mathbf{20}_{(00100)} = \mathbf{175}^{s}_{(00200)} + \mathbf{35}^{\text{Adjt},\,s}_{(10001)} + \mathbf{1}^{s}_{(00000)} + \mathbf{189}^{a}_{(01010)}$$

$$\mathbf{15}_{(01000)} \times \mathbf{20}_{(00100)} = \overline{\mathbf{210}}^{s}_{(01100)} + \overline{\mathbf{84}}^{s}_{(10010)} + \bar{\mathbf{6}}^{s}_{(00001)}$$

$$\mathbf{35}^{\text{Adjt}}_{(10001)} \times \mathbf{35}^{\text{Adjt}}_{(10001)} = \mathbf{405}^{s}_{(20002)} + \mathbf{189}^{s}_{(01010)} + \mathbf{35}^{\text{Adjt},\,s}_{(10001)} + \mathbf{1}^{s}_{(00000)} + \mathbf{35}^{\text{Adjt},\,a}_{(10001)}$$
$$+\mathbf{280}^{a}_{(20010)} + \mathbf{280}^{a}_{(01002)}$$

Embeddings

$SU(6) \supset SU(5) \times U(1)$

$$\mathbf{6} = \mathbf{5}_{-1} + \mathbf{1}_{5}; \qquad \mathbf{35} = \mathbf{24}_{0} + \mathbf{1}_{0} + \mathbf{5}_{-6} + \bar{\mathbf{5}}_{6}$$

$SU(6) \supset SU(4) \times SU(2) \times U(1)$

$$\mathbf{6} = (\mathbf{4}, \mathbf{1})_{-1} + (\mathbf{1}, \mathbf{2})_{2};$$
$$\mathbf{35} = (\mathbf{15}, \mathbf{1})_{0} + (\mathbf{1}, \mathbf{3})_{0} + (\mathbf{4}, \mathbf{2})_{-3} + (\bar{\mathbf{4}}, \mathbf{2})_{3}$$

$SU(6) \supset SU(3) \times SU(3) \times U(1)$

$$\mathbf{6} = (\mathbf{1}, \mathbf{3})_{1} + (\mathbf{3}, \mathbf{1})_{-1}$$
$$\mathbf{35} = (\mathbf{8}, \mathbf{1})_{0} + (\mathbf{1}, \mathbf{8})_{0} + (\mathbf{1}, \mathbf{1})_{0} + (\bar{\mathbf{3}}, \mathbf{3})_{2} + (\mathbf{3}, \bar{\mathbf{3}})_{-2}$$

$SU(6) \supset SU(3) \times SU(2)$

$$\mathbf{6} = (\mathbf{3}, \mathbf{2}); \qquad \mathbf{35} = (\mathbf{8}, \mathbf{1}) + (\mathbf{1}, \mathbf{3}) + (\mathbf{8}, \mathbf{3})$$

$SU(\mathbf{6}) \supset SU(\mathbf{3})$

$$\mathbf{6} = \mathbf{6}; \qquad \mathbf{35} = \mathbf{27} + \mathbf{8}$$

$SU(\mathbf{6}) \supset SU(\mathbf{4})$

$$\mathbf{6} = \mathbf{6}; \qquad \mathbf{35} = \mathbf{20}_{(020)} + \mathbf{15}$$

Representation congruence: $\mathcal{Z}_6$; **Coxeter number:** 6

Marks: (1, 1, 1, 1, 1); **Level vector:** (5, 8, 9, 8, 5)

$SO(5) \sim Sp(4)$

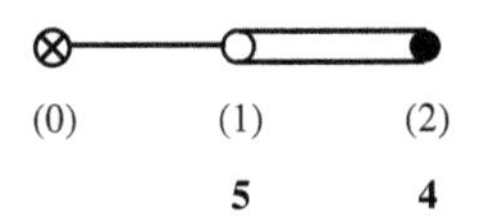

Representations & Dynkin indices

	$I^{(2)}$	$I^{(4)}$
$\mathbf{4}_{(01)}$	1	3
$\mathbf{5}_{(10)}$	2	12
$\mathbf{10}_{(02)}$	6	60
$\mathbf{14}_{(20)}$	14	252
$\mathbf{16}_{(11)}$	12	156
$\mathbf{20}_{(03)}$	21	399
$\mathbf{30}_{(30)}$	54	1836
$\mathbf{35}_{(04)}$	56	1680
$\mathbf{35}_{(12)}$	42	924

Kronecker products

$$\mathbf{5}_{(10)} \times \mathbf{5}_{(10)} = \mathbf{1}^{s}_{(00)} + \mathbf{14}^{s}_{(20)} + \mathbf{10}^{\text{Adjt}, a}_{(02)}$$

$$\mathbf{4}_{(01)} \times \mathbf{4}_{(01)} = \mathbf{1}^{a}_{(00)} + \mathbf{5}^{a}_{(10)} + \mathbf{10}^{\text{Adjt}, s}_{(02)}$$

$$\mathbf{4}_{(01)} \times \mathbf{5}_{(10)} = \mathbf{16}_{(11)} + \mathbf{4}_{(01)}$$

$$\mathbf{10}^{\text{Adjt}}_{(02)} \times \mathbf{10}^{\text{Adjt}}_{(02)} = \mathbf{1}^{s}_{(00)} + \mathbf{5}^{s}_{(10)} + \mathbf{14}^{s}_{(20)} + \mathbf{35}^{s}_{(04)} + \mathbf{10}^{\text{Adjt}, a}_{(02)} + \mathbf{35}^{a}_{(12)}$$

Embeddings

$SO(\mathbf{5}) \supset SU(\mathbf{2}) \times SU(\mathbf{2})$

$$\mathbf{5} = (\mathbf{1}, \mathbf{1}) + (\mathbf{2}, \mathbf{2}); \quad \mathbf{4} = (\mathbf{2}, \mathbf{1}) + (\mathbf{1}, \mathbf{2})$$
$$\mathbf{10} = (\mathbf{3}, \mathbf{1}) + (\mathbf{1}, \mathbf{3}) + (\mathbf{2}, \mathbf{2})$$

$SO(\mathbf{5}) \supset SU(\mathbf{2}) \times U(\mathbf{1})$

$$\mathbf{5} = \mathbf{1}_2 + \mathbf{1}_{-2} + \mathbf{3}_0; \qquad \mathbf{4} = \mathbf{2}_1 + \mathbf{2}_{-1};$$
$$\mathbf{10} = \mathbf{1}_0 + \mathbf{3}_{-2} + \mathbf{3}_2 + \mathbf{3}_0$$

$SO(\mathbf{5}) \supset SU(\mathbf{2})$

$$\mathbf{5} = \mathbf{5}; \qquad \mathbf{4} = \mathbf{4}; \qquad \mathbf{10} = \mathbf{7} + \mathbf{3}$$

Representation congruence: $\mathcal{Z}_2$; **Coxeter number:** 4

Marks: (1, 2, 2); **Co-marks:** (1, 2, 1); **Level vector:** (4, 6)

$SO(7)$

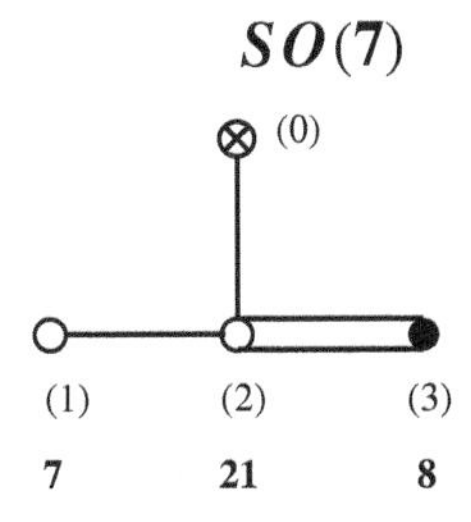

Representations & Dynkin indices

	$I^{(2)}$	$I^{(4)}$
$\mathbf{7}_{(100)}$	2	24
$\mathbf{8}_{(001)}$	2	18
$\mathbf{21}_{(010)}$	10	216
$\mathbf{27}_{(200)}$	18	600
$\mathbf{35}_{(002)}$	20	528
$\mathbf{48}_{(101)}$	28	780
$\mathbf{77}_{(300)}$	88	5280
$\mathbf{105}_{(110)}$	90	3864
$\mathbf{112}_{(003)}$	108	5148
$\mathbf{112}_{(011)}$	92	3708
$\mathbf{168}_{(020)}$	192	11136
$\mathbf{168}_{(201)}$	170	8826
$\mathbf{182}_{(400)}$	312	28704
$\mathbf{189}_{(102)}$	180	8592

Kronecker products

$$\mathbf{7}_{(100)} \times \mathbf{7}_{(100)} = \mathbf{1}^{s}_{(000)} + \mathbf{27}^{s}_{(200)} + \mathbf{21}^{\text{Adjt},\,a}_{(010)}$$

$$\mathbf{7}_{(100)} \times \mathbf{21}^{\text{Adjt}}_{(010)} = \mathbf{105}_{(110)} + \mathbf{35}_{(002)} + \mathbf{7}_{(100)}$$

$$\mathbf{8}_{(001)} \times \mathbf{8}_{(001)} = \mathbf{1}^{s}_{(000)} + \mathbf{35}^{s}_{(002)} + \mathbf{21}^{\text{Adjt},\,a}_{(010)} + \mathbf{7}^{a}_{(100)}$$

$$\mathbf{7}_{(100)} \times \mathbf{8}_{(001)} = \mathbf{48}_{(101)} + \mathbf{8}_{(001)}$$

$$\mathbf{8}_{(001)} \times \mathbf{21}^{\text{Adjt}}_{(010)} = \mathbf{112}_{(011)} + \mathbf{48}_{(101)} + \mathbf{8}_{(001)}$$

$$\mathbf{21}^{\text{Adjt}}_{(010)} \times \mathbf{21}^{\text{Adjt}}_{(010)} = \mathbf{1}^{s}_{(000)} + \mathbf{27}^{s}_{(200)} + \mathbf{35}^{s}_{(002)} + \mathbf{168}^{s}_{(020)} + \mathbf{21}^{\text{Adjt},\,a}_{(010)} + \mathbf{189}^{a}_{(102)}$$

Embeddings

$SO(7) \supset SO(5) \times U(1)$

$$\mathbf{7} = \mathbf{1}_2 + \mathbf{1}_{-2} + \mathbf{5}_0; \quad \mathbf{8} = \mathbf{4}_1 + \mathbf{4}_{-1}; \quad \mathbf{21} = \mathbf{10}_0 + \mathbf{1}_0 + \mathbf{5}_2 + \mathbf{5}_{-2}$$

$SO(7) \supset G_2$

$$\mathbf{7} = \mathbf{7}; \quad \mathbf{8} = \mathbf{1} + \mathbf{7}; \quad \mathbf{21} = \mathbf{7} + \mathbf{14}$$

$SO(7) \supset SU(4)$

$$\mathbf{7} = \mathbf{1} + \mathbf{6}; \quad \mathbf{8} = \mathbf{4} + \bar{\mathbf{4}}; \quad \mathbf{21} = \mathbf{6} + \mathbf{15}$$

$SO(7) \supset SU(2 \times SU(2) \times SU(2))$

$$\mathbf{7} = (\mathbf{1}, \mathbf{1}, \mathbf{3}) + (\mathbf{2}, \mathbf{2}, \mathbf{1}); \quad \mathbf{8} = (\mathbf{1}, \mathbf{2}, \mathbf{2}) + (\mathbf{2}, \mathbf{1}, \mathbf{2})$$
$$\mathbf{21} = (\mathbf{1}, \mathbf{1}, \mathbf{3}) + (\mathbf{1}, \mathbf{3}, \mathbf{1}) + (\mathbf{3}, \mathbf{1}, \mathbf{1}) + (\mathbf{2}, \mathbf{2}, \mathbf{3})$$

Representation congruence: $\mathcal{Z}_2$; **Coxeter number:** 6

Marks: (1, 2, 2, 2); **Co-marks:** (1, 2, 2, 1); **Level vector:** (6, 10, 6)

$SO(8)$

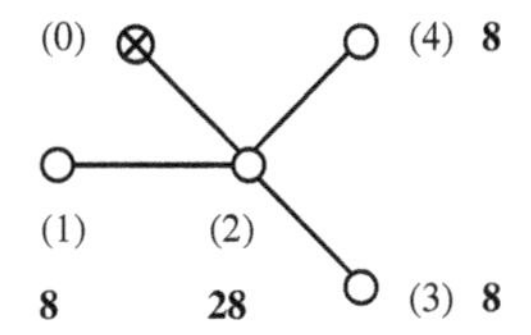

Representations & Dynkin indices

	$I^{(2)}$	$I^{(4)}$
$\mathbf{8}_{(1000)}$	2	40
$\mathbf{28}_{(0100)}$	12	480
$\mathbf{35}_{(2000)}$	20	1120
$\mathbf{56}_{(1010)}$	30	1560
$\mathbf{112}_{(1000)}$	108	10800
$\mathbf{160}_{(1100)}$	120	9120
$\mathbf{224}_{(2010)}$	200	18400
$\mathbf{294}_{(4000)}$	420	63840
$\mathbf{300}_{(0200)}$	300	31200
$\mathbf{350}_{(1011)}$	300	26400

Kronecker products

$$\mathbf{8}_{(1000)} \times \mathbf{8}_{(1000)} = \mathbf{1}^{s}_{(0000)} + \mathbf{35}^{s}_{(2000)} + \mathbf{28}^{\text{Adjt},\,a}_{(0100)}$$

$$\mathbf{8}_{(1000)} \times \mathbf{28}^{\text{Adjt}}_{(0100)} = \mathbf{8}_{(1000)} + \mathbf{56}_{(0011)} + \mathbf{160}_{(1100)}$$

$$\mathbf{8}_{(1000)} \times \mathbf{8}_{(0010)} = \mathbf{56}_{(1010)} + \mathbf{8}_{(0001)}$$

$$\mathbf{8}_{(1000)} \times \mathbf{8}_{(0001)} = \mathbf{56}_{(1001)} + \mathbf{8}_{(1000)}$$

$$\mathbf{28}^{\text{Adjt}}_{(0100)} \times \mathbf{8}_{(0010)} = \mathbf{8}_{(0010)} + \mathbf{56}_{(1001)} + \mathbf{160}_{(0110)}$$

$$\mathbf{28}^{\text{Adjt}}_{(0100)} \times \mathbf{8}_{(0001)} = \mathbf{8}_{(0001)} + \mathbf{56}_{(1010)} + \mathbf{160}_{(0101)}$$

$$\mathbf{8}_{(0010)} \times \mathbf{8}_{(0010)} = \mathbf{1}^{s}_{(0000)} + \mathbf{35}^{s}_{(0020)} + \mathbf{28}^{\text{Adjt},\,a}_{(0100)}$$

$$\mathbf{8}_{(0010)} \times \mathbf{8}_{(0001)} = \mathbf{8}_{(1000)} + \mathbf{56}_{(0011)}$$

$$\mathbf{8}_{(0001)} \times \mathbf{8}_{(0001)} = \mathbf{1}^{s}_{(0000)} + \mathbf{35}^{s}_{(0002)} + \mathbf{28}^{\text{Adjt}, a}_{(0100)}$$

$$\mathbf{28}^{\text{Adjt}}_{(0100)} \times \mathbf{28}^{\text{Adjt}}_{(0100)} = \mathbf{1}^{s}_{(0000)} + \mathbf{300}^{s}_{(0200)} + \mathbf{35}^{s}_{(2000)} + \mathbf{35}^{s}_{(0002)} + \mathbf{35}^{s}_{(0020)} + \mathbf{28}^{\text{Adjt}, a}_{(0100)} + \mathbf{350}^{a}_{(1011)}$$

Embeddings

$SO(\mathbf{8}) \supset SU(\mathbf{4}) \times U(\mathbf{1})$

$$\mathbf{8}_{(1000)} = \mathbf{6}_0 + \mathbf{1}_1 + \mathbf{1}_{-1}; \quad \mathbf{8}_{(0001)} = \mathbf{4}_{1/2} + \bar{\mathbf{4}}_{-1/2}; \quad \mathbf{8}_{(0010)} = \mathbf{4}_{-1/2} + \bar{\mathbf{4}}_{1/2}$$
$$\mathbf{28} = \mathbf{15}_0 + \mathbf{1}_0 + \mathbf{6}_1 + \mathbf{6}_{-1}$$

$SO(\mathbf{8}) \supset SU(\mathbf{2}) \times SU(\mathbf{2}) \times SU(\mathbf{2}) \times SU(\mathbf{2})$

$$\mathbf{8}_{(1000)} = (\mathbf{2}, \mathbf{2}, \mathbf{1}, \mathbf{1}) + (\mathbf{1}, \mathbf{1}, \mathbf{2}, \mathbf{2}); \quad \mathbf{8}_{(0001)} = (\mathbf{1}, \mathbf{2}, \mathbf{2}, \mathbf{1}) + (\mathbf{2}, \mathbf{1}, \mathbf{1}, \mathbf{2});$$
$$\mathbf{8}_{(0010)} = (\mathbf{2}, \mathbf{1}, \mathbf{2}, \mathbf{1}) + (\mathbf{1}, \mathbf{2}, \mathbf{1}, \mathbf{2})$$
$$\mathbf{28} = (\mathbf{3}, \mathbf{1}, \mathbf{1}, \mathbf{1}) + (\mathbf{1}, \mathbf{3}, \mathbf{1}, \mathbf{2}) + (\mathbf{1}, \mathbf{3}, \mathbf{2}) + (\mathbf{1}, \mathbf{1}, \mathbf{1}, \mathbf{3}) + (\mathbf{2}, \mathbf{2}, \mathbf{2}, \mathbf{2})$$

$SO(\mathbf{8}) \supset SU(\mathbf{3})$

$$\mathbf{8}_{(1000)} = \mathbf{8}; \quad \mathbf{8}_{(0010)} = \mathbf{8}; \quad \mathbf{8}_{(0001)} = \mathbf{8}; \quad \mathbf{28}_{(0100)} = \mathbf{8} + \mathbf{10} + \overline{\mathbf{10}}$$

$SO(\mathbf{8}) \supset Sp(\mathbf{4}) \times SU(\mathbf{2})$

$$\mathbf{8}_{(1000)} = (\mathbf{4}, \mathbf{2}); \quad \mathbf{8}_{(0010)} = (\mathbf{4}, \mathbf{2}); \quad \mathbf{8}_{(0001)} = (\mathbf{5}, \mathbf{1}) + (\mathbf{1}, \mathbf{3})$$
$$\mathbf{28} = (\mathbf{10}, \mathbf{1}) + (\mathbf{1}, \mathbf{3}) + (\mathbf{5}, \mathbf{3})$$

$SO(\mathbf{8}) \supset SO(\mathbf{7})$

$$\mathbf{8}_{(1000)} = \mathbf{1} + \mathbf{7}; \quad \mathbf{8}_{(0010)} = \mathbf{8}; \quad \mathbf{8}_{(0001)} = \mathbf{8}; \quad \mathbf{28} = \mathbf{21} + \mathbf{7}$$

Representation congruence: $\mathcal{Z}_2$; **Coxeter number:** 5

Marks: (1, 1, 2, 2, 1); **Level vector:** (6, 10, 6, 6)

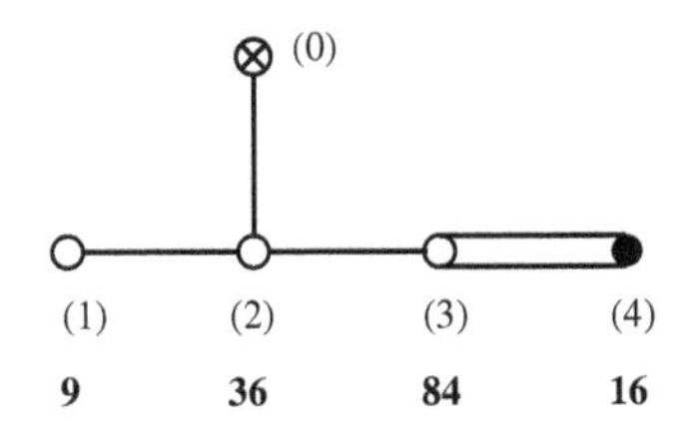

Representations & Dynkin indices

	$I^{(2)}$	$I^{(4)}$
$\mathbf{9}_{(1000)}$	2	40
$\mathbf{16}_{(0001)}$	4	80
$\mathbf{36}_{(0100)}$	14	520
$\mathbf{44}_{(2000)}$	22	1160
$\mathbf{84}_{(0010)}$	42	2040
$\mathbf{126}_{(0002)}$	70	3800
$\mathbf{128}_{(1001)}$	64	3200
$\mathbf{156}_{(3000)}$	130	11960
$\mathbf{231}_{(1100)}$	154	10760
$\mathbf{432}_{(0101)}$	300	21360
$\mathbf{450}_{(4000)}$	550	75800
$\mathbf{495}_{(0200)}$	440	41440
$\mathbf{576}_{(2001)}$	464	40000
$\mathbf{594}_{(1010)}$	462	37560
$\mathbf{672}_{(0003)}$	616	58400
$\mathbf{768}_{(0011)}$	640	55040
$\mathbf{910}_{(2100)}$	910	99320
$\mathbf{924}_{(1002)}$	770	67000
$\mathbf{1122}_{(5000)}$	1870	355640
$\mathbf{1650}_{(0110)}$	1650	174600
$\mathbf{1920}_{(3001)}$	2240	288640
$\mathbf{1980}_{(0020)}$	2310	285960

Kronecker products

$$\mathbf{9}_{(1000)} \times \mathbf{9}_{(1000)} = \mathbf{1}^{s}_{(0000)} + \mathbf{44}^{s}_{(2000)} + \mathbf{36}^{\mathrm{Adjt},\,a}_{(0100)}$$
$$\mathbf{9}_{(1000)} \times \mathbf{36}^{\mathrm{Adjt}}_{(0100)} = \mathbf{9}_{(1000)} + \mathbf{84}_{(0010)} + \mathbf{231}_{(1100)}$$
$$\mathbf{9}_{(1000)} \times \mathbf{84}_{(0010)} = \mathbf{36}^{\mathrm{Adjt}}_{(0100)} + \mathbf{126}_{(0002)} + \mathbf{594}_{(1010)}$$
$$\mathbf{9}_{(1000)} \times \mathbf{16}_{(0001)} = \mathbf{16}_{(0001)} + \mathbf{128}_{(1010)}$$
$$\mathbf{36}^{\mathrm{Adjt}}_{(0100)} \times \mathbf{36}^{\mathrm{Adjt}}_{(0100)} = \mathbf{1}^{s}_{(0000)} + \mathbf{44}^{s}_{(2000)} + \mathbf{126}^{s}_{(0002)} + \mathbf{495}^{s}_{(0200)} + \mathbf{594}^{a}_{(1010)} + \mathbf{36}^{\mathrm{Adjt},\,a}_{(0100)}$$
$$\mathbf{36}^{\mathrm{Adjt}}_{(0100)} \times \mathbf{84}_{(0010)} = \mathbf{9}_{(1000)} + \mathbf{84}_{(0010)} + \mathbf{126}_{(0002)} + \mathbf{231}_{(1100)} + \mathbf{924}_{(1002)} + \mathbf{1650}_{(0110)}$$
$$\mathbf{36}^{\mathrm{Adjt}}_{(0100)} \times \mathbf{16}_{(0001)} = \mathbf{16}_{(0001)} + \mathbf{128}_{(1001)} + \mathbf{432}_{(0101)}$$
$$\mathbf{84}_{(0010)} \times \mathbf{84}_{(0010)} = \mathbf{1}^{s}_{(0000)} + \mathbf{44}^{s}_{(2000)} + \mathbf{126}^{s}_{(0002)} + \mathbf{495}^{s}_{(0200)} + \mathbf{924}^{s}_{(1002)} + \mathbf{1980}^{s}_{(0020)} + \mathbf{36}^{\mathrm{Adjt},\,a}_{(0100)} + \mathbf{84}^{a}_{(0010)} + \mathbf{594}^{a}_{(1010)} + \mathbf{2772}^{a}_{(0102)}$$
$$\mathbf{84}_{(0010)} \times \mathbf{16}_{(0001)} = \mathbf{16}_{(0001)} + \mathbf{128}_{(1001)} + \mathbf{432}_{(0101)} + \mathbf{768}_{(0011)}$$
$$\mathbf{16}_{(0001)} \times \mathbf{16}_{(0001)} = \mathbf{1}^{s}_{(0000)} + \mathbf{9}^{s}_{(11000)} + \mathbf{126}^{s}_{(0002)} + \mathbf{36}^{\mathrm{Adjt},\,a}_{(0100)} + \mathbf{84}^{a}_{(0010)}$$

Embeddings

$SO(\mathbf{9}) \supset SU(\mathbf{4}) \times SU(\mathbf{2})$

$\mathbf{16} = (\mathbf{4}, \mathbf{2}) + (\bar{\mathbf{4}}, \mathbf{2})$; $\quad \mathbf{9} = (\mathbf{6}, \mathbf{1}) + (\mathbf{1}, \mathbf{3})$

$\mathbf{36} = (\mathbf{15}, \mathbf{1}) + (\mathbf{1}, \mathbf{3}) + (\mathbf{6}, \mathbf{3})$

$SO(\mathbf{9}) \supset SO(\mathbf{8})$

$\mathbf{16} = \mathbf{8}_{(0010)} + \mathbf{8}_{(0001)}$; $\quad \mathbf{9} = \mathbf{8}_{(1000)} + \mathbf{1}$; $\quad \mathbf{36} = \mathbf{8}_{(1000)} + \mathbf{28}$

$SO(\mathbf{9}) \supset Sp(\mathbf{4}) \times SU(\mathbf{2}) \times SU(\mathbf{2})$

$\mathbf{16} = (\mathbf{4}, \mathbf{1}, \mathbf{2}) + (\mathbf{4}, \mathbf{1}, \mathbf{2})$; $\quad \mathbf{9} = (\mathbf{5}, \mathbf{1}, \mathbf{1}) + (\mathbf{1}, \mathbf{2}, \mathbf{2})$

$\mathbf{36} = (\mathbf{10}, \mathbf{1}, \mathbf{1}) + (\mathbf{1}, \mathbf{3}, \mathbf{1}) + (\mathbf{1}, \mathbf{1}, \mathbf{3}) + (\mathbf{5}, \mathbf{2}, \mathbf{2})$

$SO(\mathbf{9}) \supset SO(\mathbf{7}) \times U(\mathbf{1})$

$\mathbf{16} = \mathbf{8}_{1} + \mathbf{8}_{-1}$; $\quad \mathbf{9} = \mathbf{1}_{2} + \mathbf{1}_{-2} + \mathbf{7}_{0}$; $\quad \mathbf{36} = \mathbf{1}_{0} + \mathbf{21}_{0} + \mathbf{7}_{2} + \mathbf{7}_{-2}$

$SO(\mathbf{9}) \supset SU(\mathbf{2}) \times SU(\mathbf{2})$

$\mathbf{16} = (\mathbf{4}, \mathbf{2}) + (\mathbf{2}, \mathbf{4})$; $\quad \mathbf{9} = (\mathbf{3}, \mathbf{3})$

$\mathbf{36} = (\mathbf{3}, \mathbf{1}) + (\mathbf{1}, \mathbf{3})) + (\mathbf{3}, \mathbf{5})) + (\mathbf{5}, \mathbf{3})$

Representation congruence: $\mathcal{Z}_2$; **Coxeter number:** 8

Marks: (1, 1, 2, 2, 2); **Co-marks:** (1, 1, 2, 2, 1);
Level vector: (8, 14, 18, 10)

$SO(10)$

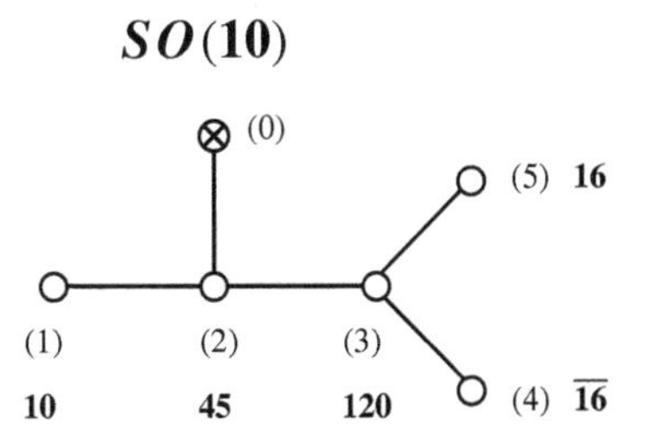

Representations & Dynkin indices

	$I^{(2)}$	$I^{(4)}$
$\mathbf{10}_{(10000)}$	2	60
$\mathbf{16}_{(10000)}$	4	150
$\mathbf{45}_{(01000)}$	16	960
$\mathbf{54}_{(20000)}$	24	1920
$\mathbf{120}_{(00100)}$	56	4560
$\mathbf{126}_{(00002)}$	70	6900
$\mathbf{144}_{(10001)}$	68	5670
$\mathbf{210}_{(10000)}$	112	10560
$\mathbf{210}_{(30000)}$	154	21420
$\mathbf{320}_{(11000)}$	192	21120
$\mathbf{560}_{(01001)}$	364	43170
$\mathbf{660}_{(40000)}$	704	145920
$\mathbf{672}_{(00003)}$	616	105180
$\mathbf{720}_{(20001)}$	532	73470
$\mathbf{770}_{(02000)}$	616	92160
$\mathbf{945}_{(10100)}$	672	88320
$\mathbf{1050}_{(10002)}$	840	124800
$\mathbf{1200}_{(00101)}$	940	136050
$\mathbf{1386}_{(21000)}$	1232	208320
$\mathbf{1440}_{(00012)}$	1256	203580
$\mathbf{1728}_{(10011)}$	1344	193920
$\mathbf{2640}_{(30001)}$	2772	559470
$\mathbf{2772}_{(00004)}$	3696	938880
$\mathbf{2970}_{(01100)}$	2706	463260

Kronecker products

$$\mathbf{10}_{(10000)} \times \mathbf{10}_{(10000)} = \mathbf{1}^{s}_{(00000)} + \mathbf{54}^{s}_{(20000)} + \mathbf{45}^{\text{Adjt},\,a}_{(01000)}$$

$$\mathbf{10}_{(10000)} \times \mathbf{45}^{\text{Adjt}}_{(01000)} = \mathbf{10}_{(10000)} + \mathbf{120}_{(00100)} + \mathbf{320}_{(11000)}$$

$$\mathbf{10}_{(10000)} \times \mathbf{120}_{(00100)} = \mathbf{45}^{\text{Adjt}}_{(01000)} + \mathbf{210}_{(00011)} + \mathbf{945}_{(10100)}$$

$$\mathbf{10}_{(10000)} \times \mathbf{16}_{(00001)} = \overline{\mathbf{16}}_{(00010)} + \overline{\mathbf{144}}_{(10001)}$$

$$\mathbf{45}^{\text{Adjt}}_{(01000)} \times \mathbf{45}^{\text{Adjt}}_{(01000)} = \mathbf{1}^{s}_{(00000)} + \mathbf{54}^{s}_{(20000)} + \mathbf{210}^{s}_{(00011)} + \mathbf{770}^{s}_{(02000)} + \mathbf{45}^{\text{Adjt},\,a}_{(01000)} + \mathbf{945}^{a}_{(10100)}$$

$$\mathbf{45}^{\text{Adjt}}_{(01000)} \times \mathbf{120}_{(00100)} = \mathbf{10}_{(10000)} + \mathbf{120}_{(00100)} + \mathbf{126}_{(00002)} + \overline{\mathbf{126}}_{(00020)} + \mathbf{320}_{(11000)} + \mathbf{1728}_{(10011)} + \mathbf{2970}_{(01100)}$$

$$\mathbf{45}^{\text{Adjt}}_{(01000)} \times \mathbf{16}_{(00001)} = \mathbf{16}_{(00001)} + \mathbf{144}_{(10010)} + \mathbf{560}_{(01001)}$$

$$\mathbf{120}_{(00100)} \times \mathbf{120}_{(00100)} = \mathbf{1}^{s}_{(00000)} + \mathbf{54}^{s}_{(20000)} + \mathbf{210}^{s}_{(00011)} + \mathbf{770}^{s}_{(02000)} + \mathbf{1050}^{s}_{(10002)} + \overline{\mathbf{1050}}^{s}_{(10020)} + \mathbf{4125}^{s}_{(00200)} + \mathbf{45}^{\text{Adjt},\,a}_{(01000)} + \mathbf{210}^{a}_{(00011)} + \mathbf{945}^{a}_{(10100)} + \mathbf{5940}^{a}_{(01011)}$$

$$\mathbf{120}_{(00100)} \times \mathbf{16}_{(00010)} = \overline{\mathbf{16}}_{(00010)} + \overline{\mathbf{144}}_{(10001)} + \overline{\mathbf{560}}_{(01010)} + \overline{\mathbf{1200}}_{(00101)}$$

$$\mathbf{16}_{(00010)} \times \mathbf{16}_{(00010)} = \mathbf{10}^{s}_{(10000)} + \mathbf{126}^{s}_{(00002)} + \mathbf{120}^{a}_{(00100)}$$

$$\mathbf{16}_{(00010)} \times \overline{\mathbf{16}}_{(00001)} = \mathbf{1}_{(00000)} + \mathbf{45}^{\text{Adjt}}_{(01000)} + \mathbf{210}_{(00011)}$$

Embeddings

$SO(\mathbf{10}) \supset SU(\mathbf{5}) \times U(\mathbf{1})$

$\mathbf{16} = \overline{\mathbf{5}}_3 + \mathbf{10}_{-1} + \mathbf{1}_{-5}$; $\quad \mathbf{10} = \overline{\mathbf{5}}_{-2} + \mathbf{5}_2$

$\mathbf{45} = \overline{\mathbf{10}}_{-4} + \mathbf{10}_4 + \mathbf{24}_0 + \mathbf{1}_0$

$SO(\mathbf{10}) \supset SU(\mathbf{4}) \times SU(\mathbf{2}) \times SU(\mathbf{2})$

$\mathbf{16} = (\mathbf{4}, \mathbf{2}, \mathbf{1}) + (\overline{\mathbf{4}}, \mathbf{1}, \mathbf{2})$; $\quad \mathbf{10} = (\mathbf{6}, \mathbf{1}, \mathbf{1}) + (\mathbf{1}, \mathbf{2}, \mathbf{2})$

$\mathbf{45} = (\mathbf{6}, \mathbf{2}, \mathbf{2}) + (\mathbf{15}, \mathbf{1}, \mathbf{1}) + (\mathbf{1}, \mathbf{3}, \mathbf{1}) + (\mathbf{1}, \mathbf{1}, \mathbf{3})$

$SO(\mathbf{10}) \supset SO(\mathbf{9})$

$\mathbf{16} = \mathbf{16}$; $\quad \mathbf{10} = \mathbf{9} + \mathbf{1}$; $\quad \mathbf{45} = \mathbf{9} + \mathbf{36}$

$SO(\mathbf{10}) \supset SO(\mathbf{7}) \times SU(\mathbf{2})$

$\mathbf{16} = (\mathbf{8}, \mathbf{2})$; $\quad \mathbf{10} = (\mathbf{7}, \mathbf{1}) + (\mathbf{1}, \mathbf{3})$

$\mathbf{45} = (\mathbf{7}, \mathbf{3}) + (\mathbf{21}, \mathbf{1}) + (\mathbf{1}, \mathbf{3})$

$SO(\mathbf{10}) \supset Sp(\mathbf{4}) \times Sp(\mathbf{4})$

$\mathbf{16} = (\mathbf{4}, \mathbf{4})$; $\quad \mathbf{10} = (\mathbf{5}, \mathbf{1}) + (\mathbf{1}, \mathbf{5})$

$\mathbf{45} = (\mathbf{5}, \mathbf{5}) + (\mathbf{10}, \mathbf{1}) + (\mathbf{1}, \mathbf{10})$

$SO(\mathbf{10}) \supset SO(\mathbf{8}) \times U(\mathbf{1})$

$\mathbf{16} = \mathbf{8}_1 + \mathbf{8}_{-1}$; $\quad \mathbf{10} = \mathbf{1}_2 + \mathbf{1}_{-2} + \mathbf{8}_0$

$\mathbf{45} = \mathbf{8}_2 + \mathbf{8}_{-2} + \mathbf{28}_0 + \mathbf{1}_0$

$SO(\mathbf{10}) \supset Sp(\mathbf{4})$

$\mathbf{16} = \mathbf{16}$; $\quad \mathbf{10} = \mathbf{10}$; $\quad \mathbf{45} = \mathbf{10} + \mathbf{35}_{(12)}$

Representation congruence: $\mathcal{Z}_2$; **Coxeter number:** 8

Marks: (1, 1, 2, 2, 1, 1); **Level vector:** (8, 14, 18, 10, 10)

G_2

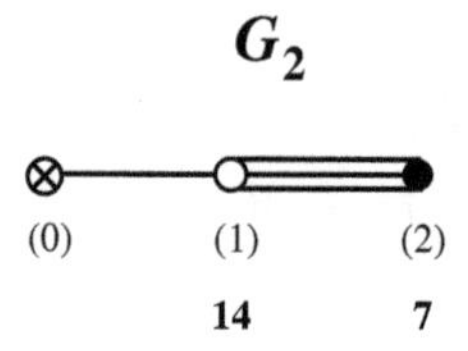

Representations & Dynkin indices

	$I^{(2)}$	$I^{(4)}$
$\mathbf{7}_{(01)}$	2	8
$\mathbf{14}_{(10)}$	8	80
$\mathbf{27}_{(02)}$	18	216
$\mathbf{64}_{(11)}$	64	1216
$\mathbf{77}_{(03)}$	88	1936
$\mathbf{77}_{(20)}$	110	3080
$\mathbf{182}_{(04)}$	312	10608
$\mathbf{189}_{(12)}$	288	8640

Kronecker products

$$\mathbf{7}_{(01)} \times \mathbf{7}_{(01)} = \mathbf{1}^{s}_{(00)} + \mathbf{27}^{s}_{(02)} + \mathbf{7}^{a}_{(01)} + \mathbf{14}^{\text{Adjt},\, a}_{(10)}$$
$$\mathbf{7}_{(01)} \times \mathbf{14}^{\text{Adjt}}_{(10)} = \mathbf{7}_{(01)} + \mathbf{27}_{(02)} + \mathbf{64}_{(11)}$$
$$\mathbf{14}^{\text{Adjt}}_{(10)} \times \mathbf{14}^{\text{Adjt}}_{(10)} = \mathbf{1}^{s}_{(00)} + \mathbf{27}^{s}_{(02)} + \mathbf{77}^{s}_{(20)} + \mathbf{14}^{\text{Adjt},\, a}_{(10)} + \mathbf{77}^{a}_{(03)}$$

Embeddings

$G_2 \supset SU(\mathbf{3})$
$\mathbf{7} = \mathbf{1} + \mathbf{3} + \bar{\mathbf{3}}$; $\quad \mathbf{14} = \mathbf{8} + \mathbf{3} + \bar{\mathbf{3}}$

$G_2 \supset SU(\mathbf{2}) \times SU(\mathbf{2})$
$\mathbf{7} = (\mathbf{2}, \mathbf{2}) + (\mathbf{1}, \mathbf{3})$; $\quad \mathbf{14} = (\mathbf{3}, \mathbf{1}) + (\mathbf{1}, \mathbf{3}) + (\mathbf{2}, \mathbf{4})$

$G_2 \supset SU(\mathbf{2})$
$\mathbf{7} = \mathbf{7}$; $\quad \mathbf{14} = \mathbf{11} + \mathbf{3}$

Representation Chang number: 0, $\pm 1 \mod(7)$; **Coxeter number:** 6

Marks: (1, 2, 3); **Co-marks:** (1, 2, 1); **Level vector:** (10, 6)

F_4

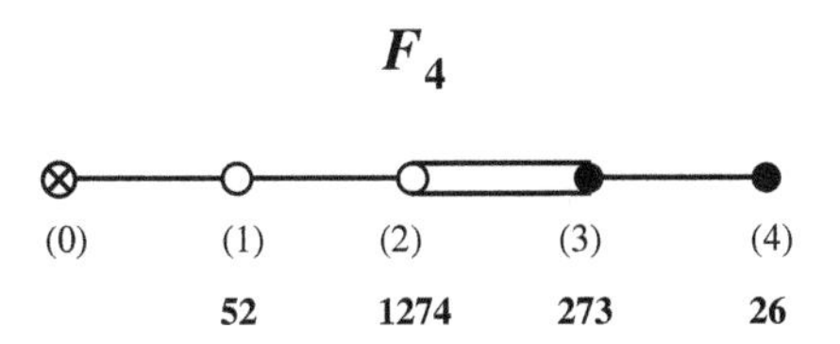

Representations & Dynkin indices

	$I^{(2)}$	$I^{(4)}$
$\mathbf{26}_{(0001)}$	6	120
$\mathbf{52}_{(1000)}$	18	600
$\mathbf{273}_{(0010)}$	126	5880
$\mathbf{324}_{(0002)}$	162	8280
$\mathbf{1053}_{(1001)}$	648	41760
$\mathbf{1274}_{(0100)}$	882	64680
$\mathbf{2652}_{(0003)}$	2142	185640
$\mathbf{4096}_{(0011)}$	3072	245760
$\mathbf{8424}_{(1010)}$	7452	712080

Kronecker products

$$\mathbf{26}_{(0001)} \times \mathbf{26}_{(0001)} = \mathbf{1}^s_{(0000)} + \mathbf{26}^s_{(0001)} + \mathbf{324}^s_{(0002)} + \mathbf{52}^{\text{Adjt},\,a}_{(1000)} + \mathbf{273}^a_{(0010)}$$

$$\mathbf{26}_{(0001)} \times \mathbf{273}_{(0010)} = \mathbf{4096}_{(0011)} + \mathbf{1274}_{(0100)} + \mathbf{1053}_{(1001)} + \mathbf{324}_{(0002)} + \mathbf{273}_{(0010)} + \mathbf{52}^{\text{Adjt}}_{(1000)} + \mathbf{26}_{(0001)}$$

$$\mathbf{26}_{(0001)} \times \mathbf{52}^{\text{Adjt}}_{(1000)} = \mathbf{26}_{(0001)} + \mathbf{273}_{(0010)} + \mathbf{1053}_{(2000)}$$

$$\mathbf{273}_{(0010)} \times \mathbf{52}^{\text{Adjt}}_{(1000)} = \mathbf{8424}_{(1010)} + \mathbf{4096}_{(0011)} + \mathbf{1053}_{(1001)} + \mathbf{324}_{(0002)} + \mathbf{273}_{(0010)} + \mathbf{26}_{(0001)}$$

$$\mathbf{52}^{\text{Adjt}}_{(1000)} \times \mathbf{52}^{\text{Adjt}}_{(1000)} = \mathbf{1}^s_{(0000)} + \mathbf{324}^s_{(0002)} + \mathbf{1053}^s_{(2000)} + \mathbf{52}^{\text{Adjt},\,a}_{(1000)} + \mathbf{1274}^a_{(0100)}$$

Embeddings

$F_4 \supset SO(\mathbf{9})$

$$\mathbf{26} = \mathbf{1} + \mathbf{9} + \mathbf{16}; \qquad \mathbf{52} = \mathbf{36} + \mathbf{16}$$

$F_4 \supset SU(\mathbf{3}) \times SU(\mathbf{3})$

$$\mathbf{26} = (\mathbf{3}, \mathbf{3}) + (\bar{\mathbf{3}}, \bar{\mathbf{3}}) + (\mathbf{1}, \mathbf{8})$$
$$\mathbf{52} = (\mathbf{6}, \bar{\mathbf{3}}) + (\bar{\mathbf{6}}, \mathbf{3}) + (\mathbf{1}, \mathbf{8}) + (\mathbf{8}, \mathbf{1})$$

$F_4 \supset Sp(\mathbf{6}) \times SU(\mathbf{2})$

$$\mathbf{26} = (\mathbf{14}, \mathbf{1}) + (\mathbf{6}, \mathbf{2})$$
$$\mathbf{52} = (\mathbf{21}, \mathbf{1}) + (\mathbf{1}, \mathbf{3}) + (\mathbf{14}, \mathbf{2})$$

$F_4 \supset G_2 \times SU(\mathbf{2})$

$$\mathbf{26} = (\mathbf{7}, \mathbf{2}) + (\mathbf{1}, \mathbf{4})$$
$$\mathbf{52} = (\mathbf{7}, \mathbf{5}) + (\mathbf{14}, \mathbf{1}) + (\mathbf{1}, \mathbf{3})$$

$F_4 \supset SU(\mathbf{2})$

$$\mathbf{26} = \mathbf{9} + \mathbf{17}$$

Representation Chang number: 0, ±1 mod(13); **Coxeter number:** 12

Marks: (1, 2, 3, 4, 2); **Co-marks:** (1, 2, 3, 2, 1);
Level vector: (22, 42, 30, 16)

E_6

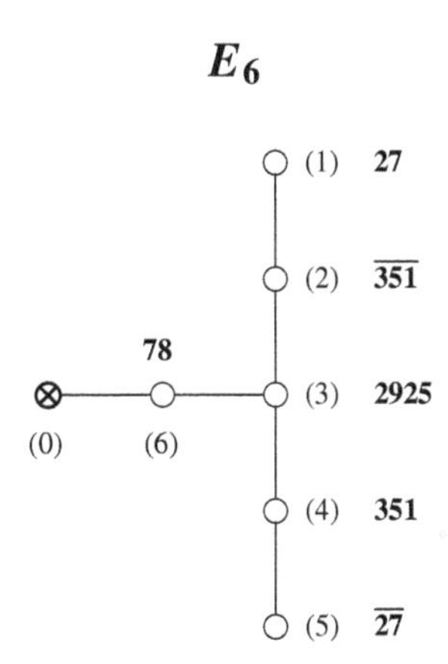

Representations & Dynkin indices

	$I^{(2)}$	$I^{(4)}$
$\mathbf{27}_{(100000)}$	6	336
$\overline{\mathbf{27}}_{(000010)}$	6	336
$\mathbf{78}_{(000001)}$	24	2016
$\mathbf{351}_{(000020)}$	168	23520
$\mathbf{351}_{(000100)}$	150	18480
$\mathbf{650}_{(100010)}$	300	40320
$\mathbf{1728}_{(000011)}$	960	158592
$\mathbf{2430}_{(000002)}$	1620	326592
$\mathbf{2925}_{(001000)}$	1800	332640
$\mathbf{3003}_{(000030)}$	2310	543312
$\mathbf{5824}_{(000110)}$	4032	846720
$\mathbf{7371}_{(010010)}$	5040	1044288
$\mathbf{7722}_{(100020)}$	5076	1271424

Kronecker products

$$\mathbf{27}_{(100000)} \times \mathbf{27}_{(100000)} = \overline{\mathbf{27}}^{s}_{(000010)} + \overline{\mathbf{351}}^{s}_{(200000)} + \overline{\mathbf{351}}^{a}_{(01000)}$$

$$\mathbf{27}_{(100000)} \times \overline{\mathbf{351}}_{(010000)} = \overline{\mathbf{5824}}_{(110000)} + \mathbf{2925}_{(001000)} + \mathbf{650}_{(100010)} + \mathbf{78}^{\mathrm{Adjt}}_{(000001)}$$

$$\mathbf{27}_{(100000)} \times \mathbf{351}_{(000100)} = \overline{\mathbf{27}}_{(100010)} + \overline{\mathbf{351}}_{(010000)} + \overline{\mathbf{1728}}_{(000011)} + \overline{\mathbf{7371}}_{(100100)}$$

$$\mathbf{27}_{(100000)} \times \overline{\mathbf{27}}_{(000010)} = \mathbf{1}_{(000000)} + \mathbf{78}^{\mathrm{Adjt}}_{(000001)} + \mathbf{650}_{(100010)}$$

$$\mathbf{27}_{(100000)} \times \mathbf{78}^{\mathrm{Adjt}}_{(000001)} = \mathbf{27}_{(100000)} + \mathbf{351}_{(000100)} + \mathbf{1728}_{(100001)}$$

$$\mathbf{78}^{\mathrm{Adjt}}_{(000001)} \times \mathbf{78}^{\mathrm{Adjt}}_{(000001)} = \mathbf{2430}^{s}_{(000002)} + \mathbf{650}^{s}_{(100010)} + \mathbf{1}^{s}_{(000000)} + \mathbf{2925}^{a}_{(001000)} + \mathbf{78}^{\mathrm{Adjt},\,a}_{(000001)}$$

Embeddings

$E_6 \supset SO(\mathbf{10}) \times U(\mathbf{1})$

$$\mathbf{27} = \mathbf{1}_4 + \mathbf{10}_{-2} + \mathbf{16}_1\,; \quad \mathbf{78} = \mathbf{1}_0 + \mathbf{45}_{-2} + \mathbf{16}_3 + \overline{\mathbf{16}}_3$$

$E_6 \supset SU(\mathbf{3}) \times SU(\mathbf{3}) \times SU(\mathbf{3})$

$$\mathbf{27} = (\mathbf{3}, \mathbf{1}, \mathbf{3}) + (\bar{\mathbf{3}}, \mathbf{3}, \mathbf{1}) + (\mathbf{1}, \bar{\mathbf{3}}, \bar{\mathbf{3}})$$

$$\mathbf{78} = (\mathbf{8}, \mathbf{1}, \mathbf{1}) + (\mathbf{1}, \mathbf{8}, \mathbf{1}) + (\mathbf{1}, \mathbf{1}, \mathbf{8}) + (\mathbf{3}, \mathbf{3}, \bar{\mathbf{3}}) + (\bar{\mathbf{3}}, \bar{\mathbf{3}}, \mathbf{3})$$

$E_6 \supset SU(\mathbf{6}) \times SU(\mathbf{2})$

$$\mathbf{27} = (\mathbf{15}, \mathbf{1}) + (\bar{\mathbf{6}}, \mathbf{2})$$

$$\mathbf{78} = (\mathbf{35}, \mathbf{1}) + (\mathbf{1}, \mathbf{3}) + (\mathbf{20}, \mathbf{2})$$

$E_6 \supset G_2 \times SU(\mathbf{3})$

$$\mathbf{27} = (\mathbf{1}, \bar{\mathbf{6}}) + (\mathbf{7}, \mathbf{3}); \qquad \mathbf{78} = (\mathbf{14}, \mathbf{1}) + (\mathbf{1}, \mathbf{8}) + (\mathbf{7}, \mathbf{8})$$

$E_6 \supset F_4$

$$\mathbf{27}_{(100000)} = \mathbf{1}_{(0000)} + \mathbf{26}_{(0001)}; \qquad \mathbf{78}_{(000001)} = \mathbf{52}_{(1000)} + \mathbf{26}_{(0001)}$$

$E_6 \supset Sp(\mathbf{8})$

$$\mathbf{27} = \mathbf{27}_{(0100)}$$

$E_6 \supset G_2$

$$\mathbf{27} = \mathbf{27}_{(02)}$$

$E_6 \supset SU(\mathbf{3})$

$$\mathbf{27} = \mathbf{27}_{(22)}$$

Representation Chang number: 0, ± 1 mod(13); **Coxeter number:** 12

Marks: (1, 2, 3, 2, 1, 2, 1); **Level vector:** (16, 30, 42, 30, 16, 22)

E_7

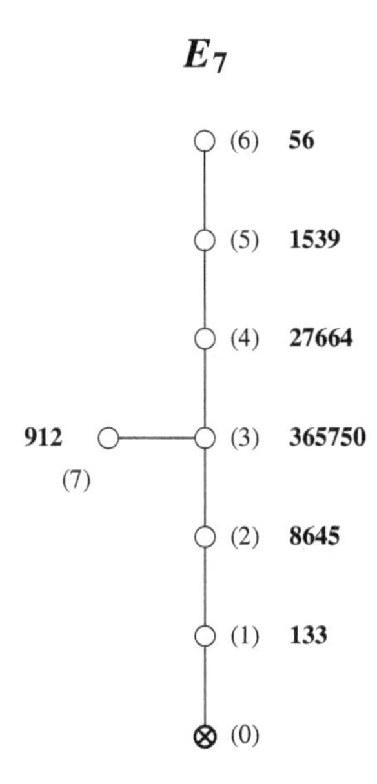

Representations & Dynkin indices

	$I^{(2)}$	$I^{(4)}$
$\mathbf{56}_{(0000010)}$	12	1008
$\overline{\mathbf{27}}_{(000010)}$	6	336
$\mathbf{133}_{(1000000)}$	36	4032
$\mathbf{912}_{(0000020)}$	660	133056
$\mathbf{1463}_{(0000100)}$	150	18480
$\mathbf{1539}_{(0000100)}$	648	120960
$\mathbf{6480}_{(1000010)}$	3240	731808
$\mathbf{7371}_{(2000000)}$	4212	1100736
$\mathbf{8645}_{(0100000)}$	4680	1153152
$\mathbf{24320}_{(0000030)}$	17280	5709312
$\mathbf{27664}_{(0001000)}$	17160	4900896
$\mathbf{40755}_{(0000011)}$	25740	7495488
……	……	……
$\mathbf{365750}_{(0010000)}$	297000	113097600

Kronecker products

$$\mathbf{56}_{(0000010)} \times \mathbf{56}_{(0000010)} = \mathbf{1}^{a}_{(0000000)} + \mathbf{1539}^{a}_{(0000100)} + \mathbf{1463}^{s}_{(0000020)} + \mathbf{133}^{\mathrm{Adjt},\, s}_{(1000000)}$$

$$\mathbf{56}_{(0000010)} \times \mathbf{133}^{\mathrm{Adjt}}_{(1000000)} = \mathbf{56}_{(0000010)} + \mathbf{912}_{(0000001)} + \mathbf{6480}_{(1000010)}$$

$$\mathbf{133}^{\mathrm{Adjt}}_{(1000000)} \times \mathbf{133}^{\mathrm{Adjt}}_{(1000000)} = \mathbf{1}^{s}_{(0000000)} + \mathbf{1539}^{s}_{(0000100)} + \mathbf{7371}^{s}_{(2000010)} + \mathbf{133}^{\mathrm{Adjt},\, a}_{(1000000)} + \mathbf{8645}^{a}_{(0100000)}$$

Embeddings

$E_7 \supset SU(\mathbf{8})$

$$\mathbf{56} = \mathbf{28} + \overline{\mathbf{28}}\,; \qquad \mathbf{133} = \mathbf{70} + \mathbf{63}$$

$E_7 \supset SO(\mathbf{12}) \times SU(\mathbf{2})$

$$\mathbf{56} = (\mathbf{32}_{(000001)}, \mathbf{1}) + (\mathbf{12}_{(100000)}, \mathbf{2})$$

$$\mathbf{133} = (\mathbf{32}_{(000010)}, \mathbf{2}) + (\mathbf{1}, \mathbf{3}) + (\mathbf{66}_{(0100000)}, \mathbf{1})$$

$E_7 \supset SU(\mathbf{6}) \times SU(\mathbf{3})$

$$\mathbf{56} = (\mathbf{20}, \mathbf{1}) + (\mathbf{6}, \mathbf{3}) + (\overline{\mathbf{6}}, \overline{\mathbf{3}})$$

$$\mathbf{133} = (\mathbf{35}, \mathbf{1}) + (\mathbf{1}, \mathbf{8}) + (\mathbf{15}, \bar{\mathbf{3}}) + (\overline{\mathbf{15}}, \mathbf{3})$$

Chang number: 0, $\pm 1 \mod(19)$; **Coxeter number:** 18

Marks: (1, 2, 3, 4, 3, 2, 1, 2); **Level vector:** (34, 66, 96, 75, 52, 27, 49)

E_8

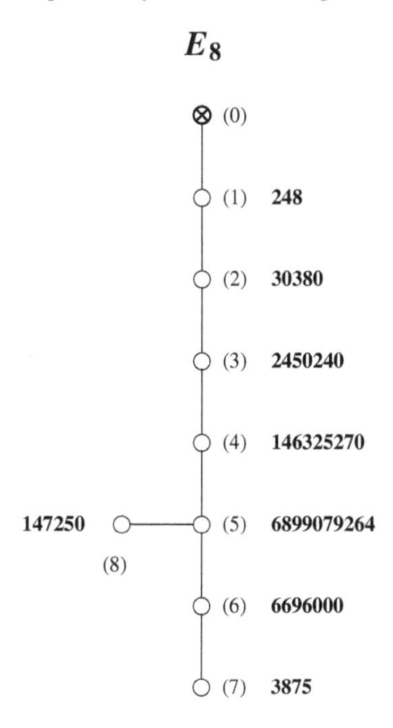

Representations & Dynkin indices

	$I^{(2)}$	$I^{(4)}$
248$_{(10000000)}$	60	8640
3875$_{(00000010)}$	1500	371520
27000$_{(20000000)}$	13500	4432320
30380$_{(01000000)}$	14700	4656960
147250$_{(00000001)}$	85500	32996160
779247$_{(10000010)}$	502740	217183680
1763125$_{(30000000)}$	1365000	715478400
2450240$_{(00100000)}$	1778400	870704640
4096000$_{(11000000)}$	3072000	1157135360
4881384$_{(00000020)}$	3936600	2154107520
6696000$_{(00000100)}$	5292000	2834818560
......	......	......
146325270$_{(00010000)}$	141605100	93799218240
......	......	......
6899079264$_{(00001000)}$	8345660400	6970295566080

Kronecker products

$$\mathbf{248}^{\mathrm{Adjt}}_{(10000000)} \times \mathbf{248}^{\mathrm{Adjt}}_{(10000000)} = \mathbf{1}^{s}_{(00000000)} + \mathbf{3875}^{s}_{(00000010)} + \mathbf{27000}^{s}_{(20000000)} + \mathbf{248}^{\mathrm{Adjt},\,a}_{(10000000)} + \mathbf{30380}^{a}_{(01000000)}$$

$$\mathbf{248}^{\mathrm{Adjt}}_{(10000000)} \times \mathbf{3875}_{(00000010)} = \mathbf{248}^{\mathrm{Adjt}}_{(10000000)} + \mathbf{3875}_{(00000010)} + \mathbf{30380}_{(01000000)} + \mathbf{147250}_{(00000001)} + \mathbf{779247}_{(10000010)}$$

Embeddings

$E_8 \supset SO(\mathbf{16})$

$$\mathbf{248} = \mathbf{120}_{(01000000)} + \mathbf{128}_{(00000001)}$$

$E_8 \supset SU(\mathbf{9})$

$$\mathbf{248} = \mathbf{80}_{(10000001)} + \mathbf{84}_{(00100000)} + \overline{\mathbf{84}}_{(00000100)}$$

$E_8 \supset E_7 \times SU(\mathbf{2})$

$$\mathbf{248} = (\mathbf{133}, \mathbf{1}) + (\mathbf{1}, \mathbf{3}) + (\mathbf{56}, \mathbf{2})$$

$E_8 \supset E_6 \times SU(\mathbf{3})$

$$\mathbf{248} = (\mathbf{78}, \mathbf{1}) + (\mathbf{1}, \mathbf{8}) + (\mathbf{27}, \mathbf{3}) + (\overline{\mathbf{27}}, \overline{\mathbf{3}})$$

$E_8 \supset SU(\mathbf{5}) \times SU(\mathbf{5})$

$$\mathbf{248} = (\mathbf{1}, \mathbf{24}) + (\mathbf{24}, \mathbf{1}) + (\mathbf{5}, \overline{\mathbf{10}}) + (\overline{\mathbf{5}}, \mathbf{10}) + (\overline{\mathbf{10}}, \overline{\mathbf{5}}) + (\mathbf{10}, \mathbf{5})$$

$E_8 \supset F_4 \times G_2$

$$\mathbf{248} = (\mathbf{52}, \mathbf{1}) + (\mathbf{1}, \mathbf{14}) + (\mathbf{26}, \mathbf{7})$$

$E_8 \supset SO(\mathbf{8}) \times SO(\mathbf{8})$

$$\mathbf{248} = (\mathbf{1}, \mathbf{28}) + (\mathbf{28}, \mathbf{1}) + (\mathbf{8}_{(1000)}, \mathbf{8}_{(0010)}) + (\mathbf{8}_{(0010)}, \mathbf{8}_{(0001)}) + (\mathbf{8}_{(0001)}, \mathbf{8}_{(1000)})$$

Chang number: 0, $\pm 1 \mod(31)$; **Coxeter number:** 30

Marks: (1, 2, 3, 4, 5, 6, 4, 2, 3);
Level vector: (92, 182, 270, 220, 168, 114, 58, 136)

References

[1] Cahn, Robert N. (1984) *Semi-Simple Lie Algebras and Their Representations* (Benjamin/Cummings, Menlo Park).

[2] Carmichael, Robert D. (1956) *Groups of Finite Order* (Dover Publications, New York).

[3] Carter, Roger W. *Simple Groups of Lie Type* (John Wiley & Sons Ltd, London).

[4] Dyson, Freeman J. (1966) *Symmetry Groups in Nuclear and Particle Physics* (W. A. Benjamin, New York).

[5] Fuchs, Jürgen (1992) *Affine Lie Algebras and Quantum Groups* (Cambridge University Press, Cambridge).

[6] Georgi, Howard (1999) *Lie Algebras in Particle Physics* (Perseus Books, Boulder).

[7] Gilmore, Robert (1974) *Lie Groups, Lie Algebras and Some of Their Applications* (John Wiley & Sons, New York).

[8] Gürsey, Feza and Tze, Chia-Hsiung (1996) *On the Role of Division, Jordan and Related Algebras in Particle Physics* (World Scientific, Singapore).

[9] Hall Jr., Marshall (1959) *The Theory of Groups* (The Macmillan Company, New York).

[10] Hammermesh, Morton (1962) *Group Theory and its Applications to Physical Problems* (Addison-Wesley, Reading).

[11] Humphreys, James E. (1972) *Introduction to Lie Algebras and Representation Theory* (Springer-Verlag, New York).

[12] Kac, V. G. (1990) *Infinite Dimensional Lie Algebras*, 3rd edn (Cambridge University Press, Cambridge).

[13] Kass, S., Moody, R. V., Patera, J., and Slansky, R. (1990) *Affine Lie Algebras, Weight Multiplicities, and Branching Rules* (University of California Press, Berkeley).

[14] Ledermann, W. (1973) *Introduction to Group Theory* (Oliver & Boyd, Edinburgh).

[15] Lomont, J. S. (1959) *Applications of Finite Groups* (Academic Press, Toronto).

[16] Miller, G. A., Blichfeldt, H. F., and Dickson, L. E. (1916) *Theory and Application of Finite Groups* (John Wiley & Sons, New York).

[17] Rose, John S. (1978) *A Course on Group Theory* (Cambridge University Press, Cambridge).

[18] Sternberg, S. (1994) *Group Theory and Physics* (Cambridge University Press, Cambridge).

[19] Tinkham, M. (1964) *Group Theory and Quantum Mechanics* (McGraw-Hill, New York).

Index

Printed in the USA
CPSIA information can be obtained
at www.ICGtesting.com
CBHW081922191124
17647CB00005B/151